The Latin American Studies Book Series

The Latin American Studies Book Series promotes quality scientific research focusing on Latin American countries. The series accepts disciplinary and interdisciplinary titles related to geographical, environmental, cultural, economic, political, urban and health-related research dedicated to Latin America. The series publishes comprehensive monographs, edited volumes and textbooks refereed by a region or country expert specialized in Latin American studies.

The series aims to raise the profile of Latin American studies, showcasing important works developed focusing on the region. It is aimed at researchers, students, and everyone interested in Latin American topics.

Submit a proposal: Proposals for the series will be considered by the Series Advisory Board. A book proposal form can be obtained from the Publisher, Andrea Brody-Barre (andrea.brody-barre@springer.com).

Romina Frontini • Paula Escosteguy
Mónica Salemme
Editors

Non-human Animals and People: Interactions, Agents and Materialities

An Overview From Southern South America

Springer

Editors
Romina Frontini
Departamento de Humanidades
Universidad Nacional del Sur
Consejo Nacional de Investigaciones Científicas y Técnicas (CONICET)
Bahía Blanca, Buenos Aires, Argentina

Paula Escosteguy
Universidad de Buenos Aires, Facultad de Filosofía y Letras, Instituto de Arqueología
Consejo Nacional de Investigaciones Científicas y Técnicas (CONICET)
Buenos Aires, Argentina

Mónica Salemme
Centro Austral de Investigaciones Científicas (CADIC) - Consejo Nacional de Investigaciones Científicas y Técnicas (CONICET)
Ushuaia, Tierra del Fuego, Argentina

ISSN 2366-3421 ISSN 2366-343X (electronic)
The Latin American Studies Book Series
ISBN 978-3-031-97494-6 ISBN 978-3-031-97495-3 (eBook)
https://doi.org/10.1007/978-3-031-97495-3

This Springer imprint is published by the registered company Springer Nature Switzerland AG
The registered company address is: Gewerbestrasse 11, 6330 Cham, Switzerland

Preface

Across time and space, there have been diverse interspecific interactions in which people and non-human animals co-modeled their relationships. Animals are agency-bearing beings and connect to people under emotional relationships (Hamilakis and Overton 2013). In these associations, the degree of human–non-human familiarity is variable (Ingold 1994). The use of animals as food is one of the human–animal interactions, and probably the one that archaeologists have more and extensively studied, traditionally from an anthropocentric perspective. On the contrary, the sensory, emotional, and memory dimensions of these interactions are less explored. The idea of this book was born with the purpose of presenting research that focuses on some topics less explored, at least in southern South America.

To encourage several other viewpoints, this volume is composed of selected papers presented and discussed during the symposium *Animals and People: Interactions, Agents and Materialities,* held during the XXI Congreso Nacional de Arqueología Argentina, that took place in Corrientes city, Argentina, in July 2023. This symposium aimed to generate a space for reflection on the interactions between non-human animals and people over the years, considering the multiplicity of aspects involved in this relationship and the diversity of the resulting materialities. During the event, 17 papers were presented, dealing with topics such as synanthropism and mutual nurturing, non-human animals depicted in rock and vessel art, historical narratives from *loberos*, historical use of non-human animal subproducts, and the social role of canids in hunter–gatherer societies, among the most represented.

This book joins some of those contributions; it contains relevant issues to comprehensively understand the complexity of the ways of interaction and the material derivatives from them. Through eight chapters, the following topics are assessed: human–camelid interactions over time comprising several ways of approach: hunting, pastoralism, domestication, and current management; human–felid interaction; the role of non-human animals in the construction of social identity; the configuration of spatial interaction networks at various scales, as well as theoretical issues from a relational approach. These topics are explored through many different materialities and perspectives, including zooarchaeological data, stable isotope information, rock and mobile art, current animal management, hides and fleece, and ethnographic information.

Thus, this compilation contains a brief overview of the state of the art on these topics in southern South America, including Northern Chile, Northwestern and Central Argentina, Pampas, and Patagonia regions. The interdisciplinary approaches and perspectives and the novel theoretical interpretations are highlighted through different case studies. Finally, we hope this book becomes a reference text for archaeologists, anthropologists, and ethnologists who deal with the interplay among people and animals.

About this Book

Since camelids are the species most considered, the eight chapters involved in the volume are ordered following three main criteria: the first two chapters have to do with an ethnographical and a geographical criterion, ordered from north to south. The following four chapters are mainly devoted to camelids and landscape, but also to the management of those animals. Carnivores are the other big group usually treated in terms of wilderness and/or domestication, though in this compilation only one chapter is devoted to the relationship between felids and people during the pre-hispanic occupations close to the Río de la Plata littoral.

The chapters are preceded by an overview of theoretical considerations about non-human and human relationships. "Non-human Animals and People in Southern South America: An Overview" is presented by Romina Frontini, Paula Escosteguy, and Mónica Salemme; by following several authors, they emphasize some examples—though it is not an exhaustive nor definitive survey—from South America.

The second chapter "*Uywaña* and *Sarawi*. Relational Approach to Camelids and Humans (Tarapacá, Northern Chile)" is authored by Francisca Urrutia Lorenzini and Josefina González Andreu. They develop ethnographic studies focusing on herding practices in Isluga highland communities (Chile) and livestock management in neighboring Precordillera agricultural communities; they establish connections with archaeological records of human–camelid relations in the Atacama Desert. Adopting a perspective that challenges conventional understandings of domestication as human dominance over nature, their approach integrates Andean ontologies and epistemologies that emphasize the multiplicity and continuing transformation of Andean worlds.

From a current perspective and through a zooarchaeological analysis of zoomorphic iconography registered in bone technology, pottery, and rock art, combined with ethnohistorical and ethnozoological approaches, Erica Colqui, Andrea Recalde, and Daniela Tamburini present "Dialogues Between Animals and People Over Time: An Interdisciplinary Perspective on the Sierras of Córdoba, Argentina"; they explore a local history in two existing rural communities located in Central Argentina from the local ecological knowledge that interacts between inherited worlds and the reinterpreted present ones.

In the chapter "Fleece and Leather Analysis in Human and Animal Archaeological Assemblages: Southern Puna of Argentina (ca. 50 BC to AD 1800)," Andrés Romano, Sara M. L. López Campeny, Sabrina Donadio, and Florencia Figuerero Minetti present an interesting point of view about camelid and human relationships through the study of materialities of difficult preservation like leathers or animal fibers, but more or less frequent in arid zones like the Puna or Andean Altiplano. They study a lapse of almost two centuries, before and after the European contact, and consider productive and residential spaces, exchange trips to far places, and ritual practices among other topics.

Eugenia Naharro, Jorge Martínez, Andrés Izeta, and Nurit Oliszewski in their chapter "Living in Quebrada de los Corrales: Camelids, Humans and Landscape within the First Village Settlements in El Infiernillo (Tucumán, Argentina)" introduce the reader into a world of occupants of dwelling and herders that interact in productive spaces and grazing areas that drive to a typical territorialization landscape in an area of ravines in Northwestern Argentina.

Further north, Hernán Muscio, Juan Pablo Orsi, and Guido Carballo worked in two different high-altitude ravines. In their chapter "Humans and Prey. Holocene Hunter Occupation of High-Altitude Spaces in the Puna of Salta, Argentina," the authors explore spatial and chronological patterns of hunting behavior of camelids since at least the end of the Early Holocene. They conclude that hunter groups lived on camelids in a changing environment of increasing aridity and spatial heterogeneity.

In southern Patagonia, the main prey has been the guanaco since the end of the Pleistocene and throughout the Holocene. Juan Dellepiane, Josefina Flores Coni, Agustín Agnolin, Gisela Cassiodoro, Rafael Goñi, Francisco Guichón, and Anahí Re discuss different types of access to individual or groups of camelids in the chapter "Communal, Collective, and Massive Guanaco Hunts: The Regional Archaeological Contribution to Its Discussion in Southern Patagonia." The authors gather information using different lines of analysis that allow them to propose the practice of collective, communal, and/or massive hunts before European contact. Besides, the authors use ecological, biological, and social information to support their hypothesis.

Fernanda Day Pilaría devotes her study to seeing the relationship of felids and humans in the littoral of Río de La Plata, in the Pampas region. Her chapter, entitled "Pumas Between the Forest and the River, Between the Domestic and the Symbolic: Prehispanic Record of Puma Concolor in the Site Las Marías (Magdalena, Buenos Aires, Argentina)" proposes that the relationship between these felids and prehispanic human societies of the riverine littoral area of Buenos Aires province involved both domestic and symbolic experiences.

We hope the journey through this book stimulates zooarchaeologists to encourage the comprehension of the record of different materialities under new theoretical approaches.

Acknowledgment

We especially thank the contributors, reviewers, and Dr. Jorge Rabassa, Series Editor, for trusting in our proposal.

References

Hamilakis H, Overton N (2013) A multi-species archaeology. Archaeol Dialogues 20:159–173. https://doi.org/10.1017/S1380203813000214

Ingold T (1994) [1988] What's an animal? Routledge, London

Bahía Blanca, Buenos Aires, Argentina — Romina Frontini
Buenos Aires, Argentina — Paula Escosteguy
Ushuaia, Tierra del Fuego, Argentina — Mónica Salemme

List of Reviewers

ÁLVAREZ, María Clara
CONICET—Grupo de Estudios Ambientales. Instituto de Matemática Aplicada. Universidad Nacional de San Luis, Argentina
BUC, Natacha
CONICET—Facultad de Filosofía y Letras, Universidad de Buenos Aires—Instituto Municipal de Investigaciones Antropológicas de Chivilcoy, Argentina
BELOTTI LÓPEZ DE MEDINA, Carlos
CONICET—Instituto de las Culturas, Museo Etnográfico J. B. Ambrosetti, Facultad de Filosofía y Letras, Universidad de Buenos Aires, Argentina
DRANSART, Penélope
University of Aberdeen, Scotland
LABARCA, Rafael
Escuela de Antropología, Facultad de Ciencias Sociales, Pontificia Universidad Católica de Chile, Chile
LÓPEZ MENDOZA, Patricio
Museo de Historia Natural y Cultural del Desierto de Atacama, Chile
MEDINA, Matías
CONICET—Facultad de Ciencias Naturales y Museo, Universidad Nacional de La Plata, Argentina
MEDRANO, Celeste
CONICET—Laboratorio de Antropología Especulativa Instituto de Ciencias Antropológicas, Universidad de Buenos Aires, Argentina
MORENO, Enrique
IRES-CONICET/Escuela de Arqueología, Universidad Nacional de Catamarca, Argentina
MIYANO, Juan Pablo
CONICET—Instituto de las Culturas, Universidad de Buenos Aires, Argentina
NIELSEN, Axel
CONICET—Instituto de Arqueología, Facultad de Filosofía y Letras, Universidad de Buenos Aires, Argentina
PAZZARELLI, Francisco

CONICET—Instituto de Antropología de Córdoba, Universidad Nacional de Córdoba, Argentina

SANTIAGO, Fernando

Centro Austral de Investigaciones Científicas (CADIC-CONICET), Laboratorio de Geomorfología y Cuaternario, Argentina

SIERPE, Víctor G.

Centro de Estudios de Historia y Arqueología, Instituto de la Patagonia, Universidad de Magallanes, Chile

WOLVERTON, Steven

Department of Geography and the Environment. University of North Texas, USA

Contents

Chapter 1
Non-human Animals and People in Southern South America: An Overview

Romina Frontini, Paula Escosteguy, and Mónica Salemme

1.1 Introduction

The human-non-human-animal relationships across time and space is a research topic in many social and humanities disciplines, including Philosophy, History, Sociology, Ethnography, Ethnobiology, Anthropology, and Archaeology (Ingold 1994 [1988]). As an example, in Art History the links between non-human and people are considered by analyzing the pieces of art throughout human history. The representation of animals is highlighted as one of the most frequent motifs in Paleolithic art in Europe. Indeed, one of the earliest depictions is the so-called lion-man figurine, recovered in Hohlenstein-Stadel, Germany. This piece of art carved in ivory was dated as early as *ca.* 30,000 years BP. It represents a lion head and a human body; it was interpreted both as a humanized lion and as a human descending from a lion (Mithen 1996). This figurine, together with the realistic non-human animal paintings in the Upper Paleolithic rock art, constitutes early materialities of

R. Frontini (✉)
Consejo Nacional de Investigaciones Científicas y Técnicas (CONICET) - Departamento de Humanidades, Universidad Nacional del Sur,
Bahía Blanca, Argentina

P. Escosteguy (✉)
Consejo Nacional de Investigaciones Científicas y Técnicas - (CONICET) Instituto de Arqueología, Facultad de Filosofía y Letras, Universidad de Buenos Aires, Buenos Aires, Argentina
e-mail: pdescosteguy@uba.ar

M. Salemme (✉)
Centro Austral de Investigaciones Científicas (CADIC) - Consejo Nacional de Investigaciones Científicas y Técnicas (CONICET), and Universidad Nacional de Tierra del Fuego,
Ushuaia, Argentina
e-mail: msalemme@conicet.gov.ar

R. Frontini et al. (eds.), *Non-human Animals and People: Interactions, Agents and Materialities*, The Latin American Studies Book Series,
https://doi.org/10.1007/978-3-031-97495-3_1

the past non-human-human relationships (Sanchidrián 2001; Mithen 1996). Furthermore, in the Middle Ages, animal images were used to express the world-wide visions strongly pushed by the ideas of Christianism prevailing at that time, as well as those rooted in previous and external traditions to those religious ideas (Museu Episcopal de Vic 2023).

The question *What's an animal?* has a long tradition in Anthropological research. In the seminal book from Ingold (1994 [1988]), derived from discussion presentations at the World Archaeology Congress in 1986, it was proven that this question cannot be solved by a singular theoretical frame or paradigm (Ingold 1994 [1988]: 15). Multiple perspectives involved in the understanding of the human-non-human-animal relationships derived into a critical approach to the perception of human uniqueness, considering that the comprehension of the interplays between human and non-human animals in different societies relates to the profound definition of what is to be human (Ingold 1994 [1988]; Segata and Lewgoy 2016). The manner in which we think about non-human animals and the questions we ask about their role in past societies implies the implementation of models of thinking and perception linked to our own realities. The choice of the title *¿Qué es un animal*? (What's an animal?), for the book edited by Medrano and Vander Velden (2020), is a tribute to the tradition of anthropological reflection on this subject and also highlights the current relevance of this question in Anthropological research including study cases from Brazil, Paraguay, Peru, Argentina, and Chile. In the prologue, Tim Ingold (2020) affirms the need to approach these issues by listening carefully to South American indigenous groups, calling for situated anthropologies.

In this regard, Laguens (2023) proposes understanding animality as the outcome of a multiplicity of relationships rather than an inherent quality of beings. This author also emphasizes the difficulty of defining the boundaries of what is an animal and what is human, even in our society (Laguens 2023:13). As a result, categories such as "animal" or "animality" can become so broad that they encompass all sorts of entities, obscuring the multiplicity of existing dimensions and relationships. In addition, Ingold (2000:9) argues that human life is not separate from nature but rather an inherent part of an ongoing process within the organic world, shaped by the ways of acting and the interactions and interwoven relationships within the environment.

The binary interpretation of human-non-human-animal relationships considered as opposition of culture vs. nature has been also confronted within Anthropology, by the so-called ontological turn, developed around the very beginning of the twenty-first century (Thomas 2002; Haber 2004; Muscio 2019). The developments from the Amazonian Anthropology are of particular interest since they propose theoretical tools raised on the basis of the study of South American native communities (Århem 1996; Viveiros de Castro 2004; Fausto 2007; Descola 2014). These different theoretical approaches consider that the set off of the human domain from the other entities' world is proper of the Western way of thinking, while non-Western societies usually make other distinctions. Then, the understanding of societies is based on how society entities define themselves within the proper community and relate to each other (Tola and dos Santos 2020). Within the framework of these theories, different concepts such as *animism* (Descola 2014), *eco-cosmology* (Århem

1996) and *perspectivism* (Viveiros de Castro 2004) have been developed. The *animism* proposed by Descola (2014) states that non-human animals and other entities are clusters of dynamic qualities and relationships that constitute "ways of making worlds," closely anchored in practices and actions in inter-agency links. The *eco-cosmology* approach followed by Århem (1996) states that there exists an integrated order, a community of beings, in which society and nature are involved equally. *Perspectivism,* on the other hand, arose from the study of the mythology of Amerindian groups, which posits the existence of an initial stage in which non-human animals and people were undifferentiated. Thus, the essence of people and non-human animals is the same while the specificity of different bodies gives rise to different perspectives; this implies distinct ways of inhabiting worlds, related to capacities and anatomical devices (Viveiros de Castro 2004).

Based on the above-mentioned ideas, we can state that nowadays there is a consensus that non-human animals are agency-bearing beings and that they have the capacity to bond in emotional relationships with people (Hamilakis and Overton 2013). In these various associations, the boundary between human and non-human is permeable according to the degree of human-non-human-animal familiarity (Ingold 1994 [1988]). In different societies, species acquire social value, replacing the Western view of humanity-animality opposition (sensu Carman 2018).

In Archaeology, animals have often been "silenced" with their agency overlooked in favor of a predominantly human-centered perspective (Ingold 1994 [1988]; Muscio 2019; Laguens 2023). This situation could be reconsidered by acknowledging non-human animals as integral members of the communities that shape human societies. In doing so, animals would be recognized for their agency, physical presence, role in relationships, and power (Nielsen and Maryañski 2018).

Across time and space, there have been diverse interactions in which people and non-human animals co-modeled their bonds. Traditional interpretations have been centered on the economic use of animals, mainly related to subsistence practices (Binford 1978; Gifford-González 2018; Hill 2013; Boyd 2017). The use of animal products by people for food purposes is one of the links that has been extensively addressed by Zooarchaeology. More recently, under the influence of ethnographic and ethnoarchaeological studies and the post-processual perspective, greater emphasis has been placed on social, symbolic, and ritual aspects. Consequently, also the transformation of non-human animals into food began to be analyzed concerning the social significance of commensality, taboos, identities, and other non-utilitarian factors (Russel 2012; Boyd 2017). In addition, eating and feasting are understood as a key moment of the interspecific bond, integrating various sensory, emotional, and memory dimensions (Le Breton 2006).

Furthermore, non-human animals play a societal role linked to group cohesion through memory or identity issues that can be assessed through the studies of faunal remains in archaeological assemblages. They comprehend not only food substances but also other corporeal elements, such as bones, feathers, teeth, and skins. These materials can be transformed into artifacts, trophies, masks, personal ornaments, and funerary trousseau. Meanwhile, the final treatment of animal remains reflects their social role in life (Hamilakis and Overton 2013). Moreover, several studies have demonstrated another aspect of the human-non-human interrelationship when an individual whose species continues to be habitual prey turns into a companion species (sensu Haraway 2003). In these cases, their status is changed and even some of their traits are humanized. At the farthest end of this relationship, certain non-human animals attain the status of subjects, a bond of affection is established and they become considered relatives (Bugallo and Tomasi 2012; Carman 2018; Nielsen 2024; Urrutia Lorenzini and González Andreu this volume; Colqui et al. this volume). A clear example is the case of female pet keepers from Amazonian groups, who nurse baby monkeys alongside their own infants (Erikson 2000; Politis 2007). This special relationship as social companion with non-human animals that are usually obtained for food purposes is interpreted as a way of compensating the destruction caused by hunting (Erikson 2000; Escosteguy 2014).

In other cases, the interrelationships are visible through the representations of animals in rock or mobiliary art (e.g., Gallardo 2018; Aschero and Schneier 2023; Yacobaccio et al. 2023) including the use of animal subproducts as blank, i.e., eggshells for decoration (Carden et al. 2024), bones for manufacturing or decorating artifacts or ornaments (Fiore 2011; Buc 2019) or gastropods used as personal adornments or containers for hallucinogens or pigments (Horta Tricallotis et al. 2022).

In Latin America, by the beginning of the twenty-first century, the study of non-human animals was mainly devoted to the zooarchaeological contexts analyzing the assemblages from a subsistence viewpoint; nevertheless, the agendas began to focus on the role of non-human species in the symbolic, political, and social issues (Mengoni Goñalons et al. 2010). As a consequence, in the last two decades, there has been a growth in academic events and publications that dealt with novel approaches towards the relationships between humans and other non-human animals (e.g., Ramos Roca and Corona-M. 2017; Corona-M. 2019, among many others). Some recent examples are: the symposium *Más allá de la representación: enfoques teóricos y metodológicos para el estudio de representaciones visuales de plantas y animales en Arqueología* (Beyond representation: theoretical and methodological approaches for the study of visual representation of plants and animals in Archaeology), leaded by Persis B. Clarkson and Daniela Valenzuela, during the XX Congreso Nacional de Arqueología Chilena held in Concepción, Chile (2015); the symposium *Entre la necesidad y el tabú. Interacciones económicas, sociales y simbólicas de humanos y animales* (Between necessity and taboo. Economic, social, and symbolic interactions of humans and animals), chaired by Laura Miotti and Eloisa García Añino, held at the IV Congreso Nacional de Zooarqueología Argentina, Ushuaia (2016); the Dossier *De océano a océano. Múltiples miradas sobre las relaciones entre humanos y animales en los Neotrópicos* (From ocean to ocean, multiple views on human-animal relations in the Neotropics) published by Muñoz et al. (2019) in the journal *Archaeofauna* (Volume 28), as a result of the meeting from the Neotropical Zooarchaeology Working Group-International Council of Archaeozoology held in San José de Mayo, Uruguay (2017); the symposium *Las arqueofaunas y sus narrativas: del Ethos cazador-recolector al pastoril* (The archaeofaunas and their narratives: from the hunter-gatherer to the herder ethos), led by Laura Miotti and Mónica Salemme and held at the III Argentine Workshop on Ethnobiology and Society (2021); the symposium *Ni Cultura, Ni Naturaleza. Ni Pasado, Ni Presente. Acercamientos Relacionales a Mundos,*

Paisajes y sus Historicidades (Neither Culture nor Nature. Neither Past nor Present. Relational Approaches to Worlds and Landscapes and their Historicities), led by Magdalena García and Andrés Troncoso and held at the XXII Congreso Nacional de Arqueología Chilena (2021).

This chapter reflects an overview of the ways in which the interactions between non-human animals and people have been assessed from a Zooarchaeological perspective. The key works highlighting theoretical frameworks in relation to selected topics are offered. Due to the enormous production of knowledge, we will focus on

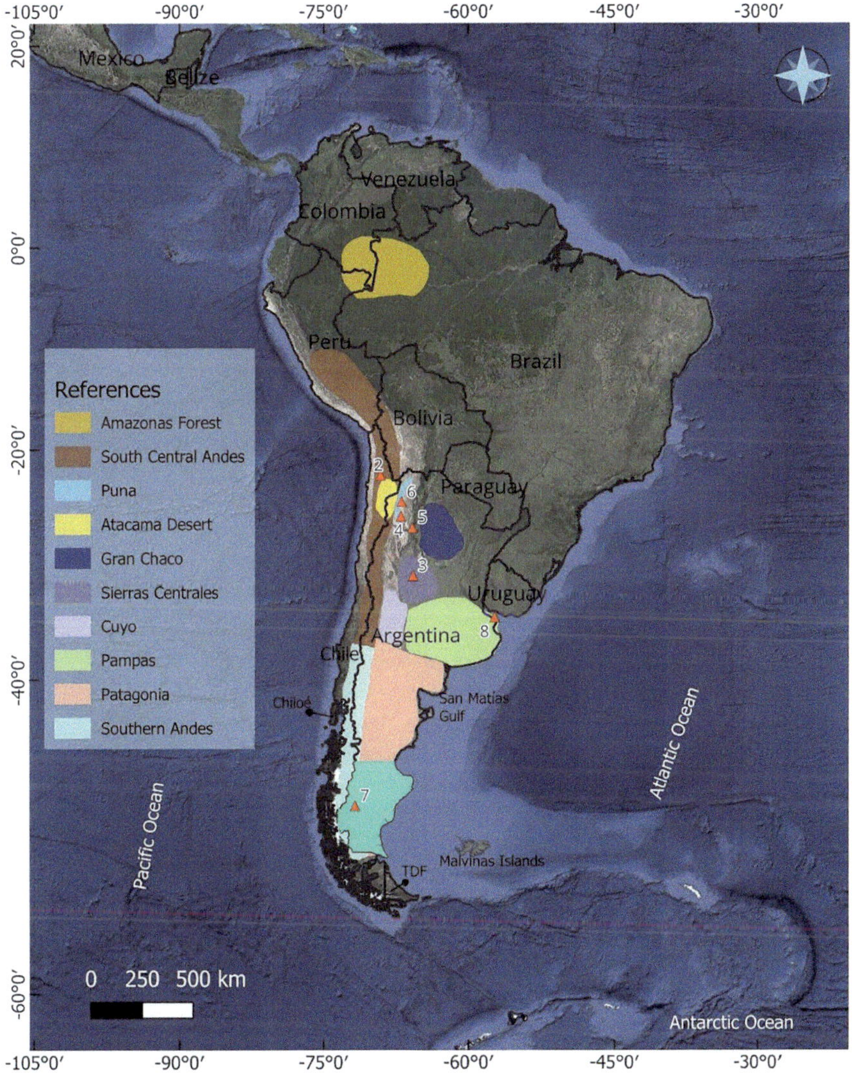

Fig. 1.1 Latin America and the Southern Cone of South America. The polygons indicate the areas mentioned in the text and the colored triangles with a number indicate the location of the sites/areas developed in each respective chapter

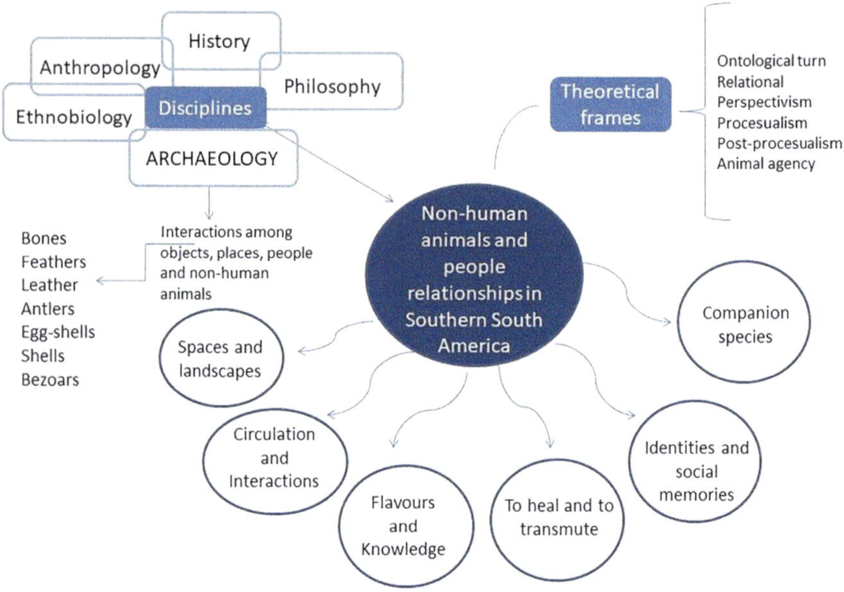

Fig. 1.2 Outline plan of the conceptual and thematic issues assessed in this chapter

southern South America, i.e., Argentina, Chile, Uruguay, and Paraguay (Fig. 1.1), even though it is not an exhaustive analysis. We additionally present some examples of other countries of South America, including Brazil, Bolivia, and Colombia, even from Central America. In the first section, a comprehensive overview of the non-human animals usually studied and a critical discussion on the use of the "species" concept are provided. In the second section, we focus on selected emergent topics and their development; for this purpose, we have elaborated major categories for grouping them and to identify the main trends (Fig. 1.2). The high diversity of subjects as well as the geographical extension prevented us from discussing the productions in-depth, and we have included essential works and their achievements. Thirdly, we reflect on the materialities that were studied within the above-mentioned issues and their links with other disciplines. Finally, a critical consideration of the theoretical issues, the materialities, the models, and the limitations of these approaches is undertaken.

1.2 Approaches to Understanding the Relationships Between Non-human Animals and People

1.2.1 Non-human Animals

Regarding non-human animal groups, there is a current debate on the suitable use of the concept of species as Biology has defined it. The labeling of non-human animals into species is seen by different authors as an anthropocentric classificatory system that is external to the specific social organizations of human groups under study (Fudge 2000; Ingold 2013; Laguens 2023). This reflection agrees with the many different roles and engagements that a non-human animal assumes or performs in a given society. Although we consider this to be a valid approach, as a practical decision, throughout the chapter we mention the animal groups as they appear in the cited original publications.

The interactions between mammals and humans are one of the most frequently analyzed. This group comprehends diverse continental but also marine species. The Artiodactyls, particularly wild and domesticated camelids are the most relevant in almost all the highlands and plains of the Southern Cone (i.e., Miotti and Marchionni 2014; Leibowicz et al. 2015; Miguez et al. 2017; Martel 2018; Cartajena et al. 2019; Lema 2019; Abbona et al. 2020; Cohen and Martinez 2020; Medina and Rivero 2020; Casanova Menéndez 2021; Moreno et al. 2021; Ottalagano 2021; del Papa et al. 2022; Erramouspe 2022; Otero and Fuchs 2022; Capriles et al. 2023; Grant and Vilá 2023; Manzano-García et al. 2023; Politis 2023; Nielsen 2024; Pazzarelli and Lema 2024, just to mention some key papers from different areas). Usually, their role is analyzed from the subsistence up to the symbolic function in the social organization. In this volume, at least four chapters considered camelids as the axis of the non-human-human relationships' studying different materialities (see Urrutia Lorenzini and González Andreu this volume; Romano et al. this volume; Muscio et al. this volume; and Dellepiane et al. this volume) while others, besides, assess these relationships between camelids and landscape as a total of daily life (see Naharro et al. this volume; Muscio et al. this volume; and Dellepiane et al. this volume). Camelids are also involved through the mobiliary art, as they appear incised or painted in domestic or sacred ceramic pots; moreover, in some cases, anthropozoomorphic vessels (funerary urns) have been interpreted as beings embodying the territory (Zotta and Acuto 2023). Very often, in the rock art, isolated animals or groups of stylized quadrupeds are interpreted as camelids (Gallardo and Yacobaccio 2005; Aschero et al. 2009; Aschero and Isasmendi 2018; Gallardo 2018; Martel 2018); likewise, they are thought as memory archives in the imaginary of returning a place, besides as territorial markers in secular and ritual times, for landscapes with an excellent supply of different resources (Aschero and Schneier 2023).

Carnivores have been related to humans from the past. The canids are a group recurrently present in archaeological assemblages concerning different contextual interpretations; particularly, dogs have played a relevant and multifaceted or multidimensional role from the past to the present (e.g., Bonomo 2006; Prates et al. 2010;

Prates 2014; Weinberg 2019; Acosta and Loponte 2023; Apesteguía 2023; González Venanzi et al. 2023a; Mendoza España 2023). Along the Southern Andes, from 2000 years ago until the Inka Period, canids (i.e., dogs and foxes) have been depicted on stone sculptures, vessels, and metal artifacts, as well as being part of offerings and funerary goods (Mendoza España 2022). Canines of canids are also frequently used to manufacture beads for necklaces or bracelets that are recovered in human burials from the Arroyo Seco Site 2 (Interserrana area, Argentina; Laporte 2014).

From an anthropological perspective, in the South Andean region dogs are perceived as bridges that connect and accompany both humans and non-humans throughout life and in their transition to the world of the dead (Weinberg 2019; Mendoza España 2022). As a result, the human-dog relationship becomes deeply intertwined, shaping and transforming the experience of being with others (Weinberg 2019). It is also frequent that dogs help in hunting other animals such as rodents, suids, rheids, tegus, among others (Álvarez and Heidler 2019; Giardina et al. 2021; Escosteguy et al. 2023; Manzano-García et al. 2023); moreover, along with other roles and functions, during Late Holocene in Central Patagonia, dogs were used by hunter-gatherers for load transport (Zilio et al. 2024). The most frequent mentions worldwide are those of canids associated as funerary goods in human burials, indicating not only the companion but also the hierarchical position of the dead person as suggested by different authors (Petrocelli 1975; Berón 2010; Prates 2014) for different archaeological sites from Pampas and Patagonia regions. Recently, through the analysis of the dental tartar, the role of canids among hunter-gatherers was considered in terms of synanthropic link during the Late Holocene in the Pampas region (Argentina) (González Venanzi et al. 2023b).

Other carnivores are related to humans in the people's imaginary; in different regions of South America, felids are frequently represented in mobiliary art (Capriles et al. 2023), as it is reflected in pottery vessels of different cultures from Mesoamerica and the Andean area (González 1961; Dillehay and Kaulicke 1984–1985), in objects made of their hides (Capriles 2002), in engraved motifs from religious architecture or in their bones or canines associated with burials. Also, in the iconography of various areas (e.g., Andean area and Sierras Centrales) expressions relating to the human/feline metamorphosis are frequent; some of them refer to the use of their skins, mainly by warriors (Pastor et al. 2015; Gordillo 2018).

Felids are animals that provoke fear; at the same time, they play a remarkable role in the social life of communities living in the plains or in the mountains, in the forest or in arid landscapes. In spite of the fact that they are carnivores and kill other animals for feeding, they represent the life symbol since they take other lives to continue living; it reflects fertility and renaissance for aboriginal people and present farmers. In the case of another species of felid, the pumas (*Puma concolor*), different reports of several areas where it inhabits, considered they are involved in the symbolic spheres (like trophies) but also as feed (the meat) or medicine (the fat) (see Day Pilaría this this volume).

Regarding another large feline species, jaguar's (*Panthera onca*) remains have been recovered, primarily related to burials from three archaeological sites in western Uruguayan territory. These occupations were probably related to the

Goya-Malabrigo complex. It should be noted that in one of the sites, possible cut marks were identified, while in another, stains that probably remind the skin of the jaguar and could be associated with pigments (iron oxide), while at the Cañada Saldaña site, perforated canines of this species were identified and interpreted as hanging ornaments. One of them also presents decorations (Ferrari et al. 2023). In the same region but on the present Argentine territory, the jaguar was modeled in vessel appendages (particularly, their heads) or other features as the claws or spots decorates the pottery (Bastourre 2021; Bonomo et al. 2021) and deciduous teeth from this feline displayed anthropic marks (Acosta et al. 2015). Another similar findings that point to the complex role of felids, are the pendants manufactured on teeth recovered in different sites along Northeastern Argentina (Píccoli et al. 2020). It is also known the relevance of jaguar on the ideology and iconography of Andean societies (González 1961; Capriles 2002; Gordillo 2018; López et al. 2018; Coloca and López 2021; among others). Finally, throughout history, written sources record the fear that people had of this feline, considering it as dangerous; nevertheless, like among the Nukak in the Amazon, the use of its skins are often mentioned, sometimes involving war events or war-related issues (Politis 2007; Pastor et al. 2015; Ferrari et al. 2023). Among the Nukak, in the Colombian Amazon, jaguars are considered ancestral spirits that use the skin of those animals to participate in daily life, so those amazonian societies never eat their meat (Politis 2007).

Studies on the interaction between people and marine species, including mammals, birds, fish, and invertebrates are usually centered on the economic use and the effects of its overexploitation (Fiore and Zangrando 2006; Torres 2009; Stoessel 2015; Beovide et al. 2017; Scartascini 2017; Bas et al. 2020; Calás et al. 2024; Cruz and Ercolano 2024; Fossile et al. 2024, among many others), but also in the ideological sphere or as mobility markers (Gordillo 2021a, b; Horta Tricallotis et al. 2022). Fishes are analyzed as food and the technology for their capture is described through historical and modern knowledge (Colonese et al. 2023). The use of white shark (*Carcharodon carcharias*) teeth as pendants and probably as tools was registered in a site close to the Atlantic coast, in the Pampas region (Cione and Bonomo 2003). Cetaceans have been assessed through ethnography, ethnoarchaeology, history, and archaeology; the narratives and perceptions of whales in the historical periods and at present are evaluated by Quiroz and Carreño (2017). The relatively big size of these species related to human size and the perception as a monstrous animal is interpreted as the base of the attraction. For native people from Tierra del Fuego associated to the sea, whales, pinnipeds, and dolphins played a relevant role in their social identity and rituals beyond food, technology, and heating (Orquera and Piana 1999; Massone and Prieto 2005), while in the Atacama Desert coast (Northern Chile), a collection of cetacean bone remains, mostly skulls, was recorded on a painted rocky block, and preliminarily interpreted as an offering (Bastías Croudo et al. 2023). Moreover, some mummies of the Chinchorro culture were wrapped with fur of pinnipeds (Standen and Santoro 2004). During historical times, until the end of the nineteenth century, commercial harvesting led to the depletion of fur seals colonies in the South Shetlands islands (Pastene and Quirós 2010); this was discussed by Cruz and Salerno (2023) and interpreted as a result of extractivist practices as part of the fur seals industry in the Antarctic seas.

Bird remains are usually part of funerary goods, including bones and feathers (Belloti López de Medina 2012). They are also represented in rock and portable art (Carden 2008; Otero et al. 2017; Lema 2019; Otero et al. 2021; Ottalagano 2021; Manzano-García et al. 2023, among many others) as complete bodies or some parts of them, for example, parrot heads, footprints, tridigits (Bonomo et al. 2021). Besides, their bones are frequently used to manufacture instruments such as beads, tubes, or points (e.g., Fiore 2011; Tívoli and Salemme 2015; Pérez Jimeno and del Papa 2016). Bird feathers have a great value in Amerindian societies, based on their colors; for Goya-Malabrigo bird remains were interpreted as derived from feathers use (Bonomo et al. 2021). Another type of bird-human relationship is that of companion animals. A significant case from the Andean area occurred in a residential enclosure at the Palo Blanco site (around 600 A.D.), where the remains of a macaw (*Ara militaris*), of allochthonous origin, were recovered. Certain bone pathologies would indicate that it was hunted young and kept in captivity (Baldini and Sempé 2011). Ethnographically, it has been documented that birds are primarily captured, cared for, and admired for their gentle behavior, aesthetic appeal, and melodious singing (Manzano-García et al. 2023). Flightless birds like ostrich-like or penguins have been considered as food, but also as symbolic animals; footprints of rheids were represented as tridigits in rock art (i.e., Piedra Museo Locality, southern Patagonia, Salemme and Miotti 2022; Carden 2022) and also represented in the musical instruments used by Patagonian natives (Oliva et al. 2018). Engraved eggshells were also recovered in Argentinean Pampas and Patagonia (e.g., Fiore and Borella 2010; Carden et al. 2024). Penguins took part in the dishes of historical Antarctic expeditions when they survived a complete winter (Schillat 2007).

1.2.2 Human and Non-human Relationships through Zooarchaeology

The broad and complex relationships between human and non-human animals in the past and also in living communities can be assessed and grouped into key topics. Even though they are not fix categories and may overlap, the following can be distinguished: the role of non-human animals in the interactions and circulations of goods, in the construction of dwelled landscapes, in the healing and transmuting practices, in building and sharing identities and memories, in the social values of food, and in the relationship as companion animals. We focus on these topics with examples from different social organizations, across several geographical locations, diverse environments, and chronologies, always referred to South America.

1.2.2.1 Circulation and Interactions

The circulation of non-human animal products, together with other goods, is a common practice among social groups and its role and significance have been of interest to Anthropology and Archaeology for decades (Lazzari 2005 and references therein). The circulation of animal objects is generally used for the reinforcement of social interactions among groups and the exchange of information, but they are not always directly or exclusively related to subsistence practices, i.e., the case of guanacos hunted on occasion of the Hain ceremony among the Selk'nam of Tierra del Fuego to introduce the young males to the adult life (Chapman 1986).

Different non-human animal remains recovered distant from their natural habitat sustained the interpretation of the intentional transportation by humans, acting as facilitators of transactions. An example of this is developed in Romano et al. (this volume) studying fleeces and leathers from archaeological contexts; they are able to interpret the relationship between people and other animals as reflected through the shared productive and residential places, the ritual areas and the exchange sustained through long trips to different environments.

Among invertebrates, shells served as exchange items in various societies across different regions. Both unmodified shells and shells transformed into personal ornaments have been transported by humans along tens to hundreds of kilometers from their natural habitat revealing the circulation, interaction, and exchange networks among populations from distinct geographical regions and diverse cultural trajectories (Orquera and Gómez Otero 2007; Bonomo and Aguirre 2009; Berón 2019; Gordillo 2021a, b; Calás et al. 2024; Hammond et al. 2024). Such exchanges have been documented in South America from very early times. This is the case of the Pacific coast, in Northern Chile, where the record of malacological remains in sectors far from the coastline dates back to 12,000 years BP and continued throughout the Holocene (Hernández Castillo et al. 2025). Indeed, in this area, there was an emphasis on the transportation of species that were difficult to capture, so it was suggested that nutritional use alone did not fully account for shell transport to inland regions (Hernández Castillo et al. 2025).

The circulation of animal products in the Andes includes not only shells but also dry fish, feathers, beaks, and animal skins (Nielsen et al. 2019). As another example, the record of the same motif decorations on rheid eggshells in Northern Patagonia and Pampas Central Dunefields was interpreted as the existence of exchange networks on a macroregional scale (Carden and Martínez 2014; Carden et al. 2024). Likewise, for Historical times, written sources from voyagers and ethnographers describe that feathers were used among natives for exchange with Europeans (Salemme and Frontini 2011; Salemme and Miotti 2022, and references therein). While at present, the procurement of feathers from some species such as the rhea continues in order to commercialize them in urban areas (Álvarez and Heidler 2019; Escosteguy et al. 2023).

Other by-products obtained from non-human animals that circulated in wide exchange and commercialization networks were hides or skins. This is easier to verify in documentary and ethnohistoric sources, as in the case of the trade of mammal skins such as vicuña, coypu, capybara, foxes, seals, among others (Escosteguy 2013, 2014; Miyano and Ratto 2020; Cruz and Salerno 2023; Escosteguy et al. 2023). However, there are also records of "overo" lizard hides being obtained for sale (Escosteguy et al. 2023). From the material record, there are major biases, although the preservation of allochthonous skins such as of jaguar in archaeological sites of the Bolivian Altiplano shows that the exchange took place over a long period of time (Capriles 2002).

1.2.2.2 Spaces and Landscapes

For decades, the construction and transformation of natural space for living have been assessed through different theoretical approaches and material information (Ingold 1993 and references therein; David and Thomas 2008). The currently accepted concept of landscape is the one that defines it as the places where the interactions between communities of people and non-human entities occur; they are geographically and historically defined (Walker 2012). This current definition of landscape is then concerned with the practices and actions occurring in a certain space (David and Thomas 2008). Ingold (1993) has created the concept of "task-scape" for naming it. In this sense, the archaeology of non-human animals integrates the discussions of the landscapes understood from different perspectives, including the use of animals as indicators of past landscapes (Deschamps 2005; Mainland 2008; Miotti and Marchionni 2023); the study of the representations of animals in art (i.e., Gallardo and Yacobaccio 2005; Aschero and Schneier 2023), the assessments of the ancient hunting and capture terrestrial landscapes (Belardi et al. 2021; Franco et al. 2021; Moreno et al. 2021; Dellepiane et al. this volume; Muscio et al. this volume); the manufacture of structures for fish capture (Torres 2009; Cardillo and Scartascini 2015; Vázquez and Zangrando 2017), among many others.

Non-human animals inhabit particular environments according to their biological and ecological adaptations including their physiology, foraging behavior, and ethology (Mainland 2008). The landscape where people and non-human animals interact is related to these particular environments. Thus, people's knowledge and expertise on inhabited landscapes are essential for capturing or hunting non-human animals. Ethnoarchaeological studies on coypu (*Myocastor coypus*) chase show that *nutrieros* perfectly know how this rodent uses different water courses and how it becomes essential in hunting strategies: whereas in the lakes coypu individuals stay long, in the rivers and creeks coypus are on the move, so people use different artifacts and capturing strategies which include the location of traps (Escosteguy 2014; Escosteguy et al. 2023). Looking at the vicuña (*Vicugna vicugna*), as a grazer and obligatory drinker adapted to live in high altitudes, its capture strategies are based on its spatial distribution (Vilá 2012).

The visual representation of the same animals or motives on different blanks along various and distant spaces may demonstrate that humans and non-human animals share a way of relating to each other (Carden and León 2022). In this regard,

Bonomo et al. (2021) propose the idea of the "domestication of the landscape" (following Hodder 1990) in spite of the fact it does not exist as a true animal domestication. Their interpretation is based on the study of zoomorphic pottery appendages from pre-Hispanic assemblages from Paraná River (Argentina), where the most frequent representations are flying and aquatic non-human species as well as nocturnal animals, showing a detailed knowledge and interplay with the environment.

The archaeological landscapes were understood as ecological, social, and symbolic by Miotti and Marchionni (2023) while analyzing hunter-gatherer zooarchaeological assemblages from the Late Pleistocene to the Holocene at the Patagonian plateau. These authors integrate different lines of evidence that include taphonomy, the spatial distribution of bone remains, rock art, and ethnohistorical and ethnographic sources in order to describe those systems in which non-human animals and other beings (i.e., lakes, hills, rivers) act as sensible agents that construct and relate socially to each other into performative interactions (Miotti and Marchionni 2023: 172).

The use of natural geoforms and the construction of rock structures for observing, stalking, and capturing non-human animals constitute the materialization of human and other animal relationships within the landscape (Santiago and Salemme 2016; Dellepiane et al. this volume). In those locations, the observation, learning, and knowledge of ecological and ethological characteristics of non-human animals are essential in the link of relationships. Some examples of these constructions are corrals, fish traps, and hunting blinds. It was proposed that corrals played an important role in the process of domestication of camelids keeping them in confinement; they varied from the enclosure of a cave to function as a corral, to its construction in stone (Yacobaccio 2021). Along marine littoral sectors, the construction of rock structures for trapping fishes at intertidal sectors was registered in Tierra del Fuego, both on the Chilean and Argentinean sides (Torres 2009; Vázquez and Zangrando 2017); in Chiloé Island, Chile (Munita et al. 2004; Álvarez et al. 2008; Torres 2009); at the San Matías Gulf, Argentina (Cardillo and Scartascini 2015), and at Brazilian Atlantic coast (Colonese et al. 2015, 2023). Moreover, at an inland wetland in Belize, Central America, there is evidence of fish-trapping facilities built by hunter-gatherer-fishers that continued to be used by their Maya descendants (Harrison-Buck et al. 2024).

In different continental areas of Patagonia, diverse kinds of hunting blinds are reported by several authors (Gradín 1996; Miotti and Terranova 2015; Goñi et al. 2016; Santiago and Salemme 2016, among others) as well as hunting strategies, as it is developed by Dellepiane et al. (this volume) who discuss strategies of communal, collective and massive hunts over the main prey (guanaco) that supplied almost everything for daily life since the end of the Pleistocene along Patagonia. In the same line, strategies of landscape use in the analysis of predator-prey are exemplified by Muscio et al. (this volume) for a comprehensive and comparative study of camelid hunter's that occupied two gorges in the Argentine Puna region since the end of the Middle Holocene. In southern Cuyo area (Argentina), Giardina et al. (2021) studied strategies for hunting rheids; they participated in the practice known as *boleada* (hunting with bola stones) and reviewed how the knowledge of the landscape and the ethology of rheids must be considered for success.

The analysis of landscapes, life cycles, and rituals are understood as a total in a relational approach between people and camelids in Tarapacá altiplano and Precordillera settings (Northern Chile), as Urrutia Lorenzini and González Andreu present in this volume. In the same sense, the study of archaeozoological assemblage from an agro-pastoral village in Northwestern Argentina reveals the interaction among herders, productive spaces, and grazing areas showing several mechanisms in the process of territorialization that builds a particular landscape (see Naharro et al. this volume).

1.2.2.3 To Heal and to Transmute

The uses of non-human animal products as healing devices for neutralizing poisons, preventing diseases, protecting from evil spirits, and promoting health and fertility are described for different societies. The bezoar stones are good examples. These gastroliths formed due to the swallowing of hair, sedimentary particles, vegetable fibers, and seeds in the gastrointestinal tract of herbivores like goat, deer, sheep, antelope, horse, tapir, and guanaco. Argentinean ethnographic and folkloric records indicate that guanaco and vicuña bezoars were greatly valued by indigenous populations for their medical powers (e.g., Furlong 1912; Musters 1964; Claraz 1988; Franklin et al. 1997; Casamiquela 2000; Miguez et al. 2017; Tomassini et al. 2019; Miyano and Ratto 2020; Pelizza et al. 2024).

Historical sources describe, as well, the use of rheid products for medical purposes; i.e., the stomach pepsin was used to cure stomach diseases (see Salemme and Frontini 2011; Salemme and Miotti 2022). Other species have been involved in medical issues, both at present and in the past. For example, the fat of various animals such as felines or lizards can be used for bone and muscle ailments (Escosteguy et al. 2023; Manzano-García et al. 2023), while guanaco fiber and feet are used for children disease (ear pain) or adult conditions (facial paralysis) (Costa et al. 2022).

The representation of armadillos in hallucinogenic tablets as well as the use of armadillo bones for manufacturing musical instruments in Northwestern Argentina has been interpreted as essential components of shamanic practices in pre-Hispanic times (Horta Tricallotis et al. 2023) mainly originated in the circulation of ideas from Gran Chaco region, where armadillos are frequent in the local biome.

Among the birds, the parrots played special social roles within the native groups from Northeastern Argentina. According to ethnographic studies, they were consumed for medical purposes and also were essential in shamanic ritual and in transmutation practices (see Bonomo et al. 2021 and references therein).

1.2.2.4 Shared Identities and Social Memories

Non-human animals played a significant role in the creation, transmission, and recreation of social memories and identities. As an example, the modern use of fibers and hides tidied and located in certain places in the Argentine Southern Puna is related to ancient practices that engaged with the shaping of ancestral territories (Cohen and Martinez 2020). In a similar way, Coloca and López (2021) identified a mining-ritual landscape in the Puna of Salta (south-central Andes) where non-human animals played an important role. This is clear from the presence of over 1000 feathers of exotic and local species, some of which were part of offerings together with bones and ceramics. In addition, other symbolic indicators are the paintings (i.e., a jaguar and caravan of llamas *-Lama glama-* tied together) and the *pusno* of camelids (a semi-digested vegetal matter from the stomachs) on the cave walls. This waka mine showed continuity from the pre-Inka and Inka periods and was used as a ritual site. It was proposed that this place likely played a significant role in the movement and circulation of information, goods, and resources (López et al. 2018).

Another relevant topic is the relationships between herders and camelids, mainly llamas. For the Southern Andes, it has been assessed through different theoretical approaches and disciplines (Bugallo and Tomasi 2012; Vilá 2015; Yacobaccio et al. 2018; Nielsen 2024 and references therein). The semiotic approach proposed by Nielsen (2024) follows the theoretical concept of "companion species" elaborated by Haraway (2003) for the interpretation of this relationship. The author evidenced that domestic camelids had been emotionally and symbolically tied to herders from ca. 3000 years BP, over generations. He also assessed the Aymará concept of *Uywaña*, in coincidence with the proposal from Urrutia Lorenzini and González Andreu (this volume) in the analyses of herding communities from Tarapacá (Northern Chile). By considering the ethnographic and zooarchaeological information, these authors make a relational approach to understanding the sociohistorical trajectories of interspecies relationships in the Andes. They also show how the mobility of the herds is related to the cycles of the seasons and to the times of sowing, harvesting, and fertility, which are celebrated in festivities. These practices were identified through the archaeological record and the ethnography approach.

The persistence of the use of ancestral non-human animals' capture techniques is part of the transmission of social memory and identity practices. In the case of wild camelids, the practices recorded from pre-Hispanic times to the present day in which animals are herded into a corral for shearing and then released or for selective hunting of individuals is widely known. The corral could be made of different materials, but the collective herding technique involved a number of people walking and holding colored ribbons. This practice is regulated by the Andean cosmovision (Vilá 2014, 2015; Miyano et al. 2017; Miyano and Ratto 2020). Also, the current use of rock traps for harvesting fish on Chiloé Island (Chile) was interpreted as a way of social memory transmission (Torres 2009). In Brazil, the study of the techniques and installations of modern artisanal fishermen also made it possible to interpret the archaeological record and to establish continuities with the past (Colonese et al. 2015).

The construction of modern rural identities based on adoption of non-human animals' names for families' identification through generations is another example of the role of these entities in the construction, definition, and transmission of social identity (Colqui et al. this volume). The oral transmission of traditional knowledge in the Sierras Centrales (Argentina) together with the assessment of pre-Hispanic archaeological materialities, mainly the rock art, led to the identification of the shared memory of the communities.

1.2.2.5 Flavors and Knowledge

The selection of non-human animals for being transformed into food is arbitrary and conventional within different human groups. Humans use only a selected portion of the great diversity of edible wild resources (Le Breton 2006). The selection of ingredients and the preparation of foods structure social relationships and impacts on daily activities as well (Montón Subías 2002; Gifford-González 2018). The non-human animal substances obtained from nature only acquire the status of food after they have been involved in processing practices and transformations that are socially meaningful (Aguirre 2017). The ways in which different substances are obtained, manipulated, and transformed into food are linked to the social memory of the groups and consist of social knowledge transmitted for generations (Pazzarelli 2008). Also, culinary activities that comprehend the performance of repeated and routine domestic activities are usually carried out within the framework of traditions, normative values, and common expectations of the group. In this sense, these practices are deeply embodied in the past, in the traditions and in the history of the group, being central in the construction and transmission of collective memory (Sutton 2001; Mintz and Dubois 2002).

Particularly, Zooarchaeology has assessed these social roles of food by examining the links between foodways and ethnicity (Colasurdo and Sartori 2011); the materialities linked to cooking processes include bone fragmentation and thermoaltered remains (Frontini 2010; Castillo et al. 2018) if the consumption of fat and meat are considered (De Nigris and Mengoni Goñalons 2004), among many others.

The consumption of unusual non-human animals may show the special interactions of animals with people. This is the case of the study by Day Pilaría (this volume) in relation to the Late Holocene puma remains recovered in Las Marías site (Pampas Region, Argentina). The author infers a differential cultural consumption by analyzing the anthropic marks and skeletal pieces selected among the puma remains together with the scarce presence of this non-human animal representation in other archaeological assemblages from the region; she relates it to special meanings and values. This agrees with the interpretation given to Calera site (Pampas region, Argentina) puma remains; this is an exceptional site that shows intentional material culture arrangement derived from hunter-gatherers' ceremonies during the Late Holocene (Politis et al. 2005).

Social taboo in consuming certain species has been assessed substantially by ethnoarchaeological research. The studies performed on Amazonian native groups by Politis and Martínez (1996) and Politis and Saunders (2002) identified that some non-human animals, including tapir, deer, and jaguar were neither hunted nor consumed by any group member. But there were food taboos exclusively for women, who did not consume peccary. This mammal was hunted, processed, and consumed only by men. Researchers make a call for stating hypotheses that include foods taboo while interpreting zooarchaeological assemblages (Politis and Martínez 1996: 280).

1.2.2.6 Companion Animals

Within the framework of ethnographic and ethnoarchaeological studies, a different bond between people and different kinds of non-human animals is frequently recorded. Thus, in many South American societies, both indigenous and rural, the care of individuals of various species was observed (Erikson 2000; Politis 2007; Escosteguy 2014; Miyano and Ratto 2020; Escosteguy et al. 2023; Manzano-García et al. 2023; Colqui et al. this volume). It usually occurs when a hunter captures and kills the mother and then takes the future pets home, where they are tamed, become to be considered a member of the family; they are even given a name and take care as if they were children of the group (Serpell and Paul 1994; Erikson 2000). According to Manzano-García et al. (2023), these so-called home species, when they are offspring and are named, trigger emotional aspects in the interactions with people, which is based on companionship, affection, or the feeling of comfort in caring for another being that is often vulnerable. In consequence, these non-human animals change their status becoming inedible, and even some of their features are humanized (Erikson 2000; Politis 2007; Escosteguy 2014; Escosteguy et al. 2023).

This process of turning wild animals into companion animals can occur with different birds, mainly parrots, but also eagles, rheids, herons, and songbirds, while among mammals, monkeys, armadillos, foxes, skunks, capybaras, coypu, agoutis, sloths, coatis, pacas, vicuñas, and guanacos were recorded as pets (Erikson 2000; Politis 2007; Baldini and Sempé 2011; Escosteguy 2014; Miyano and Ratto 2020; Bonomo et al. 2021; Escosteguy et al. 2023; Manzano-García et al. 2023). For reptiles, land tortoises are also kept as pets (Manzano-García et al. 2023).

1.3 Materiality and Its Multiple Approaches

Preservational differences of non-human animals' organic materials including bones, leather, feathers, fibers among others, undoubtedly influences the type of interpretations that one can obtain from the analysis of the zooarchaeological assemblages. The conditions of temperature, humidity, soil-pH, invertebrate, and fossorial animals activities, the water and wind erosion, among many other agents and processes, from the depositional environment directly influence the preservation of organic material (Lyman 1994; Fernández-Jalvo and Andrews 2016). Also, recovery techniques and laboratory processing and conservation strategies are key topics for a complex and diverse materiality preservation. Thus, it is highlighted that a thorough understanding of the relationships between non-human animals and humans, with their complexity, can be approached with a higher level of resolution in those environmental contexts that are conducive to organic material preservation.

Besides the preservational bias, the analyses exclusively centered on bone and other biological materials generally induce a limited interpretation and understanding, (Laguens 2023). Therefore, assessing the complex relationships between people and non-human animals in the past requires the development of research and reflection strategies that integrate diverse lines of evidence as well as theoretical perspectives that move away from the Western dualistic viewpoint. In Zooarchaeology, the research has usually been centered on the analysis of bone remains and other biological materials whereas non-material information hasn't been fully integrated into the interpretations. However, the comprehension of the past should be essentially transdisciplinary, through the study of different informative sources, such as the archaeological record, ethnographic data, written sources, among others (Corona-M. 2019).

In this regard, the combined analyses that integrate different analytical tools and diverse materialities are promising in the interpretation of the non-human animals and people complex relationships. The techniques that allow us to discuss non-human animals' past diets are also substantial tools for assessing companion and care links. They include stable isotope analyses and dental tartar content that led to defining diverse camelid-human relationship (i.e., Yacobaccio et al. 2018; Sperling 2023) as well as synanthropic human-canid relations (González Venanzi et al. 2023b). Pollen analysis from the fibers, leathers, and hairs is also a good proxy to analyze the mobility of camelid herds together with shepherds through different environments (Romano et al. this volume).

It is not a new fact that the integration of material information from archaeological assemblages with that gathered through ethnographic, ethnohistorical, ethnobiological, and ethnoarchaeological sources results in a better understanding of past linkages. Ethnoarchaeological projects in Southern South America have been performed within native communities from Amazonas (Borrero and Yacobaccio 1989; Politis 2007) and also with modern groups that perform traditional activities including coypu capture and processing (Escosteguy 2014), greater rhea hunting (Giardina et al. 2021); herding (Nielsen 2024), among others.

Considering other sources of information, the folk knowledge and the written historical records from travelers, naturalists, and militars from the 16th to 19th has been used as heuristic tools to assess different aspects of non-human-animals-human relationships (Prates 2009; Salemme and Frontini 2011; Bugallo and Tomasi 2012; Frontini 2012; Escosteguy 2013; Miyano and Ratto 2020; Escosteguy et al. 2023; Manzano-García et al. 2023).

1.4 Thoughts Towards the Future

Archaeology can contribute substantially to the study of human-non-human relationships. Following Laguens (2023), we propose assessing the archaeological record as an integral product of the interactions between people, animals, and objects. Archaeology provides time depth and a diachronic perspective; it also analyzes different social groups across diverse environments (Amstrong Oma and Birke 2013). In this overview, we have identified multiple approaches towards a transdisciplinary perspective, especially between Zooarchaeology and Ethnography. Nevertheless, we find that there is a lack of systematic projects that integrate and generate fluid dialogue between researchers from other humanities. Animal studies need the construction of networks that allow generating innovative questions about human and non-human animal relations.

Another feature that we have identified is that the relational ontologies are almost restricted to the understanding of past complex societies; and to those societies where past ancestry is recognized by contemporary groups; on the contrary, it is not so often reflected in the hunter-gatherers' studies. Social memories and transmission through action remains not only a heuristic tool, but knowledge that helps our Western interpretations to open minds to other manners of being in the world. Besides, from the perspective of relational ontologies, there is a call for archaeological research to interpret materialities in its links among them.

This chapter intends to offer a general overview, without claiming to be exhaustive, of the deep diversity of archaeological studies that goes beyond the traditional economic interpretation of animal remains. Undoubtedly, it is only a small window that allows us to enrich, under an overlook of a few examples, the view of the diverse and complex role of non-human animals with people, improving our understanding of it.

Acknowledgments We are deeply grateful to Ramiro López (CADIC-CONICET), who drew the map in Fig. 1.1. We also thank the two reviewers for their constructive comments. The research was funded by grants PIP 2020 0324 to P.E., UBACyT N° 20020220200019BA to Dr. M. I. González (Universidad de Buenos Aires), PICT 2020-0381 to Dr. M. Frère, M.I. González, C. Bayón, and two of us (P.E. and R.F.); PGI Secyt UNS 24/I307 to R.F.; PIP 0405-19 to one of us (M.S.). Finally, we thank all the non-human animals that make life enjoyable. The authors are solely responsible for the arguments discussed in this paper.

Competing Interest Declaration The authors have no conflict of interest to declare that are relevant to the content of this chapter.

References

Abbona CC, Neme G, Johnson J, Kim T, Gil AF, Wolverton S (2020) Were domestic camelids present on the prehispanic south American agricultural frontier? An ancient DNA study. PLoS One 15(11):e0240474. https://doi.org/10.1371/journal.pone.0240474

Acosta A, Loponte D (2023) Los perros precolombinos del extremo meridional de la cuenca del Paraná-Plata. In: Apesteguía S, Álvarez SM (eds) Perros y otros cánidos de las américas. Origen, evolución e historia natural. Vazquez Mazzini Editores, Fundación Félix de Azara, pp 309–339

Acosta A, Buc N, Ramírez M, Prevosti F, Loponte D (2015) Producción y uso de objetos ornamentales elaborados sobre dientes de carnívoros en contextos arqueológicos del Humedal del Paraná Inferior. Revista del Museo de Antropología 8(2):33–46

Aguirre P (2017) Una historia social de la comida. Lugar Editorial, EDUNLA Cooperativa, Buenos Aires

Álvarez MC, Heidler G (2019) Conocimiento tradicional y sus implicancias para la caza de jabalí y ñandú en comunidades campesinas del sur de la provincia de San Luis, Argentina. Rev Etnobiol 17(1):5–17

Álvarez R, Munita D, Fredes J, Mera R (2008) Corrales de pesca en Chiloé. Consejo Nacional de la Cultura y las Artes, Valdivia, Chile

Amstrong Oma K, Birke L (2013) Guest editors' introduction: archaeology and human-animal studies. Soci Anim 21:113–119

Apesteguía S (2023) Perros del sur de Sudamérica: en casas de madera y cuero. In: Apesteguía S, Álvarez SM (eds) Perros y otros cánidos de las Américas. Origen, evolución e historia natural. Vazquez Mazzini Editores, Fundación Félix de Azara, pp 285–307

Århem K (1996) The cosmic food web: human-nature relatedness in the Northwest Amazon. In: Descola P, Pálsson G (eds) Nature and society, anthropological perspectives. Routledge, pp 185–204

Aschero CA, Isasmendi MV (2018) Arte rupestre y demarcación territorial: el caso del grupo estilístico B1 en el área Río Pinturas (Santa Cruz, Argentina). Rev Mus La Plata 3:112–131

Aschero CA, Schneier P (2023) Ese asunto de volver: producción y significación en el arte rupestre de Cueva de las Manos, Patagonia meridional andina. Boletín Del Museo Chileno De Arte Precolombino 28(2):91–110. https://doi.org/10.56522/BMCHAP.0050020280003

Aschero CA, Martel AR, López Campeny SM (2009) El sonido del agua. Arte rupestre y actividades productivas. El caso de Antofagasta de la Sierra, Noroeste Argentino. In: Sepúlveda M, Briones L, Chacama J (eds) Crónicas sobre la piedra. Arte Rupestre de Las Américas. Universidad de Tarapacá, pp 257–270

Baldini M, Sempé C (2011) Evidencias funerarias de relaciones de larga distancia en el Valle de Hualfín. In: de Buenader ES, de Montiel SM (eds) El hombre, el medio y sus relaciones. Oficina de Publicaciones de la Universidad Nacional de Catamarca, San Fernando del Valle de Catamarca, pp 33–43. https://naturalis.fcnym.unlp.edu.ar/handle/628872547/39120

Bas M, Salemme M, Green EJ, Santiago F, Spelle C, Álvarez M, Godino BI, Cardona LI (2020) Predicting habitat use by the Argentine hake Merluccius hubbsi in a warmer world: inferences from the middle Holocene. Oecologia 193:461–474

Bastías Croudo F, Cabello Baettig G, Gallardo Ibáñez F (2023) Piedras pintadas en la desembocadura del río Loa (desierto de Atacama, Chile). CUHSO 33(1):63–94

Bastourre ML (2021) Las relaciones humanos/animales en las sociedades prehispánicas del Paraná Medio e Inferior: un giro ontológico. Revista Del Museo De Antropología 14(2):65–80. https://doi.org/10.31048/1852.4826.v14.n2.30066

Belardi JB, Carballo Marina F, Barrientos G (2021) Hunting landscapes in the north margin of Lake Viedma (Southern Patagonia, Argentina): prey, strategies and technology. In: Belardi JB, Bozzuto D, Fernández PM, Moreno E, Neme G (eds) Ancient hunting strategies in Southern South America, (pp. 1–32). Latin American Studies Series. https://doi.org/10.1007/978-3-030-61187-3

Belloti López de Medina CR (2012) En compañía de los muertos: Ofrendas de animales en los cementerios de La Isla (Tilcara, Jujuy). Intersecciones en Antropología 13(2):345–357

Beovide L, Martínez S, Norbis W (2017) Space use patterns and resource exploitation of shell middens from the Río de La Plata coast (ca. 6000–2000 years BP), Uruguay. In: Mondini M,

Muñoz S, Fernández M (eds) Zooarchaeology in the Neotropics environmental diversity and human-animal interactions. Springer, pp 81–101
Berón M (2010) Vínculo ritual entre el perro doméstico y el hombre en sociedades de cazadores-recolectores de la Pampa occidental. In: Gutiérrez M, De Nigris M, Fernández PM, Giardina M, Gil A, Izeta A, Neme G, Yacobaccio H (eds) Zooarqueología a principios del siglo XXI: Aportes teóricos, metodológicos y casos de estudio. Ediciones El Espinillo, pp 543–550
Berón M (2019) Artefactos malacológicos. Diferentes contextos de uso en sitios arqueológicos de la Provincia de La Pampa (Argentina). Comechingonia 23(1):53–85
Binford L (1978) Nunamiut Ethnoarchaeology. Academic Press, p 509
Bonomo M (2006) Un acercamiento a la dimensión simbólica de la cultura material en la región pampeana. Relaciones de la Sociedad Argentina de Antropología XXXI: 89–115
Bonomo M, Aguirre ML (2009) Holocene Molluscs from archaeological sites of the Pampean Region of Argentina: approaches to past human uses. Geoarchaeology 24(1):59–85
Bonomo M, Politis G, Bastourre L, Moreira G (2021) Humanized nature: symbolic representation of Fauna in pottery from the Paraná River of South America. In: Bonomo M, Archila S (eds) South American contributions to world archaeology. One world archaeology. Springer, Cham. https://doi.org/10.1007/978-3-030-73998-0_16
Borrero L, Yacobaccio H (1989) Etnoarqueología de Asentamientos Aché. J Soc Am LXXV:7–33
Boyd B (2017) Archaeology and human–animal relations: thinking through anthropocentrism. Annu Rev Anthropol 46:299–316
Buc N (2019) Stylistic variability in bone tools from the low Paraná wetland. Late Holocene. Argentina J Anthropol Archaeol 56:101–112. https://doi.org/10.1016/j.jaa.2019.101112
Bugallo L, Tomasi JME (2012) Crianzas mutuas: el trato a los animales desde las concepciones de los pastores puneños (Jujuy, Argentina). Rev Esp Antropol Am 42(1):205–224. https://doi.org/10.5209/rev_REAA.2012.v42.n1.38644
Calás E, Labarca R, Gallardo F, Pestle W (2024) Animales e Intercambio: Análisis arqueofaunístico de la aldea San Salvador, Período Formativo (3000-1000 Ap), Región de Antofagasta, Norte de Chile. Boletín de la Sociedad Chilena de Arqueología 57:146–175. https://doi.org/10.56575/BSCHA.05700240838
Capriles FJM (2002) Intercambio y uso ritual de fauna por Tiwanaku: Análisis de pelos y fibras de los conjuntos arqueológicos de Amaguaya, Bolivia. Estudios Atacameños 23:33–51. https://doi.org/10.4067/S0718-10432002002300004
Capriles JM, Delaere C, Mendoza España VV (2023) Unraveling animal sacrifice and Taphonomy: analysis of faunal remains from the underwater Khoa reef sanctuary in Lake Titicaca, Bolivia. In: Delaere C, Elkin D (eds) Underwater and coastal archaeology in Latin America, 1st edn. University Press of Florida, pp 153–164. https://doi.org/10.2307/jj.10539933.18
Carden N (2008) Imágenes a través del tiempo. Arte Rupestre y construcción social del paisaje en la Meseta Central de Santa Cruz. Sociedad Argentina de Antropología, Buenos Aires
Carden N (2022) Back to a time perspective: new insights for the study of Piedra Museo's ancient rock art. In: Miott L, Salemme M, Hermo D (eds) Archaeology of Piedra Museo locality. An open window to the early peopling of Patagonia. Latin American Studies Book Series, Springer, pp 257–290
Carden N, León C (2022) El lado oculto de la localidad Para Yacu, sur de Santiago del Estero, Argentina. Un enfoque semiótico para la interpretación de las pinturas rupestres. Chungara. Rev Antropol Chilena 54(4):697–717
Carden N, Martínez G (2014) Diseños fragmentados. Circulación social de imágenes sobre huevos de Rheidae en Pampa y Norpatagonia. Boletín Del Museo Chileno De Arte Precolombino 19(2):55–75
Carden N, Martínez G, Mitchell P, Orton J (2024) Spatial and temporal trends in the distribution of engraved eggshell fragments: a comparative view from the Holocene archaeological record of southern Africa and southern South America. J Anthropol Archaeol 73:101571. https://doi.org/10.1016/j.jaa.2024.101571

Cardillo M, Scartascini F (2015) Possible fishing structures on the west coast of san Matías gulf, Río Negro, Patagonia Argentina. J Isl Coast Archaeol 11(1):133–137. https://doi.org/10.1080/15564894.2015.1074131
Carman M (2018) La animalidad en cuestión: derechos, ontologías, moralidades y políticas. Rev Mus Antropol 11(1):195–208
Cartajena I, López P, Núñez L (2019) Una aproximación desde la fauna del templete Tulán-54 (Salar de Atacama, Chile) a los nichos en la arquitectura durante inicios del Formativo (ca. 3.300-2.400 AP). Archaeofauna 28:85–94. https://doi.org/10.15366/archaeofauna2019.28.007
Casamiquela R (2000) Temas patagónicos de interés arqueológico VI. Análisis etnográfico de la morfología del toldo tehuelche y sus derivaciones etnológicas (hacia una retroetnología). Intersecc Antropol 1:3–32
Casanova Menéndez MT (2021) Las imágenes del Formativo Tardío puneño (2000 a 1300 AP): análisis iconográfico de la cerámica de Casa Chávez Montículos (Antofagasta de la Sierra, Catamarca, Argentina). Mundo De Antes 15(2 (julio-diciembre)):65–91. https://doi.org/10.59516/mda.v15.251
Castillo L, Araujo E, Chiavazza H, Prieto-Olavarría C (2018) Cocinar y alimentarse en tiempos de conquista. Reconstruyendo paquetes culinarios a partir de análisis cerámicos y arqueofaunísticos (Mendoza, siglos XV-XVII). Arqueología 24(2):109–132. https://doi.org/10.34096/arqueologia.t24.n2.5003
Chapman A (1986) Los Selk'nam. Editorial Emecé
Cione, Bonomo M (2003) Great white shark teeth used as pendants and possible tools by early-middle Holocene terrestrial mammal hunter-gatherers in the eastern pampas (southern South America). Int J Osteoarchaeol 13:222–231
Claraz J (1988) Diario de Viaje de exploración al Chubut, 1865–1866. Ediciones Marymar, Buenos Aires
Cohen ML, Martinez MS (2020) Q'uepis, ancestros y territorios. Prácticas rituales históricas en contextos prehispánicos en Antofagasta de la Sierra, Noroeste Argentino. Chungara Revista de Antropología Chilena 54(2):291–307
Colasurdo MB, Sartori JI (2011) La conformación de la etnicidad a partir de los hábitos alimenticios: su abordaje desde la antropología y la arqueología histórica. Rev Arqueol Hist Argent Latinoam 5:125–146
Coloca FI, López GEJ (2021) Archaeology and pre-Hispanic mining rituals in the south-Central Andes (Puna of Salta, northwestern Argentina). J Anthropol Archaeol 62:101275
Colonese A, Begossi A, Brugere C, Marques Bandeira A, Brandi R, Guedes L, Wiedemann M, Azevedo P (2015) Bridging ancient and modern artisanal fisheries in Latin America: assessing the role of cultural heritage in poverty alleviation in coastal Brazil. Antiquity Proj Gall 89:344. http://www.antiquity.ac.uk/projgall/colonese344
Colonese AC, Brugere C, Ramires M, Clauzet M, Brandi R, Marques Bandeira A, Guedes L, Wiedemann M, Reyes-García V, Begossi A (2023) The legacy of pre-Columbian fisheries to food security and poverty alleviation in the modern Amazon. In: Colonese AC, Milheira RG (eds) Historical ecology and landscape archaeology in lowland South America. Interdisciplinary Contributions to Archaeology, pp 3–20. https://doi.org/10.1007/978-3-031-32284-6_1
Colqui EP, Recalde A, Tamburini M (this volume) Dialogues between animals and people over time: an interdisciplinary perspective on the sierras de Córdoba, Argentina. In: Frontini R, Escosteguy P, Salemme M (eds) Animals and people: interactions, agents and materialities. An overview from Southern South America. Latin American book series. Springer
Corona-M. E (2019) Diversas facetas de las interacciones entre los humanos y los animales: algunos registros en las Américas. Revista Etnobiología 17(2):5–10
Costa T, Weihmüller MP, Manzano-García J (2022) Cacería de guanacos en el Chaco Árido (Córdoba, Argentina): una mirada desde la zooarqueología. Chungará 54(1):127–148. https://doi.org/10.4067/s0717-73562021005002502
Cruz I, Ercolano B (2024) Negro sobre blanco. La explotación de pingüinos y cormoranes en la Patagonia Austral durante el Holoceno. Archaeofauna 33(1):63–80

Cruz MJ, Salerno MA (2023) Aullidos congelados. La mercantilización de lobos y elefantes marinos en la industria lobera antártica (Shetland del Sur, siglo XIX). Libro de Resúmenes, pp. 458, XXI Congreso Nacional de Arqueología Argentina, Corrientes. http://congresoscnaa.org/congreso/wp-content/uploads/2023/08/Libro-de-Resumenes-XXI-CNAA.pdf

David B, Thomas J (2008) Handbook of landscape archaeology. Routledge

Day Pilaría F (this volume) Pumas between the forest and the river, between the domestic and the symbolic. Prehispanic record of Puma concolor in the site Las Marías (Magdalena, Buenos Aires, Argentina), In R. Frontini, P. Escosteguy & M. Salemme (Eds.) Animals and people: interactions, agents and materialities. An overview from Southern South America. Latin American Book series, Springer

De Nigris M, Mengoni Goñalons G (2004) The guanaco as a source of meat and fat in the southern Andes. In: Mulville J, Outram A (eds) The Zooarchaeology of fats, oils and Dayring. Oxbow Books, Oxford

del Papa LM, Zarza MM, Lamenza GN, Del Papa MC (2022) Los silbatos de hueso del sitio arqueológico Palo Blanco (Berisso, Provincia de Buenos Aires). Mundo De Antes 16(2):41–68

Dellepiane J, Flores Coni J, Agnolin A, Cassiodoro G, Goñi R, Guichón F, Re A (this volume) Communal, collective and massive guanaco hunts: the regional archaeological contribution to its discussion in southern Patagonia. In: Frontini R, Escosteguy P, Salemme M (eds) Animals and people: interactions, agents and materialities. An overview from Southern South America. Latin American book series. Springer

Deschamps CM (2005) Late Cenozoic mammal bio-chronostratigraphy in southwestern Buenos Aires Province, Argentina. Ameghiniana 42:733–750

Descola P (2014) Modes of being and forms of predication. Hau: J Ethnogr Theory 4(1):271–280

Dillehay T, Kaulicke P (1984-1985) Aproximación metodológica: El Comportamiento del Jaguar y la Organización Socio-Espacial Humana. Relac Soc Argent Antropol XVI:27–36

Erikson P (2000) The social significance of pet-keeping among Amazonian Indians. In: Podberscek AL, Paul ES, Serpell JA (eds) Companion animals & us: exploring the relationships between people & pets. Cambridge University Press, Cambridge, pp 7–26

Erramouspe V (2022) Pensando la movilidad pastoril desde el registro zooarqueológico en la Puna Seca de Jujuy, Argentina. Mundo de Antes 16(1):253–267

Escosteguy P (2013) El uso de fuentes documentales y etnográficas para la interpretación del registro arqueofaunístico de coipo. Rev Arqueol Hist Argent Latinoam 7:41–65

Escosteguy PD (2014) Estudios etnoarqueológicos con cazadores de coipo de Argentina. Antípoda. Revista de Antropología y Arqueología 1(20):145–165. https://doi.org/10.7440/antipoda20.2014.07

Escosteguy P, Salerno V, González MI, Frère MM (2023) Usos de la fauna en el río Salado bonaerense: saberes locales actuales. Mundo de Antes 17(1):109–140. https://doi.org/10.59516/mda.v17.272

Fausto C (2007) Feasting on people. Eating animals and humans in Amazonia. Curr Anthropol 48(4):497–530

Fernández-Jalvo Y, Andrews P (2016) Atlas of taphonomic identifications. Springer

Ferrari A, González EM, Manzuetti A (2023) Repaso integrador de la presencia del jaguar (Panthera onca) en Uruguay con nuevos datos arqueozoológicos. Rev Mus Antrop 16(2):97–118. https://doi.org/10.31048/1852.4826.v16.n2.40595

Fiore D (2011) Art in time. Diachronic rates of change in the decoration of bone artefacts from the Beagle Channel region (Tierra del Fuego, Southern South America). J Anthropol Archaeol 30(4):484–501. https://doi.org/10.1016/j.jaa.2011.07.002

Fiore D, Borella F (2010) Geometrías delicadas. Diseños grabados en cáscaras de huevo de rheidae recuperados en la costa norte del golfo San Matías, Río Negro. Intersecciones Revista de Antropología 11:277–293

Fiore D, Zangrando AF (2006) Painted fish, eaten fish: artistic and archaeofaunal representations in Tierra del Fuego, Southern South America. J Anthropol Archaeol 25(3):371–389. https://doi.org/10.1016/j.jaa.2006.01.001

Fossile T, McGrath K, Comes P, Villanueva J, Sayle KL, Gilson SP, Haimovici M, Alves MC, Carrion Bartz M, da Rocha Bandeira D, Borba FM, Ferreira J, Colonese AC (2024) The historical ecology of subsistence and early commercial fisheries in mangrove systems in Brazil. J Archaeol Sci 166:105986

Franco N, Vetrisano L, Gilio BL, Cirigliano NA, Bianchi P (2021) Hunting blinds in the southern end of the Deseado Massi: collective hunting strategies during the late Holocene. In: Belardi JB, Bozzuto D, Fernández PM, Moreno E, Neme G (eds) Ancient hunting strategies in Southern South America. Latin American Studies Series, pp 1–32. https://doi.org/10.1007/978-3-030-61187-3

Franklin W, Bas MF, Bonacic CF, Cunazza PC, Soto VN (1997) Striving to manage Patagonia guanacos for sustained use in the grazing agroecosystems of Southern Chile. Wildl Soc Bull 25(1):65–73

Frontini R (2010) Las arqueofaunas en la cocina. Actividades culinarias en el sitio 2 de la localidad arqueológica el guanaco. Arqueología 16:191–208

Frontini R (2012) El aprovechamiento de animales en valles fluviales y lagunas del sur bonaerense durante el Holoceno. Unpublished PhD. Facultad de Filosofía y Letras, Universidad de Buenos Aires

Fudge E (2000) Perceiving animals, humans, and beasts in early modern English culture. Macmillan Press, Londres

Furlong G (1912) Hunting the Guanaco. The Outing Magazine 61(1):3–20

Gallardo F (2018) Estilos de arte rupestre e interacción social en el Desierto de Atacama (Norte de Chile). Mundo de Antes 1:13–78

Gallardo F, Yacobaccio H (2005) Wild or domesticated? Camelids in early formative rock art of the Atacama Desert (northern Chile). Lat Am Antiq 16:115–130

García M, Troncoso A (2021) Ni cultura, ni naturaleza. Ni pasado, ni presente. Acercamientos relacionales a mundos, paisajes y sus historicidades. Libro de resúmenes del XXII Congreso Nacional de Arqueología Chilena, p. 2. Universidad Austral de Chile, Sociedad Chilena de Arqueología. Puerto Montt, Chile

Giardina M, Otaola C, Franchetti F (2021) Hunting, butchering and consumption of Rheidae in the South of South America: an actualistic study. In: Belardi IEJB, Bozzuto DL, Fernández PM, Moreno EA, Neme GA (eds) Ancient hunting strategies in Southern South America. Springer, pp 159–174). The Latin American Studies Book Series

Gifford-González D (2018) An introduction to Zooarchaeology. Springer

Goñi R, Cassiodoro G, Flores Coni J, Dellepiane J, Agnolin A, Guichón R (2016) Estrategias de caza y movilidad. Parapetos del sitio K116 (Meseta del Strobel, Santa Cruz). In: Mena F (ed) Arqueología de Patagonia: de mar a mar. CIEP-Ñire Negro, Coyhaique, pp 441–449

González AR (1961) [U3]. La Cultura de la Aguada del NO Argentino. Revista del Instituto de Antropología 2–3:205–2053 -

González Venanzi L, Romano V, Saghessi D, Serna A, Vercellini MR, Prates L (2023a) Special pet, special care. Diet, provenance, and health analyses of a dog reveal strong ties with humans in Patagonia. Archaeol Anthropol Sci 15:168. https://doi.org/10.1007/s12520-023-01877-3

González Venanzi L, Bonomo M, Prevosti F (2023b) Sinantropismo de zorros (*Dusicyon avus*) en Nutria Mansa 1 (Holoceno tardío, provincia de Buenos Aires) inferido por análisis de tártaro dental. Libro de Resúmenes, pp. 461–462, XXI Congreso Nacional de Arqueología Argentina, Corrientes. http://congresoscnaa.org/congreso/wp-content/uploads/2023/08/Libro-de-Resumenes-XXI-CNAA.pdf

Gordillo I (2018) Los pueblos de la Aguada: Vida y arte. Academia Nacional de la Historia, Ciudad Autónoma de Buenos Aires

Gordillo S (2021a) Caracoles marinos en contextos arqueológicos en el centro de Argentina (32° LS–64° LO): Materialidades simbólicas extendidas en el territorio. Chungara 53:379–398

Gordillo S (2021b) Registros arqueomalacológicos en el centro de Argentina: El uso de Megalobulimus y otros moluscos en cuentas y adornos personales. Intersecciones en Antropología 22:83–95

Gradín C (1996) Grabados y parapetos de la zona sur de la meseta del lago Buenos Aires (Prov. de Santa Cruz). In: Goméz Otero J (ed) Arqueología. Sólo Patagonia. CENPAT-CONICET, Puerto Madryn, pp 173–184

Grant J, Vilá B (2023) Las llamas como componentes bioculturales de la Puna de Catamarca. Mundo de Antes Mundo de Antes 17(1):247–278

Haber AF (2004) Arqueología de la naturaleza, naturaleza de la arqueología. In: Haber AF (ed) Hacia una arqueología de las arqueologías sudamericanas. Universidad de los Andes, Facultad de Ciencias Sociales, Centro de Estudios Socioculturales e Internacionales, pp 15–32

Hamilakis H, Overton N (2013) A multi-species archaeology. Arch Dial 20:159–173. https://doi.org/10.1017/S1380203813000214

Hammond H, Zilio L, Nuevo-Delaunay A, Méndez C (2024) Middle through late Holocene long-distance transport of exotic shell personal adornments in Central West Patagonia (southern South America). The archaeomalacological assemblage of Baño Nuevo 1. PLoS One 19(5):e0304454. https://doi.org/10.1371/journal.pone.0304454

Haraway DJ (2003) The companion species manifesto. Dogs, people and significant others. Prickly Paradigm Press

Harrison-Buck E, Krause SM, Brouwer Burg M, Willis M, Perrotti A, Bailey K (2024) Late archaic large-scale fisheries in the wetlands of the pre-Columbian Maya Lowlands. Sci Adv 10(47):eadq1444. https://doi.org/10.1126/sciadv.adq1444

Hernández Castillo D, Troncoso A, Méndez C, Pascual D, Armstrong F, Nuevo-Delaunay A, Grasset S, Pérez I, Vera F, Delgado M, Escudero A, Pino M, Larach P (2025) 12,000 years away from the sea: long-term circulation of Pacific shells in the semiarid north of Chile (South America). Quat Sci Rev 348:109112. https://doi.org/10.1016/j.quascirev.2024.109112

Hill E (2013) Archaeology and animal persons: toward a prehistory of human-animal relations. Environ Soc Adv Res 4:117–136

Hodder I (1990) The domestication of Europe: structure and contingency in neolithic societies. Basil, Blackwell

Horta Tricallotis H, Gordillo S, Yáñez J (2022) El caracol terrestre Megalobulimus en contextos funerarios de San Pedro de Atacama (Chile) y su rol en el intercambio ritual circumpuneño (400-1400 dC). Lat Am Antiq 33(4):753–772. https://doi.org/10.1017/laq.2022.4

Horta Tricallotis H, Gudemos M, Lema V, Peña-Villalobos I (2023) Circulación interregional de armadillos (orden Cingulata) y su vínculo con artefactos del complejo alucinógeno en el sur andino (400-1400 DC). Chungara Revista de Antropología Chilena 55(1):99–124

Ingold T (1993) The temporality of the landscape. World Archaeol 25(2):152–174

Ingold T (1994) [1988] What's an animal? Routledge, London

Ingold T (2000) The perception of the environment: essays on livelihood, dwelling and skill. Routledge, London

Ingold T (2013) Anthropology beyond humanity. Soumen Antropologi/J Finnish Anthropol Soc 38(3):15–23

Ingold T (2020) Prefacio ¿Qué es un animal en la América del Sur indígena? In: Medrano C, Velden FV (eds) ¿Qué es un animal? 1st edn. Rumbo Sur, pp 9–13

Laguens A (2023) Zoocializando con personas no humanas: de llamas y otras especies en la vida cotidiana de sociedades Aguada del Noroeste de Argentina (s. VI-XI d.c.). Fragmentos del Pasado Revista de Arqueología 7:11–47

Laporte L (2014) Estudio del ajuar funerario de los entierros humanos. In: Politis G, Gutiérrez MA, Scabuzzo C (eds) Estado actual de las investigaciones en el sitio arqueológico Arroyo Seco 2 (Partido de Tres Arroyos, Provincia de Buenos Aires, Argentina), (pp: 393–416). Serie Monográfica Número 5. INCUAPA-CONICET-UNICEN

Lazzari M (2005) Traveling objects and spatial images: exchange relationships and the production of social space. In: Funari P, Zarankin A, Stovel E (eds) Global archaeological theory. Contextual voices and contemporary thoughts. Springer, pp 191–210

Le Breton D (2006) El Sabor del Mundo. Una Antropología de los Sentidos. Nueva Visión, Buenos Aires

Leibowicz I, Ferrari A, Jacob C, Acuto F (2015) Petroglifos en el Valle Calchaquí Norte (Salta, Argentina): camélidos, montañas y apropiación inkaica del paisaje local. Chungará Rev Antropol Chil 47(4):575–587
Lema V (2019) Contenedores, cuerpos y topologías: un análisis integral de la colección arqueológica de Pampa Grande (Salta, Argentina). Antípoda. Revista de Antropología y Arqueología 37:95–118
López GEJ, Coloca FI, Rosenbusch ML, Solá P (2018) Mining, macro-regional interaction and ritual practices in the South-Central Andes: the first evidence for turquoise exploitation from the late Prehispanic and Inca periods in North-Western Argentina (Cueva Inca Viejo, Puna de Salta). J Archaeol Sci Rep 17(2):81–92
Lyman RL (1994) Vertebrate Taphonomy. Cambridge University Press
Mainland I (2008) The uses of archaeological faunal landscape in archaeology. In: David B, Thomas J (eds) Handbook of landscape archaeology. Routledge, London, pp 544–550
Manzano-García J, Costa T, Mignino J (2023) ¿El bicho se caza o es de casa?: dicotomía entre el consumo y mascotismo de fauna silvestre en el Chaco Seco de la Provincia de Córdoba. Mundo De Antes 17(1):215–246. https://doi.org/10.59516/mda.v17.273
Martel A (2018) Caravanas de llamas: tecnología clave para la interacción social prehispánica. In: Grau HR, Babot MJ, Izquierdo AE, Grau A (eds) La Puna Argentina: Naturaleza y Cultura. Fundación Miguel Lillo, Tucumán, pp 319–320
Massone M, Prieto A (2005) Ballenas y delfines en el mundo Selk'nam: una aproximación etnográfica. Magallania 33(1):25–35
Medina M, Rivero D (2020) Hunting and skeletal element abundance of guanaco during the Holocene of sierras of Córdoba, Argentina. Journal of archaeological. Sci Rep 29:02074. https://doi.org/10.1016/j.jasrep.2019.102074
Medrano C, Vander Velden F (2020) ¿Qué es un animal? Buenos Aires, Rumbo Sur
Mendoza España V (2022) Más allá de la amistad: ofrendas y ajuares funerarios de perros en sitios arqueológicos del Altiplano boliviano. Revista Archaeobios 16(1):89–109
Mendoza España V (2023) Cánidos de la altura: el perro y el zorro andino prehispánico del altiplano boliviano. In: Apesteguía S, Álvarez SM (eds) Perros y otros cánidos de las américas. Origen, evolución e historia natural. Vazquez Mazzini Editores, Fundación Félix de Azara, pp 285–307
Mengoni Goñalons G, Arroyo-Cabrales J, Polaco O, Aguilar F (eds) (2010) Estado actual de la Arqueozoología Latinoamericana/ Current advances for the Latin -American Archaeozoology. Instituto Nacional de Antropología e Historia, México, p 176
Miguez G, Nasif N, Vides ME, Caria MA, Gudemos M (2017) Piedras bezoares en contexto: primer estudio de su relevancia en comunidades Prehispánicas del Noroeste de Argentina. Chungará Rev Antropol Chil 49(3):343–357
Mintz SW, Dubois CD (2002) The anthropology of food and eating. Annu Rev Anthropol 31:99–119
Miotti L, García Añino E (2016) Entre la necesidad y el tabú. Interacciones económicas, sociales y simbólicas de humanos y animales. In: En IV Congreso Nacional de Zooarqueología Argentina. Universidad Nacional de Tierra del Fuego, Ushuaia, p 27
Miotti L, Marchionni L (2014) Autopodios de guanacos en sitios arqueológicos. Equifinalidad entre los palatable y lo preservado. Revista Chilena de Antropología 21: 122–129. Universidad de Chile
Miotti L, Marchionni L (2023) La agencia de los cazadores-recolectores y de los animales en la construcción de los paisajes arqueológicos de Patagonia. Mundo de Antes 17(1):169–197
Miotti L, Terranova E (2015) A hill full of points in terra incognita from Patagonia: notes and reflections for discussing the way and tempo of initial peopling. PaleoAmerica 1(2):181–196. https://doi.org/10.1179/2055556315Z.00000000019
Mithen S (1996) Arqueología de la mente. Orígenes del arte, de la religión y de la ciencia. Crítica, Barcelona

Miyano JP, Ratto N (2020) Vicuñas y humanos en el pasado reciente (siglos XIX y XX) del oeste tinogasteño (Catamarca, Argentina). Rev Colomb Antropol 56(2):289–312. https://doi.org/10.22380/2539472X.632

Miyano JP, Lantos I, Ratto N, Orgaz M (2017) Animales e Incas en el Oeste Tinogasteño (Catamarca, Argentina). Lat Am Antiq 28(1):28–45. https://doi.org/10.1017/laq.2016.7

Montón Subías S (2002) Cooking in zooarchaeology: is this issue still raw? In: Milner M, Milner N (eds) Consuming passions and patterns of consumption. McDonald Institute Monographs, pp 7–15

Moreno E, Martínez J, Aschero C (2021) Ancient hunting strategies of wild camelids through the study of multiple lines of archaeological evidences at southern Argentine Puna. In: Belardi JB, Bozzuto D, Fernández PM, Moreno E, Neme G (eds) Ancient hunting strategies in Southern South America. Springer, pp 1–32. Latin American Studies Series. https://doi.org/10.1007/978-3-030-61187-3

Munita D, Álvarez R, Ocampo C (2004) Corrales de piedra, pesca pasiva en la costa interior de Chiloé. Bol Soc Chil Arqueol 37:61–74

Muñoz AS, Ramos Roca E, Borges C, Luz C, de Souza RC, Fernández PM, Beovide L (2019) Introducción. Archaeofauna 28:7–11

Muscio H (2019) Animalidad/humanidad: una dicotomía estratégicamente persistente que entorpece el avance científico en arqueología. Revista Arqueología 19 Dossier: 81–101. Facultad de Filosofía y Letras, UBA

Muscio H, Orsi JP, Carballo G (this volume) Humans and prey. Holocene hunter occupation of high altitude spaces in the Puna of Salta, Argentina. In R. Frontini, P. Escosteguy & M. Salemme (Eds.) Animals and people: interactions, agents and materialities. An overview from Southern South America. Latin American Book Series, Springer

Museu Episcopal de Vic (2023) Besties. Els animals en l'art medieval del MEV. Go Books! Editorial, Vic, Spain. 101 pages

Musters G (1964) Vida entre los patagones. Un año de excursiones por tierras no frecuentadas, desde el Estrecho de Magallanes hasta el Río Negro. Ediciones Solar/Hachette

Naharro ME, Izeta A, Martínez J (this volume) Living in Quebrada de los Corrales: camelids, humans and landscape within the first village settlements in El Infiernillo (Tucumán, Argentina). In: Frontini R, Escosteguy P, Salemme M (eds) Animals and people: interactions, agents and materialities. An overview from Southern South America. Latin American Book Series. Springer

Nielsen A (2024) Herders and llamas, companion species in the southern Andes during the last three Millenia. In: Pochettino ML, Capparelli A, Stampella PC, Andreoni D (eds) Nature(s) in construction ethnobiology in the confluence of actors, territories and disciplines. Springer, pp 501–518. Latin American Book Series

Nielsen AE, Maryañski JM (2018) Los pastores y sus caravanas en la era del Tawantinsuyu. In: Shimada I (ed) El Imperio Inka. Pontificia Universidad Católica del Perú, Lima, pp 473–512

Nielsen A, Berenguer RJ, Pimentel G (2019) Inter-nodal archaeology, mobility, and circulation in the Andes of Capricorn during the late intermediate period (AD 1000–1450). Quat Int 533:48–65

Oliva F, Paniza MC, Morales N (2018) Relaciones simbólicas entre sociedades indígenas y el mundo animal en Ventania (Provincia de Buenos Aires, Argentina): el caso de los Rheidae. Archaeofauna 27:233–252

Orquera LA, Piana E (1999) Arqueología de la región del canal Beagle. Publicaciones de la Sociedad Argentina de Antropología, Buenos Aires

Orquera L, Gómez Otero J (2007) Los cazadores-recolectores de las costas de Pampa, Patagonia y Tierra del Fuego. Relac Soc Argent Antropol 32:75–99

Otero C, Fuchs ML (2022) Análisis en clave de género de contextos mortuorios en una capital incaica del Collasuyu (Tilcara, Argentina). Antípod Rev Antropol Arqueol 49:37–65

Otero C, Bordach MA, Mendonça OJ (2017) Las prácticas funerarias en el Pucará de Tilcara (Jujuy, Argentina). Nuevos aportes para su conocimiento a partir del caso de la Unidad

Habitacional 1. Antípoda. Revista De Antropología Y Arqueología 1:27. https://doi.org/10.7440/antipoda27.2017.06

Otero C, Akmentins MS, Quinteros SA (2021) Animales en acción: usos rituales de fauna silvestre y de representaciones zoomorfas en contextos incaicos del Pucará de Tilcara (Quebrada de Humahuaca, Argentina). Estudios Atacameños Arqueología y Antropología Surandinas 67:e3926

Ottalagano F (2021) Animales representados y animales cazados: aportes para el estudio arqueológico de la interacción simbólica humanos-fauna entre los cazadores-recolectores complejos de las Tierras Bajas del Paraná (Argentina). Chungara Rev Antropol Chil 53(2):237–260

Pastene L, Quirós D (2010) An outline of the history of whaling in Chile. In: Human culture from the perspective of traditional maritime communities, international symposium report no. 1. Kanagawa Shimbun Press, Kanagawa, pp 73–98

Pastor S, Recalde A, Tissera L, Ocampo M, Truyol G, Chiavassa-Arias S (2015) Chamanes, Guerreros, Felinos: Iconografía de Transmutación en el Noroeste de Córdoba (Argentina). Boletín SIARB 29:71–85. https://siarb-bolivia.org/wp-content/uploads/2019/08/bol29c.pdf

Pazzarelli F (2008) Notas acerca de una arqueología de la comida. La Zaranda De Ideas 4:157–162

Pazzarelli F, Lema VS (2024) La geografía del estómago / el estómago de la geografía: texturas, cuerpos y ofrendas de arrieros en los Andes argentinos. Antípoda Revista De Antropología Y Arqueología 54:3–25

Pelizza F, Pastor S, Sario G, Medina ME (2024) Artefactos en piedras bezoares y su rol simbólico en el Período Prehispánico Tardío (Sierras de Córdoba, Argentina). Revista del Museo de La Plata 9(1):22–35. https://doi.org/10.24215/25456377e187

Pérez Jimeno L, del Papa L (2016) Presencia del grupo morfológico definido como tubo-ave e hipótesis de uso. El humedal del Paraná Medio y el Chaco Seco como caso de estudio. Revista De Antropología Del Museo De Entre Ríos 2(1):100–118. Recuperado a partir de https://ramer.ar/revista/index.php/ramer/article/view/91

Petrocelli J (1975) Nota preliminar sobre hallazgos arqueológicos en el valle del río Luján (Población Río Luján, Campana, Pcia. De Buenos Aires). Actas del Primer Congreso de Arqueología Argentina, pp. 251–270. Rosario

Píccoli CV, Barboza MC, Ávila JD, Gallego OF, Carvallo M, Pusterla S, Torri M, Contreras FI, Zilli F, Martínez SA, Cuaranta P, Contreras S, Saucedo ON, Monferran MD, Fernandez-Pacella L, Cabaleri NG, Weibel F, Mujica JI (2020) El sitio Isla El Disparito (Iberá, Corrientes). Aportes a la arqueología del Chaco húmedo (Argentina). Rev Mus La Plata 5(2):716–744

Politis G (2007) Nukak. Ethnoarchaeology of an Amazonian People. Walnut Creek, Left Coast Press

Politis G (2023) Cuarenta años después de La distribución del guanaco (Mammalia, Camelidae) en la provincia de Buenos Aires durante el Pleistoceno Tardío y Holoceno. Los factores climáticos como causa de su retracción. Publ Electrón Asoc Paleontol Argent 23(1):331–359

Politis G, Martínez G (1996) La cacería, el procesamiento de las presas y los tabúes alimenticios. In: Politis G (ed) Nukak. Editorial Kinopia Bolivar, pp 231–280

Politis G, Saunders N (2002) Archaeological correlates to Ideological Activities. In: Miracle P, Milner N (eds) Consuming passions and patterns of consumption. McDonald Institute Monographs, Cambridge, pp 113–130

Politis G, Messineo P, Kaufmann C, Barros MP, Álvarez MC, Di Prado V, Scalise R (2005) Persistencia ritual entre cazadores-recolectores de la llanura pampeana. Bol Arqueol PUCP 9:67–90

Prates L (2009) El uso de los recursos por los cazadores-recolectores posthispánicos de Patagonia continental y su importancia arqueológica. Relac Soc Argent Antropol XXXIV:201–229

Prates L (2014) Crossing the boundary between humans and animals. The extinct fox (Dusicyon avus) in a mortuary hunter-gatherer context of Patagonia (Río Negro, Argentina). Antiquity 88:1–13

Prates L, Prevosti FJ, Beron M (2010) First records of prehispanic dogs in Southern South America (Pampa-Patagonia, Argentina). Curr Anthropol 51:273–280

Quiroz D, Carreño G (2017) Narrativas etnográficas sobre ballenas y balleneros en las costas de Chile. Antípod Rev Antropol Arqueol 1(28):35–55
Ramos Roca E, Corona-M. E (2017) La importancia de diversas, complementarias y comparativas miradas en la investigación sobre las interacciones entre los humanos y la fauna en América Latina. Antípod Rev Antropol Arqueol 1(28):13–29. https://doi.org/10.7440/antipoda28.2017.01
Romano AS, López Campeny SM, Donadio S, Figuerero Minetti F (this volume) Fleece and leather analysis in human and animal archaeological assemblages. Southern Puna of Argentina (ca. 50 BC to AD 1800). In: Frontini R, Escosteguy P, Salemme M (eds) Animals and people: interactions, agents and materialities. An overview from Southern South America. Latin American Book Series. Springer
Russel N (2012) Social Zooarchaeology. Humans and animals in prehistory. Cambridge University Press
Salemme M, Frontini R (2011) The exploitation of rheidae in Pampa and Patagonia (Argentina) as recorded by chroniclers, naturalists and voyagers. J Anthropol Archaeol 30:473–483
Salemme M, Miotti L (2022) The rheids as palaeoenvironmental and consumption indicators during the latest Pleistocene and the Middle Holocene. In: Miott L, Salemme M, Hermo D (eds) Archaeology of Piedra Museo locality. An open window to the early peopling of Patagonia. Springer, Switzerland, pp 257–290). Latin American Studies Book Series
Sanchidrián JL (2001) Manual de arte prehistórico. Ariel Prehistoria
Santiago F, Salemme M (2016) Guanaco hunting strategies in the northern plains of Tierra del Fuego, Argentina. J Anthropol Archaeol 43:110–127. https://doi.org/10.1016/j.jaa.2016.07.002
Scartascini F (2017) The role of ancient fishing on the desert coast of Patagonia, Argentina. J Island Coast Archaeol 12(1):115–132
Schillat M (2007) Pemmican and penguin-breast, but no pie: daily problems of polar explorers during the Heoic age of Antarctic exploration. In: Rabassa J, Borla ML (eds) Antarctic peninsula and Tierra del Fuego: 100 years of Swedish-Argentine scientific cooperation at the end of the world. Taylor and Francis, pp 169–178
Segata J, Lewgoy B (2016) Animals in anthropology. VIBRANT—Vibrant Virtual Braz Anthropol 13(2):27–37
Serpell JA, Paul ES (1994) Pets and the development of positive attitudes to animals. In: Manning A, Serpell JA (eds) Animals & human society: changing perspectives. Routledge, London, pp 127–144
Sperling A (2023) Mesa para tres. Una aproximación a las relaciones entre humanos, camélidos y perros del Cementerio de pPca 8 (900 d.c. 1450 d.c.) a partir de la dieta. Libro de Resúmenes. XXI Congreso Nacional de Arqueología Argentina, Corrientes. http://congresoscnaa.org/congreso/wp-content/uploads/2023/07/Programa-Congreso-Arqueologia-2023.pdf
Standen VG, Santoro C (2004) Patrón funerario arcaico temprano del sitio Acha-3 y su relación con Chinchorro: Cazadores, pescadores y recolectores de la costa norte de Chile. Lat Am Antiq:89–109
Stoessel L (2015) Tendencias preliminares sobre el consumo de peces durante el Holoceno medio en el área de transición Pampeano-patagónica oriental (provincia de Buenos Aires). Archaeofauna Int J Archaeozool 24:103–117
Sutton D (2001) Remembrance of repasts: an anthropology of food and memory. Berg, Oxford
Thomas J (2002) Archaeology's humanism and the materiality of the body. In: Hamilakis Y, Pluciennik M, Tarlow S (eds) Thinking through the body. Archaeologies of corporeality. Kluwer Academic/Plenum Publishers, pp 29–45
Tívoli AM, Salemme MC (2015) Diversidad faunística en La Arcillosa 2: el aprovechamiento de aves por casadores –recolectores del norte de Tierra del Fuego. Archaeofauna 24:253–249
Tola F, dos Santos A (2020) Ontology and ontologies. Theoretical, political, and methodological debates. América Crítica 4(2):163–172. https://doi.org/10.13125/americacritica/4478[R6]
Tomassini R, Kaufmann C, Frontini R (2019) Modern guanaco (Lama guanicoe, Camelidae) bezoars: an approach towards identification in the fossil record. Int J Paleopathol 26:111–121

Torres J (2009) La Pesca Entre Los Cazadores Recolectores Terrestres De La Isla Grande De Tierra Del Fuego, Desde La Prehistoria A Tiempos Etnográficos. Magallania 37(2):109–138

Urrutia Lorenzini F, González Andreu J (this volume) Uywaña and Sarawi. Relational approach to camelids and humans (Tarapcá, Northern Chile). In: Frontini R, Escosteguy P, Salemme M (eds) Animals and people: interactions, agents and materialities. An overview from Southern South America. Latin American Book Series. Springer

Vázquez M, Zangrando AF (2017) Estructuras de Pesca en el Canal Beagle. Magallania 45(1):101–122

Vilá B (2012) Camélidos Sudamericanos. Colección Ciencia Joven, Eudeba

Vilá B (2014) Una aproximación a la etnozoología de los camélidos andinos. Etnoecológica 10(5):1–16

Vilá B (2015) Camélidos en Santa Catalina (Jujuy, Argentina): Manejo de vicuñas y caravanas de llamas. Etnobiología 13(3):19–37

Viveiros de Castro E (2004) Exchanging perspectives. The transformation of objects into subjects in Amerindian ontologies. Common Knowledge 10(3):463–484

Walker J (2012) Recent landscape archaeology in South America. J Archaeol Res 20:309–355. https://doi.org/10.1007/s10814-012-9057-6

Weinberg M (2019) Especies compañeras después de la vida: pensando relaciones humano-perro desde la región surandina. Antípod Rev Antropol Arqueol 1(36):139–161. https://doi.org/10.7440/antipoda36.2019.07

Yacobaccio HD (2021) The domestication of South American camelids: a review. Anim Front 11(3):43–51

Yacobaccio H, Samec C, Marcoppido G, Vilá B (2018) Domesticación De Camélidos Y Patologías Óseas: Un Caso De Estudio De La Puna Argentina. Cuadernos del Instituto Nacional de Antropología y Pensamiento Latinoamericano—Series Especiales 6:71–79

Yacobaccio HD, Gallardo F, Sepúlveda M (2023) Interacción humano-camélido en la Puna de Atacama (área Surandina): Caza y demografía humana desde el arte rupestre. Lat Am Antiq 34(2):276–294. https://doi.org/10.1017/laq.2022.2

Zilio L, Hammond H, Peralta González S, Parolin ML, Montes A, Ocampo SM, Moreno JE (2024) Huachen: first record of use of dogs for carrying loads among hunter-gatherers in South America. Human-dog interactions on the late Holocene in Colhué Huapi lake, Argentine Patagonia. J Archaeol Sci Rep 57, september 2024. https://doi.org/10.1016/j.jasrep.2024.104621

Zotta ML, Acuto FA (2023) Corporalidades entrelazadas: humanos, animales y entidades no-humanas en un mundo permeable. Libro de Resúmenes, pp. 469, XXI Congreso Nacional de Arqueología Argentina, Corrientes. http://congresoscnaa.org/congreso/wp-content/uploads/2023/08/Libro-de-Resumenes-XXI-CNAA.pdf

Chapter 2
Uywaña and *Sarawi.* Relational Approach to Camelids and Humans (Tarapacá, Northern Chile)

Francisca Urrutia Lorenzini and Josefina González Andreu

2.1 Introduction

This work seeks to position in terms of social life the different forms of dwelling of both human groups and camelids, thus, we are compelled to pay attention to the multiplicity of organizations and comparative approaches. Ethnographic and zooarchaeological records, carried out over 20 years in the Tarapacá region through various investigations (García et al. 2023; Santana-Sagredo et al. 2024; Uribe et al. 2020a, b; Urrutia et al. 2023; Vidal-Elgueta et al. 2024), are brought into account. Although animal and human behaviors are equated, this does not proceed in reductionist terms. On the contrary, it involves exploring other types of social reality understandings and narratives of change, recognizing the way in which other entities and forces are bearers of attention and accounting for the plurality of ways of being in the environment (Despret 2022; Stengers 2005).

The Tarapacá territory emerged through a long history of behaviors and habits, not necessarily linked to optimal adaptation, but relating at once both time and space to certain attachment conditions by means of long reproductive cycles and social processes. Therefore, the relation between humans and camelids entails certain ecological practices (Stengers 2005) or ecological entanglements (Despret 2022), within which all organisms play a crucial role in joint becoming. This coexistence deployed a variety of dense interactions, where local solutions occurred at different historical moments. Andean experience put in motion the attendance to the many ways of dwelling that multiply worlds, challenging our ability to describe and act with them.

F. Urrutia Lorenzini (✉)
Doctorado Antropología Universidad de Tarapacá, Arica, Chile

J. González Andreu
Independent researcher, Santiago, Chile

R. Frontini et al. (eds.), *Non-human Animals and People: Interactions, Agents and Materialities*, The Latin American Studies Book Series,
https://doi.org/10.1007/978-3-031-97495-3_2

2.2 Material and Methods

The ethnographies comprehend extended works in communities of the *Precordillera* and the highlands (Urrutia 2011, 2019, 2022), specifically, in the Camiña (2500 m.a.s.l.) and Nama (3000 m.a.s.l.) *quebradas* (i.e., ravines), and in the Isluga altiplano (4000 m.a.s.l.) (Fig. 2.1). The living spaces and the deployment of socio-productive strategies are described from complementary cycles formed between both landscapes, stipulating the multiple interrelationships among the organization of human groups and the organization of the herd, also incorporating agricultural practices. The main productive orientation depends on the environmental dispositions, but always supposes the combination of both animals and plants bred by human families.

The archaeological analysis includes records corresponding to those sites (Camiña-1, Nama-1, and Pukarqollo), as well as other settlements (Pariqollo, Kuico, Chusmiza, and Jamajuga) localized in the surrounding *quebradas* (Gonzalez 2010, 2020). All archaeological sites are dated to the Late Intermediate period (AD 900–1450), when herding and agriculture crystallized local processes in compliance

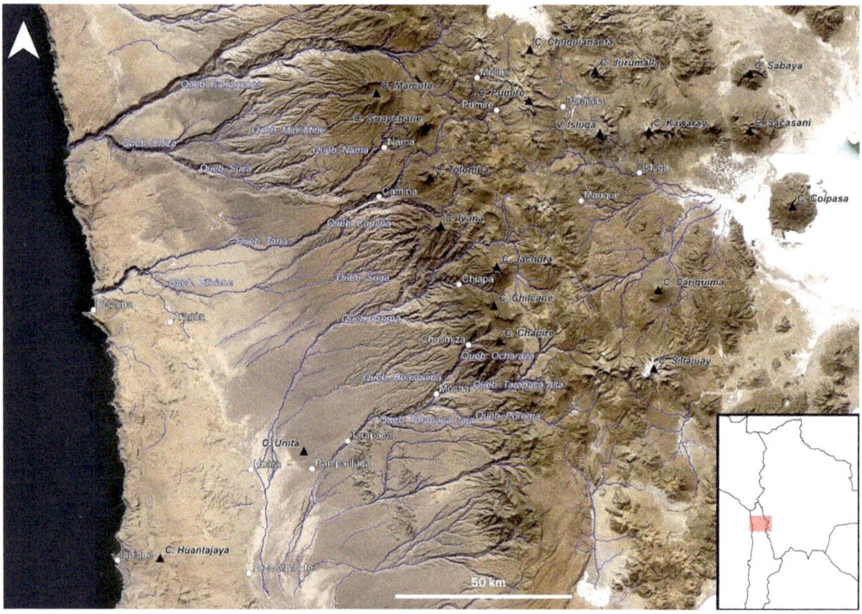

Fig. 2.1 The section specified on the white box at the bottom, corresponds to the northern Chile section shown in the main map. From north to south, the principal ravines or *quebradas* from Camarones to Tarapacá are specified. These *quebradas* and their names are underlined in blue. To the west, the map includes the coast of the Pacific Ocean and to the east, the hills located in the altiplano region, identified in black. In white letters, the names of actual villages appear where the ethnographies were produced, which coincide with the archaeological sites named in the text. Image by P. Mendez-Quiros

with regional developments that anchored its dynamics in early times (Santana-Sagredo et al. 2024; Uribe et al. 2020a, b; Vidal-Elgueta et al. 2024). The identification and quantification processes followed Mengoni Goñalons (1999) and Gifford-González (2018), while the specimens were identified using osteological manuals (Sierpe 2015; Reise 1973) and bone collections. Because studies were performed at different times, quantification results presented here are summarized as NISP, meaning the total anatomically and taxonomically identified elements, not including minimal fragments and unidentified specimens.

We are both part of a larger research team who has been studying Tarapacá's archaeology over a long time of collective work. Simultaneously, each one of us have carried out our own deepening joint understandings and interpretations. Ethnography allows us to pay attention and give importance to Andean thoughts and practices, opening our perception and comprehension of social reality and eventually, its archaeological record. In the Andean communities of Tarapacá, the triad of camelids, humans, and pastures is lived having the ancestors of the landscape, *Pachamama*, and tutelary mountains, as main owners and breeders. Thus, we show the cyclical interrelation between rituals and commemorations with socio-productive strategies and activities that link a series of interconnected calendars, where the Andean way of life and the historical dimension of their cultural traditions can be seen.

The ethnographic results are distinguished within historical contexts and contingent relationships, while pre-Hispanic evidence enhances material memories recalling and enacting ancestral traditions. The camelid-human relationship is approached from the mutual breeding and the diplomacy related to their Andean cosmopolitics (Stengers 2005), seeking to illustrate archaeological results not only in terms of domestication and consumption, but attempting to go further into the relations embodied. We aim to describe its flourishing in local, fragile, and changing environments such as Tarapacá, identifying solid actors with their own power to affect and being affected in a myriad of settings.

2.3 The Social Life of Camelids

Camelids are the largest mammal in South America with two wild species, vicuña (*Vicugna vicugna*) and guanaco (*Lama guanicoe*), along with two domesticated species, llama (*Lama glama*) and alpaca (*Lama pacos*). Archaeological evidence shows that the transition from hunting to breeding occurred around 6000 years ago in the Central Andes (Lavallée 1995) and, in parallel, in other parts of the Andes such as southern Peru, northern Chile and northwestern Argentina (Yacobaccio 2003), establishing forms of coexistence that influenced both the agricultural process (Langlie and Capriles 2021) and the shaping of the landscape (Lane 2006).

Ethological studies (Franklin 1983) carried out on vicuñas and guanacos in their natural habitat, in a sort of animal ethnography, show that they organize themselves for access to pastures, grouping into three basic forms of societies: family groups,

groups of males, and solitary individuals. Among the vicuñas, which are today circumscribed between 3700 and 4900 m.a.s.l. (González and Donoso 2020), this organization is stricter than in guanacos, which live from sea level to 4000 m.a.s.l., allowing for greater flexibility and mobility in the composition of the groups. In particular, the leading figure among guanacos is not necessarily a male, since this role may fall on one or more females, given that the establishment of ranks is the result of the interactions experienced during the formation of the group, as events in permanent remodeling and change (Correa et al. 2013).

Within the basic scheme established by Franklin (1983) and other authors (for instance, Vilá 2002; Vilá and Arzamendia 2020), among vicuñas the family group consists of a male and three or four females with their young, occupying a pasture and resting area where they spend their days, in some cases moving away from the resting area, generally in foothills, during the first half of the day, and then returning to it during the second half of the day to spend the night (Franklin 1983) or, in other environments, feeding throughout the morning, moving to drink water and resting around midday, and then returning to graze until the end of the day (Vilá and Arzamendia 2020).

In this sense, the study by Arzamendia et al. (2018) found that the high nutritional requirements of females, who gestate and nurse at the same time, acts "as an organizing resource for the polygynous dynamics of the groups" (Vilá and Arzamendia 2020: 90), determining the formation and distribution of family groups. This basic family group structure is governed and maintained by one of the components, generally a male, who uses expulsion strategies, both of intruders and of young 1-year-old males in order not to overload the use of pastures and to avoid crossbreeding with females of the group. These expelled males form herds that move through annexed territories and eventually establish new family groups. Solitary males, either adults in search of territory or expelled senile individuals, constitute the third group.

It is most likely that the observation and experiencing of this distinctive territorial and social organization enabled hunter-gatherers to become breeders by adapting these practices to their own needs and managing the groups of animals now under human care, the herd. Thus, through the castration of the annoying and crafty young and their derivation to transport or meat for consumption, the intervention in crossbreeding, and the separation of exclusive groups of females, among other strategies, human families organize the groups of camelids at their convenience, entrusting the care of each group to different people, according to the size of their herd. For this, the strong territoriality that characterizes them helps in pastoral management, granting in some cases an independence that, at least in small herds, allows humans to release the different groups unsupervised to pasture areas, sometimes for an entire season (Tomka 1992), leaving the territories and pastures to be traveled daily to their discretion, although always within the framework previously defined by the human, who maps out the areas and movements according to the seasons and availability.

Breeding, with its repetitive habits, is a three-way arrangement between animals, humans, and pastures, all cared for by deities, in which animals and humans teach and learn from each other. Shepherds replicate the animal's social organization, so that camelid's territorial and social wisdom results in a way of inhabiting spaces, while dealing with changes and multiple reconfigurations (Despret and Meuret 2016).

2.4 Highland Herding

Broadly speaking, today's herd management in the altiplano is articulated around different vegetational formations and in correspondence with different altitudinal zones (Gundermann 1984). The *estancias (*i.e., altiplanic villages*)*, composed by groups of houses and corrals, belonging to one or two extended families, are scattered along the edge of *bofedales* or high Andean wetlands. This location is conceived as a nucleus of concentric rings through which social life takes place (Villagrán and Castro 2004), materializing certain territorial and kinship logics that intertwine different beings and modes of bonding (Urrutia 2022). Social life is material and spiritual, between the public and the private, distinguishing and relating human populations *jaqi* with the world of sacred entities or *wak'as* and the wild environment of the *sallqa*, populated by powers and interconnections that configure their own socio-natural groups.

In zonal terms, the *pampas tolares* are distinguished, characterized by shrub species generically called *tolas*, hence its name; together with the *pajonales* that are located in a higher altitudinal gradient and whose designation also refers to the existence of various types of straw or *wichu* that camelids also eat. Likewise, the *bofedales* are distinguished as *uma*, as a humid place, as opposed to *waña* or dry zones that correspond to *tolares* and *pajonales* (Gundermann 1984; Urrutia 2019; Villagrán and Castro 2004).

Taking this into consideration, we can still appreciate the organization of the herd activities around the different formations of vegetation, which involves concentration movements in *bofedales* and *pampas* surrounding the estancias, together with gradual dispersion movements in more distant *waña* lands. These herd movements are also orchestrated in relation to the ordering of the annual cycle into three epochs or seasons with their respective activities and transitional moments (Urrutia 2019). In addition to the customary use rights on each estancia, generally patrilocal, most families also maintain other grazing lands corresponding to them through matrilineal shares. On these lands, they typically have smaller houses and corrals located near the *estancia* or locality to which they are affiliated and that they occupy during certain times of the year.

The cold and windy period (*thaya pacha*) takes place from April to August, when the animals are in excellent conditions after the exuberance of pastures; therefore, it is oriented fundamentally to slaughter those individuals that are between 2 and 3 years old; in most of the cases this work is carried out in the most distant corrals of the family *estancia*, while the rest of the herd grazes in neighboring *pampas*. In August, after the "meat harvest," *pawas* or *kollas* are held to Pachamama and the hills, ceremonies of thanksgiving for fertility and rituals to ask permission to begin the new cycle. Then come the dry months (*junt'u pacha*) from September to November when shearing, or "wool harvesting," takes place according to the availability of pastures; either at the beginning of the dry season if a great shortage is expected so as not to reduce the quality of the fiber, or towards the end of this season so that it grows nicely in the wet months of greater abundance and quality of pastures. All in all, it is easier to herd because the animals stay together and are appeased by the decrease in food. The rainy season (*jallu pacha*) from December to March is the busiest time as calving and mating take place, along with the ceremonies of livestock branding (*Wayño* in Isluga or *K'illpa* in Visviri, both Aymara territories in the extreme north of Chile) and carnivals. Despite the continuous work and attention required by the herd, it is a time that is lived with joy and gratitude, as it is lavished in community and ceremonial festivities that enhance the intensity of activities around the herds.

In the altiplano, during the dry season, animals are kept in the vicinity of the *bofedales* and despite all the care taken, mortality is extremely high, especially between October and November, which is the critical season, decimating a third and up to two thirds of the herds (Villagrán et al. 1999). Maintenance throughout the dry months is made possible by complementing, and often supplementing, their feed with baled pasture from the low valleys. Care is intensified with deworming to avoid weakening of the animals. On the other hand, herding at this time requires little work, so the families try to clean the channels that cross their respective *bofedales*, even artificially expanding the watercourses to irrigate a larger area (Urrutia 2019). They are also busy preparing or digging the land for cultivation, cleaning it of roots and leveling it, and then fertilizing the soil before sowing quinoa and potatoes, sometimes beans and garlic, where they also accompany their respective rituals.

The transition to the wet season takes place ideally in December, when the first raindrops begin to fall and the vegetation covers are renewed. The arrival of the rains in January is received with great rejoicing and agitation, as the herders must be extremely vigilant with their animals. Towards the end of the month the calving season begins, and between January and February it is followed by mating or crossbreeding, which is known as "sowing of the (camelid) cattle." Since the gestation period of these animals is 11 to 12 months, production is organized by coordinating births and breeding to coincide more or less on the same date, choosing for this purpose the beginning of the wet season, when the time of greater availability of pastures begins, which ensures the good development of the herd and the optimum growth of both the offspring and the fiber. The same occurs in wild herds, which calve between December and February.

Each year only half of the group of females is destined to mating, the rest graze with the *maltones* (young animal that has not reached sexual maturity) from the previous crossbreeding and is destined for wool harvesting. Pregnant females are not sheared since this work affects and can spoil the gestation process. During breeding, the herders take the opportunity to make a genetic selection of their animals, placing as stud those males considered stronger, larger, or of the preferred color of the moment, with abundant and good-quality wool. The females intended for mating are grazed alongside the males in the *bofedales*, and sometimes the breeding occurs in the corrals of the *estancias*. These corrals, typically circular and smaller, are situated near the main quadrangular corral. Meanwhile, the remaining females are kept in the *pampas*. Although the distance between these sectors is not great, the herder primarily tends to the troop of mothers in the *pampas*. Nevertheless, at least once a day, he checks that everything is in order with the other animals (Urrutia 2011, 2019).

Also in February, the livestock branding ceremonies or *floreos* (*Wayño* or *K'illpa*) take place, where the animals are marked with cuts on their ears and by pinning wool flowers on them, through a ritual assembly of registers and materialities, signaling at the same time family and community relationships that connect with the environment and the sacred entities that lord it over. Although access to grazing lands is communal, animals are always individually owned, regardless of who may be in charge of them (Dransart 2002; Urrutia 2011). During these ceremonies, some camelids are given to daughters and sons, who in the past actively participated with their parents in the care of the herds. In Aymara, while bred animals are distinguished as *uywa* (or *yuya* for cultivated plants), it is the tutelary hills, called *uywiris* or *achachilas* as the primal ancestors who foster everything around, that are the true owners of the surrounding landscape and who lend the raising of certain herds to human groups (Arnold 2016; Grebe 1984). Thus, each generation of animals must be reintroduced to the care of human families during the marking or *enfloramiento* ceremonies of the herd (Dransart 2011).

In the highlands, herding shapes much of the calendar, routines, and movements across various landscapes. Similarly, in the *quebradas* or ravines, the care and development of agricultural fields concentrate most activities and energies. However, in both ecozones, agropastoral economies flourish amidst profound transformations and with distinctive emphases.

2.5 Camelid Herds in Precordillera and Low Valleys

Until the 1970s, alongside periods of concentrating the family herd in *bofedales* and dispersing them in the surrounding altiplano pastures, there was a transhumance of livestock to the highlands of the adjacent *Precordillera*, known as *costeo* (Gundermann 1984; Martínez 1989; van Kessel 1996). From April to September, herders and their animals descended from their ranches in the Cordillera to the *paskanas* or smaller settlements in the upper parts of the *quebradas*: the *pattja qherwa* or headwaters of ravines and *suni* or the lower plateau. There, they took advantage of mature rainy pastures after the wet season, which fattened llamas; it also allowed them to reserve the high Andean pastures and meadows for times of greater fodder shortage. By March, the hills of the *pattja qherwa* turn green and the *suni* plains resemble a sea of grass; from May onwards, the pastures become less tender, and with the frosts of July, they turn almost woody. Thus, through *costeo,* herders annually occupied lands located across transitional territories, with each Altiplanic family having its respective places established through kinship ties, godparenting, and longstanding customs, with families residing in the *Precordillera*.

Moreover, the main towns of Camiña in the *quebrada* and Isluga in the altiplano share Santo Tomás as their main patron saint, celebrating this advocation at different times: in Camiña, it coincides with the feast of Corpus Christi (which moves between May and June), and in Isluga, it is commemorated on its Catholic saint's day (December 21). The *costeo* routes along the once intense network of *tropero* (drover) roads were interwoven with pilgrimages during the patron saint festivities in both directions, ritually connecting social and cosmopolitical links between the two landscapes with their respective productive calendars. Since the end of twentieth century, the upper part of the Camiña *quebrada* has been mostly inhabited by families from Isluga, who have bought and own a large part of the agricultural fields, without losing their rights to the Altiplanic *estancias* or ceasing to participate actively in their festivals and ceremonies. There is an interesting tension between native families from the *Precordillera*, whose residents are a minority and still own much of the land, and families with altiplano roots who have a long history of interrelations with these landscapes (Urrutia 2011).

Although no longer practiced systematically, as beings, spaces, and times involved have reduced their warp, some Altiplanic families who moved their main residence to the *quebradas* and turned to agriculture still maintain minor circuits of transhumance or mini-*costeo* (Urrutia 2011; García 2018;). The cessation of widespread *costeo* practices coincided with significant migratory flows of the Altiplano population to the agricultural valleys and cities, framed by a series of prolonged droughts between the 1960s and 1970s that dramatically reduced the number of livestock in the Cordillera. In the past, it was common for families to have 500 or more animals; today, herds barely exceed a hundred (Dransart 2002; Urrutia 2019).

In parallel to the high Andean grazing practices, their seasonal movements and historicities, permanent camelid herds have existed and still exist in the *Precordillera* and lowlands, in a smaller proportion because of the predominant agricultural orientation. In *quebradas*, the livestock is kept and foraged in corrals, near the houses or arranged in the vicinity of threshing floors and cultivation terraces. The agricultural sectors are called *chacras*, while *campo* refers to the surrounding hillsides and hills. In the upper Camiña *quebrada*, most families have two or more circular stone corrals that are located near a set of *chacras*, where some are reserved for growing alfalfa for fodder. In Nama, which is higher, at the head of the *quebrada*, the corrals tend to be quadrangular and large, arranged on agricultural sectors, as well as circular and grouped on the hillside near the houses. In the Tarapacá *quebrada* and the lower part of the Camiña *quebrada*, in addition to stone corrals, there are constructions made of cane, like *quinchos*, and combinations of both materials. In Mocha, for example, a corral consists of a space delimited by wooden fences with a roof that protects from the sun, located on the edge of crop fields.

The animals live most of the time in the corrals, where they are fed and watered. Only when the harvest is finished and before preparing the land for sowing again are the animals taken to the fields to consume the leftover vegetable matter. They are often left tied to a stake, with the length of the rope controlling how far each one can graze. In the corrals, the camelids are given *chala* (stalk, leaves, and cover) of corn, along with alfalfa, when possible, supplemented with grass and *sorona* (*Tessaria absinthioides*). The first task at dawn is to tend to the livestock; then the farmers have breakfast and go to work in the fields. At the end of the day, the last task is also to feed and care for the animals after the crops labor.

Just like in the altiplano, the males that mate with the females are changed every 2 or 3 years to improve genetics and preserve diversity. However, in the *precordillera*, the herds are very small and are almost entirely dedicated to meat production; also their fiber is considered of lower quality than that produced in the altiplano. Additionally, depending on the family, rituals are performed during mating and marking ceremonies although they are simpler compared to the grand *floreos* of the altiplano. Most farmers in the *quebradas* have 5 to 10 animals in one or two corrals; those with the most barely exceed 20 with three to four corrals; although in some villages in the upper parts, there are still families that maintain 50 to 80 animals, which was the norm in the past.

It is said that there used to be more animals in the *quebradas*, and the grazing lands in the surrounding hills were utilized. Occasionally, they were taken to places with abundant pastures close to springs. Often, families had farms or fruit orchards there, benefiting from all the circuits around more distant agricultural lands. It was common to have horses and mules for travel and cargo transport, along with cattle, sheep, and camelids. Troops of Altiplanic groups also arrived with their products to exchange for those from the *quebradas*.

> I grew up in Berenguela [estancia in Isluga] with purely natural things. When I was a child, my father traded with people from Apamilca [a town in the upper Camiña quebrada]. They would load up with llamos [male llamas] and donkeys. Some llamos were called quintaleros, special for carrying, they were castrated and carried charqui [dried and salted meat] and wool, sometimes quinoa, cheese and ropes, fabrics, also salt. He exchanged for corn, pumpkin, pears and quinces. The exchange was conducted either by arroba or by sack. Four arrobas make a sack and it is a quintal, about twelve and a half kilos today. An arroba of charqui for one of corn, that's how it was, by cambalache [trading]. In the old days they didn't talk about a kilo, they talked about a quintal and one arroba. They used to go on the troperos roads and they used to trade (Isidora Ramos, farmer and shepherdess, community of Apamilca, 2022. Personal communication).

2.6 Zooarchaeological Records in the Tarapacá Region

Evidence of a longstanding relationship between camelids and humans is apparent in the pre-Hispanic refuse found at different archaeological sites across Tarapacá region. At Mocha, two *pukaras*, or hillside settlements, are situated in close proximity to each other and to the present-day town. The Pariqollo *pukara* (2170 m.a.s.l.), located at the confluence of the Tarapacá and Ocharaza *quebradas* (Fig. 2.2), represents a settlement consisting of approximately 200 terraced dwellings that were occupied during the Late Intermediate period (AD 900–1450). Among the eight excavated enclosures, skeletal remains of consumed animals are found in the refuse,

Fig. 2.2 Pariqollo *pukara* showing the enclosures distributed on summit and slopes. Red colored earth in the back section of the hill also bears enclosures but with less frequency due to minor space availability. Image by P. Mendez-Quiros

with 24.21% of the specimens, equivalent to at least seven individuals, belonging to the Camelidae family. The complete skeleton of the animal is represented, with an age range that includes sub-adults and adults (Kaufmann 2009).

Currently, llama meat is part of the *quebrada* diet, which is combined with that of other animals, both raised in corrals or purchased in the cities. In Mocha, as in other villages in the *Precordillera*, its inhabitants are mainly dedicated to agriculture, keeping in use many *chacras* of probably pre-Hispanic fabric, and giving vitality to the local Andean traditions. The fulfillment of ceremonies is an important effort within the productive and social strategies, likewise, the most elaborate rituals involve the *wilancha* within their protocols, consisting in the sacrifice of a bred animal (*llama* or lamb) to thank the *Pachamama*, the tutelary hills and the ancestors, as well as god, saints, and virgins, for the fostering of life they accomplished, also to bring about good luck and good health in all social relations, either among humans and with other beings and forces.

The remaining percentage of refuse originates from the possible hunting of vizcachas (*L. viscacia*), along with a small number of Guinea pig (*Cavia* sp.) feces, indicating the breeding of this rodent, which holds deep and multifaceted significance in Andean culture (Morales 1994; Rofes and Wheeler 2003; Castro 2009; Urrutia 2011). Today, vizcacha is considered a wild animal, often kept as a pet and only sometimes consumed; a very skillful procedure is required to cut off the tail before it sours the meat. As for Guinea pigs, they are still raised by some families today, mainly for meat consumption, as well as for medicinal and ritual purposes. Additionally, in most of the excavated units, a small number of camelid feces are present, suggesting their possible use as fuel, which would have been transported from the corrals located in the shaded areas along the river or near the cultivation areas, as is still the case today. At times, families have *llamos* or *alpacos* (male alpacas) that are *guachitos* (camelid offspring not nursed by their mothers and cared for by humans); these *guachitos* share the space around the house with other domestic animals and are highly cherished within the family. Furthermore, it is plausible to consider the presence of camelid feces as part of offerings in domestic rituals, as it has been documented in the Jujuy Puna. Similarly, remnants of natural-colored fleece suggest the use of their hides in dwellings, as well as the production of yarns.

Located on the opposite bank is the Kuico *pukara* (2350 m.a.s.l.), a settlement built on a long and narrow plateau of difficult access, at the confluence of the Tarapacá and Poroma *quebradas*, where the bases of approximately 700 stone-walled enclosures are preserved (Fig. 2.3). Occupied for a short period of time, parallel to the Pariqollo occupation, the skeletal remains found here, recovered from

Fig. 2.3 Kuico *pukara* with the enclosures distributed in the upper meseta. Pariqollo *pukara* is seen on the opposite river bank, to the left, as indicated. Between the hills, the vegetation shows the agricultural fields watered by the Tarapacá river, and the current Mocha town above. Image by P. Mendez-Quiros

seven enclosures, are predominantly from the Camelidae family (43.62%), with the entire skeleton represented and age classes restricted to juveniles up to 2 years old (Kaufmann 2009) for seven individuals. As in Pariqollo, there are also camelid feces and their possible association with combustion for culinary, ritual, or other purposes. Fleece and fur remnants and *Cavia* sp. feces are absent, probably due to the shorter time it was occupied.

Other five *pukaras*, four of them located in different *quebradas*, and one in the altiplano are also included (Uribe et al. 2007; García and Uribe 2012). Camiña (2400 m.a.s.l.), in the Camiña or Tana *quebrada*, composed of 588 densely clustered enclosures, of which 13 were excavated; Nama (3000 m.a.s.l.), situated at the confluence of the Nama and Guayaquil *quebradas*, with 558, also densely clustered enclosures, 13 of them excavated; Jamajuga (2880 m a.s.l.), in the headwaters of the Mamiña *quebrada*, upstream from the present town, with 135 enclosures of which five were excavated. Finally, Chusmiza (3240 m.a.s.l.), located in one of the main tributaries that feed the water system of the Tarapacá *quebrada* itself, where nine of the 147 enclosures were studied. In all four *pukaras*, situated close to the present-day villages, living spaces and storage structures were functionally distinguished (Urbina and Adan 2006). Also, all of them are surrounded by agricultural fields that optimize agricultural space by the construction of hydraulic management systems and extensive terraces.

The associated zooarchaeological results show that camelid remains represent 24.21% in Pariqollo, 43.62% in Kuico, 29.82% in Camiña, 54.25% in Nama, 64.86% in Jamajuga, and 85.14% in Chusmiza, with camelid refuse shared, in different amounts, with rodents that include the vizcacha (*Lagidium viscacia*), chinchilla (*Chinchilla brevicaudata*), and cururo (*Ctenomys* sp.), along with fish, reptiles, and birds in minor proportion (Table 2.1). Reptilia (*Liolaemus* sp.) are possibly the result of natural processes, but there are known medicinal uses that justify their incorporation in the quantification. In Mocha, reptiles were described as used for wound treatments, where lizards, opened in halves, are applied on the affected body member, secured by a bandage and left for several days until the drying of the wound.

Higher up in the altiplano, the Pukarqollo *pukara* (3900 m a.s.l.), is composed of 586 enclosures built on the slopes of an isolated hill surrounded by pastures and *bofedales*, near the present-day Isluga village, whose current occupation is only available for community ritual commemorations, established by ancestral rules and practices, or *costumbres*, along the calendar cycle. In the archaeological site, seven structures were studied, showing that 57.4% of the bone remains correspond to camelid elements, with the same rodent species identified in the *quebradas* among the accompanying fauna (Gonzalez 2010).

Vegetable refuse associated to the *quebrada pukaras* include *Zea mays* and *Chenopodium quinoa* as the most significant plants cultivated, with a less significant presence of *Lagenaria* sp., *Cucurbita* sp. and *Gossypium* sp., all of them cultivated at low yields, eventually oriented to self-consumption (García and Uribe 2012). Compared to former periods, local plants gathering in Camiña and Nama is diminished (although present as tree species such as q'ero (*Escallonia angustifolia*)

Table 2.1 Zooarchaeological refuse in *quebrada* and altiplano *pukaras*, where NISP represents the number of taxa elements identified

Pukara	Taxa	NISP	% NISP
PARIQOLLO	Birds?	9	3.16
	Camelidae	69	24.21
	Rodents	18	6.32
	Mammalia	188	65.96
	Reptilia	1	0.35
	Total	*285*	
KUICO	Birds?	2	2.13
	Camelidae	41	43.62
	Mammalia	51	54.26
	Total	*94*	
CAMIÑA	Birds	3	0.49
	Camelidae	181	29.82
	Rodents	400	65.90
	Mammalia	4	0.66
	Fish	19	3.13
	Total	*607*	
NAMA	Birds	1	0.27
	Camelidae	198	54.25
	Rodents	82	22.47
	Mammalia	84	23.01
	Total	*365*	
JAMAJUGA	Birds	2	2.70
	Camelidae	48	64.86
	Rodents	23	31.08
	Canidae	1	1.35
	Total	*74*	
CHUSMIZA	Camelidae	63	85.14
	Rodents	6	8,11
	Mammalia	5	6.76
	Total	*74*	
PUKARQOLLO	Birds	5	0.89
	Camelidae	322	57.40
	Rodents	233	41.53
	Reptilia	1	0.18
	Total	*561*	

and algarrobo (*Neltuma* sp.) (García and Uribe 2012), representing a shift in Late Intermediate communities' management, when extended and intensified interaction took place, while the *quebrada* sites were consolidating their main position on Tarapacá developments. On the other hand, the exclusive presence of quinoa (*Chenopodium quinoa*) in the Altiplanic Pukarqollo advises a different relevance on agricultural activities, where herding fosters their main socio-productive strategy, incorporating the management of their own cultivated plants, of which only some species remain through time (today, along with quinoa, potatoes, and garlic are sometimes cultivated). In highland landscapes, one can see at first sight the grazing herd fluency with big corrals surrounding the *bofedales* near each *estancia*, while agricultural fields generally appear after a second view. In comparison, in *quebradas* landscapes the prominence of *chacras* is conspicuous and cultivated terraces spread to foothills with an intricate irrigation system, giving corrals and herds a secondary role.

Relationship between camelids and humans includes the presence of other animal species, as much as botanical species, as it is observed in archaeological sites covering these two Andean ecozones, showing the combined use of cultivated and collected plants, together with raised and hunted animals. The appearance of sea shell (*Oliva peruviana*) in Pukarqollo, very common in the *quebrada pukaras*, especially as an offering in both public and domestic contexts, suggests the ritualization of a coastal component in the Tarapacá archaeological complex. The time when agropastoral strategies flourished on inland territories and *Precordillera* played a central role, connecting highlands and lowlands (Uribe 2006; Valenzuela 2010). Distinctive practices were used, according to environment and to different decisions that involved cultural and political aspects among segmentary and autonomous groups with social diversity, who may have managed their herds in different and intertwined ways (Uribe 2006; Santana-Sagredo et al. 2024).

2.7 Calendar and *Costumbres*

The annual climatic cycle of dry, wet, and cold seasons, so evident in the foothills and the altiplano, gradually loses its influence as one descends to the more arid pampas. However, during the rainy season, its effects are also noticeable in the lower valleys, pampas, and coast when the river channels swell and the spring water recharge allows for the regeneration of an otherwise seemingly desert landscape. Today, Andean farmers in the lowlands describe the pampas as fertile land, a productive desert that fascinated the early Spanish chroniclers who traveled through the Tarapacá region (Urrutia et al. 2023). This also implies the need to adapt to certain changes in the climatic cycle, such as long periods of drought or devastating flood events.

Thus, the rainy season is a time of great rejoicing everywhere and involves a lot of hustle and bustle with the camelid herds, with calving and crossbreeding in December, flowering in January, and Carnivals in February. During Carnivals, homage is paid to the Pachamama and tutelary hills, thanking them for their abundance and fertility with offerings of the first fruits and crops. *Carnivals is when Pachamama gives birth* (Lupercina Chuquillijta, farmer, community of Mocha, 2023. Personal communication). In Camiña and Isluga, these celebrations are often joked about as times of mating in the community, with songs featuring picaresque lyrics while people dance in lively rounds. The similarities of mating and crossbreeding among different types of beings (*Pachamama,* camelids, and humans) reveal the poetics of the efforts to maintain and promote balance. This balance is understood as a genetic force in permanent movement, a constant flow of reproduction, and the circular forms that materially embody good and fertile relations (Pazzarelli 2022).

The transition to the cold season is perceived as the last attendance for the livestock harvest, along with the quinoa and potato maturation in the altiplano, as is the corn development in the *quebradas*. Therefore, if Carnival marks the first fruits of production, its abundance is realized with the arrival of the cold months when the proper meat and crop harvest deploys (Table 2.2). At that time, the livestock was fat and in excellent condition because they ate good-quality pasture during the rainy season, which also boosted crop growth. The meat from the animals ensures the family's consumption, and sometimes a portion is also allocated for sale; some pieces are prepared as *charqui*, dried meat.

In August, at the end of the colder months, the Pachamama is hungry, and *pawas* or *kollas* are officiated to ask for permission and for the new cycle to be favorable for everyone, bringing health and luck. In September, after the most intense cold has passed, it is ideally the turn of the animals to be "shaved." Camelids take 2 to 3 years to develop an adequate volume of fiber; however, pregnant females can miscarry due to the effort of catching and keeping them still; consequently, these tasks are coordinated so that the shepherds have new fleeces annually and distinguishing some females only to maintain reproductive flows. In the dry season, troop care is intensified due to the decrease in pastures, while also taking advantage of the time to clean channels, prepare, and sow the land.

In both the altiplano and the *quebradas*, everyone agrees that the meat of animals raised in the foothills has a better flavor than that of the high plateau and that alpaca is tastier than llama. In fact, it is a highly valued gift when relatives from the ravines give it to the couple of *alférez*, who are in charge of the festivities of the *marka* of Isluga or main town. *Charqui* can be made both in the altiplano and in the *quebradas*. It is preferably taken from the back (*ulla*) of a *llamo* or *alpaco* although any piece can be used. *The time for making charqui is in June, July, or August, depending on how cold the season is [between June during the cold season, or July and August during the transition to the dry season]. The meat is salted and dried by hanging or placing it on top, turning it over repeatedly. Once it is a little dried, it is pounded and stored in the hide of a llamo* (Emiliano Chamaca, farmer and shepherd, community of Alto Camiña, 2022. Personal communication).

Regarding the differences in flavor between *charqui* from the *Precordillera* and the altiplano, as with the meat, it is often commented that "in the quebradas it is richer, it is grassier, and not so much hill [pastures]." However, in reference to its nutritional value and durability, opinions are the opposite. "In the quebradas, the charqui lasts about a year, then it turns yellow and gives off an odor. In the altiplano, the charqui lasts longer because it is colder there, the meat stays white; here it is not as nice" (Teodora Calle, farmer and shepherdess, community of Nama, 2022. Personal communication).

According to the amount of wool, whose higher quality and quantity blossom in the altiplano, an adult llama gives 2 kilos at a time and a young llama does about 1

Table 2.2 Complementary cycles between the altiplano and the Tarapacá *quebradas*

Annual cycle		Activity calendar: Altiplano	Activity calendar: Quebradas	Festive cycle
Transition	August	Troop maintenance in meadows and high Andean plains	Soil preparation *(Covar)*	Pawas o Kollas to Pachamama San Lorenzo
Dry season *(junt'u pacha)*	September	Deworming, Wool Harvesting	Cleaning of farm agricultural channels	San Nicolás, Exaltation Chilean National Holidays
	October	Cleaning of wetland channels Quinoa and potato plantations	Corn, potato, and quinoa crops sowing	San Francisco
Transition	November	Livestock concentration on farms	Crop growth	All Saints San Andrés
	December	Appearances Breeding	Reserve Hilling	Immaculate Conception, St. Thomas Christmas/Making Nativity
Wet season *(jallu pacha)*	January	Livestock Branding (marking)	Fertilize *(Guanear)*	Shepherds, kings*Wayños* o *Killpas*
	February	Troop care in wetlands	Crop development	Virgin Candelaria, Lourdes Carnivals
Transition	March	Dispersal of livestock in *waña* areas	Grass removal Composting	San José
	April	*Traveling along the coast to the headwaters of ravines*	Crop ripening	Easter
Cold season *(thaya pacha)*	May	Troop care on hills and headwaters	Early crop harvesting	Santa Cruz, San Isidro Holy Spirit
	June	Crop harvesting Meat harvesting	Late crop harvesting	Corpus Christi, San Antonio, *Willka Kuti*, San Juan
Transition	July	Selecting seeds Preparing *charqui* and *chuño*	Land fallowing Selecting seeds Preparing *charqui (charqueo)*	Saint James Virgin of La Tirana

kilo. "We get more from the back, ulla; that part is used because the legs and the neck are not used. Before, the grandparents knew how to make slippers with the neck of big llamos." In contrast, the alpaca yields more, "three to four kilos depending on the size of the animal; a baby alpaca yields up to two kilos. Wool is also taken from the feet and the neck." (Isidora Ramos, farmer and shepherdess, community of Apamilca, 2022. Personal communication).

Guano is used as fertilizer for crops. In the corrals of the *quebradas*, it is cleaned periodically and piled up in fallow or unused fields. It is left covered for a couple of months so that the seeds ingested by the animals die and do not grow as weeds in the fields but enriches the selected crops. Afterwards, it is stored in sacks and mixed into the soil during preparation before planting. Another way to access guano is by taking advantage of the resting areas in the high Andean *pampas*; it is also collected in sacks and left for a period of time before being used as fertilizer, airing it in the sun.

In addition to meat, wool, and guano, camelid hides are used for flooring, chairs, and beds; hides from the neck were used to make sandals that were tied with ligaments, and they were also used in the construction of houses to tie beams and posts. Textile tools, such as *wichuñas* that tighten the weft of traditional weavings, are made from the bones of the legs. Similarly, camelid hides and urine are also used for medicinal purposes, as well as fat. Along with the *wilancha* and the divination processes involved in folds and guts during the ritual feast of main conmemorations, *llamo* fat or *untu* is a vital element in sowing ceremonies, and stuffed *llamo* fetuses also participate as special offerings to thanks and petitions for good luck.

2.8 Ethnographic Assembly

The current parallel and complementary experiences between the Altiplano and the *quebradas* imply the importance of collective and familial management, the relevance of the ritual as an ethic of sociability that unfolds in a rich calendar to negotiate and update agreements among different human and non-human beings, along with the multiplicity of modes of cooperation within various affinity groups grounded not only in kinship, but also in ceremonials and cults, or simply in technical and productive justifications. This also translates into mechanisms of multiple residency and complex work arrangements that enable both dispersion and the maintenance of property (Urrutia 2011, 2022). All intertwined by profound transformations and deep temporalities.

For Andean groups, the *mallku* and *t'alla* hills stand as their first and primal ancestors (Grebe 1984; Martínez 1989; Arnold 2016). They are referred to as tutelary deities, precisely because they unite and sustain a network of links that have a cosmological rather than merely territorial meaning. In Tarapacá, they are generically referred to as *achachilas*, *uywiris*, or *aviadores*, responsible for nurturing life and arranging things in the world. Although hills and ancestors of the landscape can correspond indistinctly to feminine and masculine figures, the *Pachamama* is their feminine counterpart, representing a receptive and generative force, while the *uywiris* embody the realm of creative powers. Both are configured according to terms that link agriculture and pastoralism with hunting and gathering, husbandry and the wild. In short, both biological and cultural reproduction ultimately depend on these entities. Therefore, the connection with them is an unavoidable component of productive and social efforts.

Based on traditional dynamics, we can affirm that Andean sociality develops through a reciprocity that invokes all living beings that populate the world, both visible and invisible. In the human labors sphere, the exchange of joint effort is instituted by the *ayni* or *torna* (reciprocal work between people) and the *mit'a* or turns (community tasks in which a representative of each domestic unit takes part). However, reciprocity practices are not only related to cooperative forms of Andean work, nor are they only related to the economic-productive spectrum; they also encompass the cultural and religious spheres, as well as each of the components of social life (Dransart 2002; Urrutia 2011).

Thus, the direction and extent of reciprocal exchange circuits define what is inside (or what is potentially included) and what is outside (or what is potentially excluded) of each social unit (family, community, *ayllu*). In this permanent and contextual relationship of interiority/exteriority, the mutual expenditure of power and energy (*ch'ama*) also describes the complementarity that must be taken into consideration to configure as a unit (Cavalcanti-Schiel 2007; Urrutia 2022). In other words, community does not constitute an inherent condition or a fixed belonging; instead, it is orchestrated in the multiplicity and fluidity of continuously ritualized links that allow the sharing of the burden of life in common, understood simultaneously as gift, debt, and duty (Esposito 2003).

"We say in Aymara ayniña kutiyaña, which means to do, to return the torna... Thus, we live, always reciprocating" (Bernabé Vilches, shepher, community of Chapicollo, 2008. Personal communication; in Urrutia 2011:198). It is, therefore, an Andean ideal signified as *uywaña* or mutual husbandry as co-breeding, "knowing how to nurture the upbringing of life" (Martínez 1989; van Kessel 1989), which entails a deep knowledge of the different forces in constant mutation that intervene in the living world, as well as the capacity to propitiate creative flows and avoid stagnation. Reciprocity is then to respectfully engage the other to act for common benefit, where ritual constitutes its textual regime or preferred communication device that stipulates and promotes the accurate social ethic, relating the many beings and the different worlds (Urrutia 2022).

It is known that the greatest enhancement of reciprocity and deployment of community ties is expressed in the festivities and political-religious positions, whose ritual calendar establishes and renews the agreements of sociality, in time with the changes that are incorporated and reformulated in the significant multiplicity of tradition. The *costumbres* or *sarawi*, as the Andeans call ceremonies, practices and beliefs linked to Andeanness, both from "*uywir* part" and "god part," are not only anchored in immemorial times and remote pasts, but above all, they embody the distinctiveness and give meaning to new situations, transforming change into tradition and leaving their mark on it (Ortega 1998). These *costumbres* evoke the ritual work of offering food to sacred beings to thank and foster its nourishing powers, as well as exhibiting the historical dimension of their own cultural traditions. In the distinction between devil's and saint's celebrations, god part and *uywir* part, the tradition not only operates in the pre-Hispanic and ancient times but also incorporates, according to its own contents, the colonial and the contemporary. *Sarawi* refers to a certain ethic of sociability that advocates Andean ideals on acting, thinking and feeling.

The festivals of the devils or the *uywir* part eloquently materialize vital and prosper links between camelids and humans. In August, both in the altiplano and in valleys and *quebradas*, *pawas* are conducted for *Pachamama* and her counterpart, the tutelary hills, to whom the offerings are directed as a generic expression of source and foundation for the living. It involves feeding her or being consumed by her, as during August, *Pachamama* is especially hungry (Bugallo and Tomasi 2012; Arnold 2016). Coca, tobacco, and *untu* (llamo fat) are given, along with sprinklings of *chicha* and alcohol, as well as *copal* and *koa* incense. It is a family ritual conducted by the owner of the house and not by a specialist. The August rituals do not distinguish between grazing or hunting animals, nor agricultural activities; instead, they are about thanking and bringing joy to tutelary deities for having raised and nurturing all the organisms that populate the landscape.

In the Isluga altiplano, the Christmas celebration at the end of December is called *Hacer Nacimiento* (to give birth), blending rituals of the god part and *uywiri* part. It is not only to commemorate the arrival of the Child God, but it is also an occasion to ask the Lord and Lady of the Animals (hills and *Pachamama*) that the births of the flock soon to come are blessed (van Kessel 1996; Urrutia 2011). Families craft small clay figures of camelids with desired shapes and attributes, as well as collect small stones with colors of their choice. All of this is offered to the Child God in the church and then buried in a place near the corral of their respective *estancia*, devoted to their own ancestors.

Carnivals during February are framed around the first fruits of the harvest, when crops and herds are in full growth and require special care to ensure the desired production for May and June. It is a community rite that combines gratitude and merriment, bringing together families and the community in a massive commemoration led by indigenous authorities (*caciques*, *mayordomos*, *yatiris*). These celebrations combine life analogies and fertile relationships between the calving of livestock and Pachamama, along with the mating of animals and humans.

The corrals stand for a fractal representation of the world; they are not merely enclosed fences but rather relational and porous settings. It is there that the rituals of livestock branding or *enfloramiento* take place, *K'illpa* in Visviri and *Wayño* in Isluga, organized throughout January, after Christmas and before Carnival (Urrutia 2019). This marks the reintroduction of herds into the care of human families and the consecration of the corral as a vital space for social reproduction (Dransart 2002, 2011). Early on, a *wilancha* (animal sacrifice and ritual feast) begins with an Andean *mesa* inside the corral, displaying a woven blanket and other minor textiles with offerings such coca leaves, alcohol, flowers, tobacco and a fire aside, where copal, incense and certain herbs or *koa* keep on smoking.

The *wilancha* is a cherished gift, led by a ritual specialist who knows how to communicate with the sacred. The reunited community delivers one of its finest animals to the sacred forces, personified in Pachamama, Achachilas and ancestors, as the true breeders of all the living. The blood of the animal is given to them in a hole in the ground, specially designed for that purpose. Part of the blood is scattered around the corners, naming the main deities loved by the community and mapping their territory. The leather and bones of the animal, and the best portion of its meat are also put into this hole, a mouth or liminal gate which connect different worlds. After the ritual feast, the hole is carefully covered until the next calendar experience (Urrutia 2011).

In the *Wayño*, after the *wilancha*, the animals are marked by cutting off their ears and pinning woolen earrings, as well as adorning them with brightly colored fleeces on their backs (Fig. 2.4). The first to pass through are the front pair, also adorned with colorful braided collars. Constant smudging and libations with alcohol are deployed to the ancestors and sacred entities. Families dance and sing in lively rounds around their flock, often keeping stuffed *tites* (wildcat, probably *Leopardus jacobita*), considered as the shepherd that takes care of the livestock of the tutelary hills.

The performance of these rituals, also known as *Señaladas* in the Jujuy Puna of northwestern Argentina, is oriented towards circular forms and forces that generate constant reproductive flows, as well as ensuring good relations between different beings and dimensions (Pazzarelli 2022). This spiral and enveloping logic is materially embodied through signs, either as evidence or announcements, which reveal the

Fig. 2.4 Camelids and humans are shown during *Killpa* festivity in Visviri village of northern Chile, in the year 2019. The alliance between them is ritually reenacted, so corrals are the center of social life, consecrating the animal attachment to human families through ear marks and colored wool flowers, honoring fertile relations with tutelary deities

similarities between interconnected beings and things with their own centrifugal and centripetal forces. For example, a herd of fat and healthy animals reflects well-conducted linkages and interactions that augur a prosperous future, or the topology of the skin and viscera of animals, especially folds or pens of guts, embodies omens that entail certain processes of consultation. It is also possible to intervene and fabricate resonances; as in the case of corrals of wool and people, built from rolls and turns, rounds and chants during the stages of the *Señalada* (Pazzarelli 2022).

All of these ceremonies, *pawas* in August at the beginning of the annual productive cycle, "Hacer Nacimiento" in Christmas, *enfloramientos* and Carnivals during the wet season, entangle the *costumbres* or *sarawi* institutionalized by the ritual calendar, which commemorate different stages of the coexistence among the *uywaña* ideal. There are other ceremonies that belong to the catholic sphere also incorporated in the *sarawi* calendar, such as patronal festivities dedicated to the saint or virgin who each Andean town was advocated since colonial times. The distinction between *uywiri* part and god part rituals, more than separate both spheres, intends to integrate multiple worlds and sacred beings into the own Andean social ethic and life rules.

The Andean *mesa* is deployed in every ceremony, even in every event of social and political importance; the *wilancha* is also dedicated to saints and virgins during the patronal festivities. That is what *sarawi* means in terms of "naturalization of Christianism" (Mills 2007); that is why *costumbres* know how to deal with changes. Sometimes Andean tradition must resist violent historical and political transformations, but mostly it has to extend its own creativeness to reenact the new into the local understandings and current experiences of sociality, authority, and sacredness.

2.9 Opening Reflections

Ethnographic uses of the materials that enhance a contact with pre-Hispanic times, such as camelid's primary and secondary products, rodent, fish and bird's bones, together with the botanical counterparts, broaden the understanding of the archaeological faunal refuse in a way that avoid the merely consumption approach (Sykes 2014). This experimental collaboration sought to shake us off habits of thoughts that place the human at the center of all explanations and open our own comprehension to other experiences.

Tarapacá archaeological results show that, after a long period of formative achievements, around AD 1000, Late Intermediate communities of Tarapacá moved from the pampa dwellings to the *quebradas* (García and Uribe 2012; Uribe et al. 2020b; Vidal-Elgueta et al. 2024), giving more prevalence to camelid herding and expanding the previous agricultural wisdom from low valleys to *Precordillera* settings. This period is also known as Regional Developments, when Tarapacá tradition was consolidated throughout a long history of different local processes, which built a common experience respecting the territory and social groups heterogeneity (Uribe 2006).

As it is seen here in the light of the ethnographies proposed, maintenance of a camelid herd in *quebradas* and altiplano slightly differs, but their remains seem to be similar. While the livestock graze freely in the highland pastures and only at night goes to big corrals of each *estancia*, the animals in the *Precordillera* spend most of their time in generally smaller corrals located near the crop fields and are foraged daily by farmers. So far, at least vizcacha and reptile's bones, as well as camelid and *Cavia* sp. feces acquire new meanings that could lead to more complex interpretations where refuse materials get closer to past functionalities, despite taphonomic issues.

The archaeological correlations must be done considering the concurrence of the other species' networks and ways of inhabiting that leave a mark in material records. The *quebrada* settlements articulate Tarapacá developments from local solutions based on agropastoralist strategies depending on each ecological entanglement, also dealing with other regional traditions. Especially from AD 1250, the coexistence with Altiplanic groups expanded the interaction circuits and densified forms of authority along the territory and between the former populations (Uribe 2006).

Interspecies relationships are grounded in the idea of reciprocity, which can be both brutal and affectionate, understood as a "strict and continuous process of mutual consumption" (Dransart 2022). As animals consume grass, and humans consume animals, the *uywiris* and Pachamama are also hungry and eat; they are the last links in the trophic chain and, at the same time, the primordial basis of life. Grazing, despite referring to animal husbandry, is named alluding to the pasture, which is the base of the food chain, where tutelary deities are the source and sustenance of life.

The analogies of sowing for crossbreeding, as well as harvesting for meat and wool production, the similarity in the mating times of camelids and humans, the calving between animals and Pachamama, as well as the resonances between the role of the leader within the herd and the role of the cacique or *mallku* of the community, are eloquent. Many agricultural metaphors are lavished, with the corrals as spirals of life, movements, and wrappings that breed life, fostering forces and shaping forms whose folds share notions of seed and nest.

The Andean way of life, embodied in *uywaña* and *sarawi* conceptualizations, points on the importance of appearances, of seeing and doing things well, as material expressions to provoke desired effects and enter into new relationships of forces. The emphasis on material metaphors enrich the archaeological interpretation. The ritual is not only at the heart of social relations and authority ideas, as well as technology and productive efforts, but mainly is present in everyday life. The power and magic of appearances to materialize agency and reinvent traditions, challenge and refresh both the ethnographic and archaeological approach. Andean communities teach us how to let ourselves be raised by the many ways of being in the environment and be aware of our impact on them.

Acknowledgments To the research team of Proyect FONDECYT 1221166, to Tarapacá Andean community members. Last but not least, to our editors and reviewers.

Competing Interest Declaration The authors have no conflict of interest to declare that are relevant to the content of this chapter.

References

Arnold D (2016) Territorios animados. Los ritos del Señor de los Animales como una base ética para el desarrollo productivo en los Andes. In: Román-López A, Galarza H (eds) Símbolos, desarrollo y espiritualidades. El papel de las subjetividades en la transformación social. ISEAT, La Paz, pp 111–159

Arzamendia Y, Carbajo A, Vilá B (2018) Social group dynamics and composition of managed wild vicuñas (*Vicugna vicugna vicugna*) in Jujuy, Argentina. J Ethol. https://doi.org/10.1007/s10164-018-0542-3

Bugallo L, Tomasi J (2012) Crianzas mutuas. El trato a los animales desde las concepciones de los pastores puneños (Jujuy, Argentina). Rev Esp Antropol Am 42(1):205–224

Castro V (2009) De ídolos a santos. Evangelización y religión andina en los Andes del sur. Colección de Antropología. Ediciones de la Dirección de Bibliotecas, archivos y museos. Santiago, Chile

Cavalcanti-Schiel R (2007) Las muchas naturalezas en los Andes. Perifèria Revista d'investigació i formació en Antropologia 7(2):1–11

Correa L, Zapata B, Samaniego H, Soto-Gamboa M (2013) Social structure in a family group of Guanaco (*Lama guanicoe*, ungulate): is female hierarchy based on "prior attributes" or "social dynamics"? Behav Process 98:92–97. https://doi.org/10.1016/j.beproc.2013.05.003

Despret V (2022) Habitar como un pájaro. Modos de hacer y de pensar los territorios. Editorial Cactus, Buenos Aires

Despret V, Meuret M (2016) Composer avec les moutons. Cardère éditeur, France

Dransart P (2002) Earth, water, fleece and fabric: an ethnography and archaeology of Andean camelid herding. Routledge, London

Dransart P (2011) Social principles of Andean camelid pastoralism and archaeological interpretations. In: Albarella U, Trentacoste A (eds) Ethnozooarchaeology: the present and past of human-animal relationships. Oxbow Books, Oxford & Oakville, pp 123–130

Dransart P (2022) Comer y ser comido: fisiologías de la práctica en Isluga, norte de Chile. In: Bugallo L, Dransart P, Pazzarelli F (eds) Animales humanos y humanos animales. Relaciones y transformaciones en mundos indígenas sudamericanos. Antropofagia, Buenos Aires, pp 355–386

Esposito R (2003) Communitas. Origen y destino de la comunidad. Amorrortu Editores, Buenos Aires

Franklin WL (1983) Contrasting socio-ecologies of South America's wild camelids: the vicuña and the guanaco. In: Eisenberg JF, Kleiman DG (eds) Advances in the study of mammalian behavior, vol 7. Special Publication American Society of Mammals, pp 573–628

García M (2018) Otra "costa" hay en la puna. Memorias y materialidad de un espacio pastoril en la sierra de Arica-Tarapacá (ca. 2.600-4.000 msnm), Andes del norte de Chile. PhD Dissertation. Departamento de Antropología, Universidad de Tarapacá and Universidad Católica del Norte, Arica. Unpublished

García M, Uribe M (2012) Contextos de uso de las plantas vinculadas al Complejo Pica Tarapacá, Andes centro-Sur: Arqueobotánica y agricultura en el periodo Intermedio tardío (ca. 1250–1450 DC). Estudios Atacameños 44:107–122

García M, Urrutia F, Uribe M, Mendez-Quiros P, Izaurieta R, Maldonado A, Mandakovic V, Saintenoy T, Sánchez T, Vidal-Elgueta A (2023) Pampa Iluga. Tecnología, trabajo y persistencia de un paisaje agrícola prehispánico en el desierto de Atacama (50 AC-1800 DC). Revista Norte Grande 86:1–26

Gifford-González D (2018) An introduction to zooarchaeology. Springer
Gonzalez J (2010) Arqueofauna del complejo Pica-Tarapacá (950-1450 DC), I región de Tarapacá, norte de Chile. Actas XVII Congreso Nacional de Arqueología Chilena, vol I. Sociedad Chilena de Arqueología, Ediciones Kultrún, pp 59–69
Gonzalez J (2020) Informe de análisis del material zooarqueológico superficial del sitio Iluga Túmulos, Región de Tarapacá. Technical report, Project Fondecyt 1181829. Unpublished
González B, Donoso D (2020) Distribución geográfica de la vicuña. In: González BA (ed) La Vicuña Austral. Facultad de Ciencias Forestales y de la Conservación de la Naturaleza, Corporación Nacional Forestal y Grupo Especialista en Camélidos Sudamericanos, Santiago, Chile, pp 52–59
Grebe ME (1984) Etnozoología andina: concepciones e interacciones del hombre andino con la fauna altiplánica. Estudios Atacameños 7:335–347
Gundermann H (1984) Ganadería Aymara, ecología y forrajes: evaluación regional de una actividad productiva andina. Chungara 12:99–123
Kaufmann C (2009) Estructura de edad y sexo en guanaco. In: Estudios actualísticos y arqueológicos en Pampa y Patagonia. Sociedad Argentina de Antropología, Buenos Aires, Argentina
Lane K (2006) Through the looking glass: re-assessing the role of agropastoralism in the north-central Andean highlands. World Archaeol 38:439–510. https://doi.org/10.1080/004382406008.3086
Langlie B, Capriles J (2021) Paleoethnobotanical evidence points to agricultural mutualism among early camelid pastoralists of the Andean central Altiplano. Archaeol Anthropol Sci 13(7):article 107. https://doi.org/10.1007/s12520-021-01343
Lavallée D (1995) Telarmachay. Cazadores y pastores prehispánicos en los Andes. IFEA
Martínez G (1989) Espacio y pensamiento I. Andes Meridionales. Hisbol, La Paz
Mengoni Goñalons G (1999) Cazadores de guanacos de la estepa patagónica. Sociedad Argentina de Antropología, Buenos Aires, Argentina
Mills K (2007) The naturalization of Andean Christianism. In: Po-Shia Hsia R (ed) The Cambridge history of Christianism. Cambridge University Press, New York, pp 504–534
Morales E (1994) The Guinea pig in the andean economy: from household animal to market commodity. Lat Am Res Rev 29(3):129–142
Ortega M (1998) By reason or by force: lslugueño ldentity and Chilean Nationalism. PhD. Dissertation in Social Anthropology, Department of Social Anthropology, University of Cambridge, England
Pazzarelli F (2022) Viscera/Corral. Pliegues, envoltorios y relaciones en los Andes Meridionales. In Bacchiddu G, Tola F Mundos relacionales amerindios. Modos del ser en la América del Sur Indígena. Editorial Las cuarenta, Buenos Aires, pp. 119-136
Reise (1973) Clave para la determinación de los cráneos de marsupiales y roedores chilenos. Gayana Zoología 27, Universidad de Concepción
Rofes J, Wheeler J (2003) Sacrificio de cuyes en los Andes. El caso de El Yaral y una revisión biológica, arqueológica y etnográfica de la especie *Cavia porcellus*. Archaeofauna 12:29–45
Santana-Sagredo F, Maturana A, Lemp C, le Roux P, Harrod C, Uribe M (2024) Camelids in the oasis: new evidence from the Pica 8 Cemetery, Atacama Desert, Northern Chile (AD 900-1450). In: Foodways of the ancient Andes: transforming diet, cuisine, and society. The University of Arizona Press, p 259. https://doi.org/10.2307/j.ctv37xg2bh.16
Sierpe V (2015) Atlas osteológico del guanaco. Ediciones de la Universidad de Magallanes
Stengers I (2005) Introductory notes on an ecology of practices. Cult Stud Rev 11(1):183–196
Sykes N (2014) Beastly questions: animal answers to archaeological issues. Bloomsbury Academic
Tomka S (1992) Vicuñas and llamas: parallels in behavioral ecology and implications for the domestication of Andean camelids. Hum Ecol 20(4):407–433
Urbina M, Adan L (2006) Construcciones de uso público y su distribución en las quebradas tarapaqueñas durante el periodo Intermedio Tardío (900-1450 años DC). Boletín de la Sociedad Chilena de Arqueología 39:19–34
Uribe M (2006) Acerca de la complejidad, desigualdad social y el complejo cultural Pica-Tarapacá en los Andes Centro-Sur (1000-1450 d.C.). Estudios Atacameños 31:91–114

Uribe M, Sanhueza L, Bahamondes F (2007) La cerámica prehispánica tardía de Tarapacá en valles y costa desértica. Chungara 39(2):143–170

Uribe M, Angelo D, Capriles J, Castro V, de Porras ME, García M, Gayo E, Gonzalez J, Herrera MJ, Izaurieta R, Maldonado A, Mandakovic V, McRostie V, Razeto J, Santana F, Santoro C, Valenzuela J, Vidal A (2020a) El Formativo en Tarapacá (3000-1000 AP): Arqueología, naturaleza y cultura en la Pampa del Tamarugal, Desierto de Atacama, norte de Chile. Lat Am Antiq 31(1):81–102. https://doi.org/10.1017/laq.2019.92

Uribe M, Agüero C, García M, Herrera MJ, Izaurieta R, Maldonado A, Mandakovic V, Saintenoy T, Santana F, Urrutia F, Vidal A (2020b) Pampa Iluga y las chacras de los ancestros (Tarapacá, norte de Chile): tensionando materialidades y ontologías desde la arqueología. Revista Chilena de Antropología 42:371–398. https://doi.org/10.5354/0719.2020.60497

Urrutia F (2011) Participación social en la quebrada de Camiña: Entre los avatares de la tradición y la dinámica municipal. Undergraduate dissertation for the degree in Social Anthropology. Departamento de Antropología, Universidad de Chile, Santiago. Unpublished

Urrutia F (2019) Memoria, paisaje y pastoreo en Visviri. Publicación Ocasional N° 70, Museo Nacional de Historia Natural, Santiago

Urrutia F (2022) Cacofonía de voces en Nama (Tarapacá, norte de Chile): caminos que andan y relaciones entrelazadas. Boletín de la Sociedad Chilena de Arqueología 51:5–40

Urrutia F, Uribe M, González V (2023) Representaciones e historicidades sobre la Pampa del Tamarugal: descripciones relacionales de un paisaje persistente. Revista TEFROS 21(1):10–37

Valenzuela J (2010) El material malacológico y el complejo cultural Pica-Tarapacá: uso y simbolismo de las conchas en la prehistoria tardía del norte de Chile. Undergraduate dissertation for the degree in Archaeology. Departamento de Antropología, Universidad de Chile, Santiago. Unpublished

van Kessel J (1989) Ritual de producción y discurso tecnológico. Chungara 23:73–92

van Kessel J (1996) Los aymara contemporáneos de Chile. In: Hidalgo J, Schiappacasse V, Niemeyer H, Aldunate C, Mege P (eds) Etnografía. Sociedades indígenas contemporáneas y su ideología. Editorial Andrés Bello, Santiago, pp 47–68

Vidal-Elgueta A, Lûcke A, Pérez F, Wissel H, Maldonado A, Uribe M (2024) Challenging the Atacama Desert: agronomic and water conditions for pre-Hispanic maize agriculture in hyper arid environments inferred by $\delta^{18}O$ isotopes. J Archaeol Sci Rep 59:104775. https://doi.org/10.1016/j.jasrep.2024.104775

Vilá B (2002) La silvestría de las vicuñas, una característica esencial para su conservación y manejo. Ecol Austral 12:79–82

Vilá B, Arzamendia Y (2020) Comportamiento de la vicuña austral. In: González BA (ed) La Vicuña Austral. Facultad de Ciencias Forestales y de la Conservación de la Naturaleza, Corporación Nacional Forestal y Grupo Especialista en Camélidos Sudamericanos, Santiago, Chile, pp 86–94

Villagrán C, Castro V (2004) Ciencia Indígena de los Andes del Norte de Chile. Editorial Universitaria, Santiago

Villagrán C, Castro V, Sánchez G, Hinojosa F, Latorre C (1999) La tradición altiplánica: estudio etnobotánico en los Andes de Iquique, Primera Región. Chile Chungara 31(1):81–186

Yacobaccio D (2003) Procesos de intensificación y de domesticación de camélidos en los Andes Centro Sur. Memorias del Tercer Congreso Mundial sobre Camélidos I:211–216

Chapter 3
Dialogues Between Animals and People Over Time: An Interdisciplinary Perspective on the Sierras of Córdoba, Argentina

Erica Paola Colqui, Andrea Recalde, and Daniela Tamburini

3.1 Introduction

In the winter of 2022, Marianela, a rural inhabitant of Los Gigantes (LG) (Córdoba, Argentina), revealed to us a very peculiar custom: families who have lived in the region for a long time have the name of a non-human animal as another form of designation besides their given name. To help us to understand this practice, she likened it to a nickname, but unlike nicknames, it does not carry a mocking connotation, such as referring to the physical attributes of individuals. Some families share the same surname without having close (or any) blood ties, so we asked, "If surnames are repeated, do the nicknames also repeat?" She replied, "No, there are surnames that are the same but the nicknames are not[...].We, on my father's side,

E. P. Colqui (✉)
Consejo Nacional de Investigaciones Científicas y Técnicas (CONICET), Instituto de Estudios Históricos (IEH-CEH), Córdoba, Argentina
e-mail: erica.colqui@mi.unc.edu.ar

A. Recalde
Consejo Nacional de Investigaciones Científicas y Técnicas (CONICET), Instituto de Estudios Históricos (IEH-CEH), Universidad Nacional de Córdoba, Córdoba, Argentina
e-mail: andrea.recalde@unc.edu.ar

D. Tamburini
Centro de Ecología y Recursos Naturales Renovables, Universidad Nacional de Córdoba, Córdoba, Argentina
e-mail: daniela.tamburini@unc.edu.ar

R. Frontini et al. (eds.), *Non-human Animals and People: Interactions, Agents and Materialities*, The Latin American Studies Book Series,
https://doi.org/10.1007/978-3-031-97495-3_3

are 'The Rams,' and on my mother's side, we are 'The Calandra' (i.e., a beautiful bird). She wasn't carried[1] (laughter)."

How does the ram integrate into the construction of rural families' identity in the Central Sierras of Córdoba, Argentina? This animal is part of the livestock that lives in the everyday environment of these families. They are an indispensable part of their capital, essential for their subsistence. It is this situation that leads families to include it in their identity construction. What other species of animals play this role in rural communities? How is this dialogue constructed? The analysis of this practice provides a pathway to understanding the various ways in which humans and animals construct their relationships with each other and through each other (Medrano 2013). This leads us to investigate whether these forms of interaction between these beings are constituted as inherited practices with significance over time, and how they are transmitted to new generations.

This social practice involving the association of families with names of non-human animals extends across central-western Córdoba, Argentina (Nores and D'Andrea 1996; Lorca 2018; Colqui et al. 2023). In this regard, our initial focus is on the analysis of ethnozoological knowledge present in the rural communities of Los Gigantes and La Playa (LP), located in different mountainous environments, but where rural families carry out similar practices. This work aims to understand how non-human animals are integrated into the everyday practices of these communities, the bonds that are formed in these specific contexts, and in this scenario, which species are associated with local families.

This ethnozoological knowledge is transmitted orally from generation to generation, persisting across several of them, thus becoming part of the collective memory of the families inhabiting the study areas. Accordingly, a study is proposed to gather information on different moments of the local historical process. To this end, in addition to investigating the present with families from these rural areas, the analysis of historical documents such as the 1921 National Folklore Survey is incorporated. This survey gathers testimonies about traditions, beliefs, and customs, where non-human animals have a prominent role. In the search for roots that allow us to understand the role of non-human animals in the process of shaping socially significant practices, the analysis of iconography present in the archaeological record of the communities that occupied La Playa and Los Gigantes before the arrival of the Spaniards to the region is incorporated. The analysis focuses on zoomorphic iconography present in various material supports, such as clay figurines, bone instruments, and especially rock art, from the Late Period (ca. 700 to 1544 AD) of the Central Sierras of Córdoba, Argentina. We consider this to be an analytical approach that allows us to glimpse how human communities constructed their worlds through integration with other beings, especially non-human animals, over time.

[1] When the flock is moved from one place to another, it is carried.

3.2 Relationships Between Humans and Non-humans, Landscapes, and Objects

The incorporation of other beings into the constitution of human worlds enables the consideration of relational frameworks wherein animals play a significant role in social development (Medrano 2018). They are conceived as beings endowed with sentiments and autonomy (Ingold 2000), possessing the capacity to interact, negotiate, dialogue, and affect individuals with whom they share everyday life (Bugallo and Tomasi 2012; Lema 2013; Bastourre 2021; among others). In this sense, the categories of domestic and wild animals have lost their analytical properties for comprehending the ways of encompassing the investigated world. Along this line, animals do not emerge in a passive relationship where people exert control and dominance over them; rather, these connections can be varied as they occupy a symmetrical position in the configuration of social practices (Descola 2004; Bugallo and Tomasi 2012; Lema 2013; Vander Velden 2021). The concept of nurturing (*crianza*) becomes relevant as it grants agency to animals, considering them as beings that engage in dialogue, negotiate, and establish agreements with the people who inhabit the same space (sensu Lema 2013). Therefore, if non-human animals can act in the reproduction of the social world, it is important to consider their agency in the construction of social and identity memory within human communities. As proposed by Haber (2022), human sociability should be conceived within the relationship among humans, landscapes, and animals.

Social memory is an act that also involves the encounter between people and objects, in a process of continual negotiation of meanings and belongings, wherein it is further established what is intended to be remembered and forgotten (Augé 1998; Ingold 2000; Jones 2007; Meskell 2008). In all acts of remembrance, the original meaning is evoked, but under different circumstances, implying changes since remembering is more an act of construction than reproduction (Connerton 1989).

As a consequence, there exists a corpus of knowledge, practices, and beliefs that are shared, accepted, negotiated, and transmitted by a human group over time. In the process of constructing a shared memory, orality becomes an important mechanism for knowledge transmission, through formats predetermined by the same human group (Severi 2010). The main is that, although this knowledge is not immune to the fluctuations they undergo, it is the "typical scenarios" or, in other terms, the ritualized spaces where forms of building bonds are reiterated over time, which provide a stable framework for these narratives to oscillate without losing the identity of a practice (Severi 2010). This knowledge consists of fields that emerge as instances of social memory formation, among which myths and oral traditions stand out. Typically transgenerational, these hold a significant influence on the identity formation of a social group (Thompson 2004).

In this process of identity construction, the utilization of symbolic resources in everyday practices constitutes the most powerful means for its reproduction (Bauman 2005; Lucy 2005). Consequently, material culture, through its intervention and mediation in these practices, constitutes one of these resources. Thus, it is in the relationships established between individuals and objects where people negotiate different senses of belonging and strengthen the bonds among group members (Hastorf 2003; Lucy 2005; Jones 2007; among others).

Dialogue thus involves not only human and non-human animals but also those that mediate and contain that dialogue (landscape, material culture, and orality) and all the historical, social, and cultural contingencies that permeate it. This work aims to analyze the various ways in which people and animals in rural communities in the central-west region of the province of Córdoba, Argentina, are related, and their role in the construction of rural identity. This analysis enables us to reconsider and investigate the diverse forms that relationships between people and animals may have taken in the pre-Hispanic past of indigenous communities in Córdoba, through the study of their materiality. As a complementary perspective, historical data from the early twentieth century are incorporated through the analysis of the National Folklore Survey (1921), which provides significant information for identifying the social processes of the signification of animals alongside humans in social practices.

3.3 Common Dialogues, Intertwined Methodologies

The rural communities under study are located in the locality of Los Gigantes and its surroundings and La Playa and the surrounding areas (Fig. 3.1). The families inhabiting these landscapes have particular ways of connecting with the environment and the beings that inhabit it. The current contexts in which these families

Fig. 3.1 Map of the study areas and the Departments of modern Córdoba province (Argentina), selected for the National Folklore Survey analysis. Spatial data https://datosestadistica.cba.gov.ar/dataset/ciudad-de-cordoba. QGIS software

move and operate are permeated by market and globalization logic through cattle and sheep farming (Ryan 2004; Cingolani et al. 2004). However, getting involved in the frameworks of reality objectification that they construct allows us to identify how they oscillate between different worlds (Descola and Tola 2018).

To advance in this study, methodological tools from ethnozoology, history, and archaeology are employed. They enable us to assess the ways in which the relationships between animals and people are constructed in different historical contexts. An ethnozoological approach enables the analysis of the present, focusing on the study of the rural communities Local Knowledge of animals. The Local Zoological Knowledge (LZK) encompasses the body of knowledge, practices, and beliefs held by traditional human communities (peasants, and indigenous peoples, among others), specifically concerning the relationship between people and animals. This knowledge is cumulative and dynamic, transmitted through several generations (adapted from Berkes 1999). To understand the LZK, open and semi-structured interviews were conducted with key informants using the "snowball" technique (Guber 2005). The set of questions was organized into two groups: one is related strictly to familiar aspects (time residing in the area, productive activities undertaken, lineage, ancestral origins, among others), and the other is focused on the perception of local fauna. However, other pertinent considerations were also incorporated during the course of the interview. Visual aids, such as bird, reptile, and mammal guides (Narosky and Yzurieta 2010; Torres and Tamburini 2019; Cabrera 2021), were used to verify the identification of the non-human animal types mentioned in the interviews. Throughout the entire process, the Code of Ethics of the Latin American Society of Ethnobiology was followed (Medinaceli et al. 2018).

To organize the universe of non-human animal species and avoid biased classification by academic categories, the classification was arranged according to the criteria of the interviewed (Medrano 2018), considering not only which animals were mentioned first but also the appreciations associated with that order (Sutrop 2001; Quinlan 2005). It is necessary to clarify that in those cases in which it was not possible to arrive at the identity of the species because they were mentioned in very general categories (i.e., lizards, snakes, vipers, birds) in the interviews and surveys, they were included in the descriptions of their uses and relationships.

In recent decades, and according to various academic fields, oral history has been revalued for its capacity to construct oral sources as cultural productions that reflect conflicting connections between collective histories and memories (Thompson 1988; Portelli 1991; Thompson 2004; Camarena and Necochea Gracia 2008).

Furthermore, when these oral memories are recorded on paper, it is necessary to resort to methodological tools of the historical discipline for their analysis. This is the case with the information collected in the National Folklore Survey from 1921, which was incorporated into the present study as it provides narratives of the oral memory of rural communities from the early twentieth century. This survey represents a unique documentary collection that gathers knowledge, oral traditions, and popular beliefs of rural populations in the country. It was a project of the National Council of Education and was carried out by teachers across the national territory. They were summoned to collect and organize the material according to the guidelines provided by an "Instruction Manual" prepared by the Council (Espósito and Di Croce 2013).

Among the 209 surveys conducted in the province of Córdoba, we selected 60 from the localities and surrounding areas studied. The surveys cover various narrative topics. In this work, we will focus on analyzing superstition, traditional or folk medicine, and omens. The instructional guide prepared for the teachers in 1921 does not offer a specific definition of these concepts. However, they are grouped under the category of beliefs and customs of human communities and adhere to the same regulations outlined above for their surveying.

Superstition is understood as the belief in spiritual forces or non-natural, religious, and magical things (Rodríguez Becerra 2021). Meanwhile, folk or traditional medicine (*curanderismo*) refers to those practices involving knowledge about the treatment of diseases and ailments where representations about the power of nature are at play, in this study, animals in the healing process (Pen et al. 2013). Finally, omens refer to the announcement or anticipation of an event or occurrence through the presence or behavior of an animal.

The organization and implementation of these surveys are significantly biased for two reasons. The first reason is that the National Council of Education selected the information to be gathered based on what was deemed useful for the strengthening of nationalism. To this initial filter is added the selection made by the teachers in the field regarding what to ask to each respondent. Thus, not all surveys have the same level or depth of information (Rosalía 2018), which was subjected to whoever was conducting the interview. However, as will be developed in the following section, it is valuable information that provides insights into the non-human animal species appearing in the selected narrative contexts for this study.

The material culture also constitutes points of reference for the communities with their past, as the repeated interaction of people and objects in specific places activates the reference to a shared memory (Lucas 2005; Jones 2007; Meskell 2008). These lines form the foundation for analyzing non-human animals present in various media (e.g., bone tools, ceramic, lithic), focusing on rock art, which was integrated into different social spheres (settlement-type occupations, transitory occupations, and small-scale occupations, among others). All this materiality is integrated into the historical process that begins in the so-called Late Period (ca. 700 to 1544 AD). The iconographic repertoire integrated into this study is part of a symbolic system through which identity ties are strengthened or negotiated (Nanoglou 2008). In this sense, the incorporation and preference for certain species of non-human animals could constitute a trait of metaphorical identification within the groups, with their ancestors, and with specific places and experiences of the landscape (Jones 1998; Meskell 2015).

To refine species identification, emphasis was placed on the recognition of diagnostic bodily attributes, namely specific morphological details that account for some particular taxa (Martin and Meskell 2012). Classifying the repertoire based on resemblance to a real referent establishes a bias difficult to overcome. Footprints or "tridigits" indicating a bird have been recorded, but this cannot be applied to other motifs. Are the points made together metonymic signs resembling the ocelli of felines? Faced with the empirical impossibility of answering affirmatively, only those animals that "resemble" non-human animals should be considered.

In essence, social memory is constructed and reproduced through the interaction of language, bodily practice, and material culture (Nielsen 2008). Thus, the dialogue between materiality and oral history allows us to access objects, their characteristics, their uses, and production contexts, which constitute expressions of aspects related to a way of thinking and living by the communities that produce them (Ingold 2012). In this sense, the bonds forged in the becoming of "being human" are built with the other, in a process where other beings have agency and autonomy (Ingold 2000).

3.4 Dialogues Between People and Animals in the Present

Currently, the study areas maintain cultural, environmental, and demographic peculiarities that both resemble and differentiate them from each other, as they are the result of their historical contingencies (Fig. 3.2). In general terms, both are strongly influenced by rurality, as they have low population density and an economic activity linked to agro-livestock exploitation, largely oriented towards the local market. According to the National Population Census conducted in 2010, the LP area has a total of 220 inhabitants with a 25% increase towards the end of the past decade (Plan Estratégico Territorial de La Playa 2018), while LG records a total of 602 inhabitants (Instituto Nacional de Estadística y Censo 2010). However, this low density in the regions has been a constant that has persisted from the early twentieth century to the present day (Censo Nacional de Población de 1914).

In both areas, agricultural and livestock activities are developed by the rhythm of the environmental characteristics of each. The municipality of LP, located at 820 m above sea level, is part of Sierra Chaco, characterized by an arid and dry climate, with annual precipitation reaching 400 mm (Demaio et al. 2002), enabling the development of agriculture for household consumption. Although rainfall is more abundant in LG, reaching 900 mm in the warmer months (Cingolani et al. 2004, 2015), it has environmental conditions typical of high-altitude environments (1800 m.a.s.l), which hinder the availability of suitable surfaces for cultivation practices. Therefore, economic activities linked to these rural spaces are primarily driven by the husbandry of Euroasiatic-origin animals (which were brought to the region after the Spanish Conquest over 500 years ago). Currently, families residing in these environments are engaged in the breeding of cattle, goats, and sheep, which respond to local market demand and cultural values assigned to them.

For the communities in both areas, in this context, the husbandry of non-human animals is a significant activity. Therefore, paying attention to the relationships woven with these beings allows us to observe the coexistence of various forms of perception and dialogue among species. This specific bond between humans and animals engenders a dialogue wherein everyday life unfolds within the temporal rhythms forged by their interaction.

Fig. 3.2 (**a**) The local environment of Los Gigantes and its surroundings. (**b**) The local environment of La Playa and its surroundings

As will be seen, the interviewed individuals categorize the universe of animals according to their terms, or in other words, what can be analyzed as specific categories. The majority of the interviewed individuals were born and lived in LG and LP all along their lives, and the occupation of these environments dates back to at least three generations ago. Regarding the question of their possible indigenous ancestry, their responses reveal differences. This recognition is more pronounced in LG, where 25% affirmed it, while 50% indicated not knowing. This allows for the possibility of indigenous ancestry to remain open, as they highlighted in their narratives that during their childhood, it was not socially acceptable to descend from indigenous communities, thus it was uncommon to inquire with their parents and/or grandparents about this topic (Colqui et al. 2023).

Much of the knowledge concerning the nurturing, care, production, reproduction, and other values assigned to non-human animals are transmitted and reproduced over time, meaning they are sustained through the transmission of CZL (Cumulative Zoological Knowledge).

When asked about the animals they were familiar with, the majority of interviewees from both areas led the list of mentions with species that are part of the productive activities on their farms (e.g., horses, sheep, and pigs, among others). However, this dominance of domestic-nurturing animals in oral discourse may be due to the extensive network of values, meanings, and emotions that characterize the relationships between humans and non-human animals, which are not solely limited to their utilitarian and/or economic value.

The term "animal" is associated with those that are accessible, easily manageable, and belong to the domestic realm, namely, nurturing animals. They thus make a distinction between these and "bichos del monte," a term referring to species that solely inhabit the wilderness, that are within and of the wilderness (Mogni 2015).

Both the nurtured and the wild animals are integrated into social life through a particular practice, which is to name families after a non-human animal species. These designations, in most cases, are not repeated and differentiate families that share the same surname but do not share genealogical lineage. This association of families with non-human animals constitutes an intra-social code, as it only holds meaning within the community (Colqui et al. 2023). This situation posed limits or barriers when formulating questions for the interviewed individuals, as we were outsiders to their tradition. In parallel, this situation becomes more complex since many of the interviewees indicated being unaware of their origin, meaning that the sense of this practice in oral memory is being lost. This social custom is part of a world of representations that are inherited, maintained in the present, and shape the relationship between people and animals (Bugallo and Tomasi 2012).

Among the non-human animals that form part of these designations, significant variability is observed, including birds, mammals, insects, and reptiles, in that order of importance for the case of LG (see Table 3.1). The majority of the mammals are defined by the livestock raised by families (e.g., lambs, sheep, bulls, among others). In contrast, the birds are those that predominantly inhabit the grassland landscape (e.g., condors) (see Table 3.1). Conversely, in LP, the mammals that inhabit the Chaco forest predominate (e.g., skunks), while birds, insects, and reptiles are present in lesser proportions.

The families are within the community as they are recognized by the non-human animal that distinguishes and identifies them. An interviewee indicated that on the paternal side, she is "Colcón," and on the maternal side, she is the "Chuña"; both are birds from the area. She related the following:

> "I am Colcón," why do they call you that? I don't know, it's a tradition from many years ago. [...] but Does it come from your dad or your mom? From my dad's side, my mom I think she was Chuña. It's a bird too [...] What characteristics does the Chuña have? I don't

Table 3.1 Non-human animals in the surnames of families from Los Gigantes, La Playa, and surrounding areas

Los Gigantes and surrounding areas						
N°		Scientific name	Mammals	Birds	Reptiles	Insects
1	Does not have (father)					
	Gallos (mother)	*Gallus gallus*		X		
2	Jotes (father)	*Coragyps atratus*		X		
	Toros (mother)	*Bos taurus*	X			
3	Carnero (father)	*Ovis orientalis aries*	X			
	Calandria (mother)	*Mimus saturninus*		X		
4	Chicharrita (father)	*Ammodramus savannarum*		X		
	Unknown (mother)					
5	Does not have (father)					
	Los zorros grises (mother)	*Lycalopex gymnocercus*	X			
6	Lagarto (father)	Lacertilia			X	
	Unknown (mother)					
7	Colcón	Unidentified bird		X		
	Chuña	Cariamidae		X		
8	Los Pavos (mother)	*Meleagris gallopavo domesticus*		X		
	Does not have (father)					
9	Does not have (mother)					
	Los Gatos (father)	*Felis* sp.	X			
10	Unknown (mother)					
	Does not have (father)					
11	Los corderos (father)	*Ovis orientalis aries*	X			
	Does not have (mother)					
12	Los mosquitos (mother)	Culicidae				X
	La Calandria (father)	*Mimus saturninus*		X		
13	Los champi (father)	Unidentified insect				X
	Unknown (mother)					

(continued)

Table 3.1 (continued)

Los Gigantes and surrounding areas						
N°		Scientific name	Mammals	Birds	Reptiles	Insects
14	Los Rupachico (father)	Unidentified worm				X
	Los Tortero (mother)	Object				
15	Las Urracas (father)	Probably *Guira guira*		X		
	Los corderos (mother)	*Ovis orientalis aries*	X			
16	Lampalagua (father)	*Boa constrictor occidentalis*			X	
	Los Librillos (mother)	Associated with the stomach cavity of ruminants	X			
17	Los Cacholotes (father)	*Pseudoseisura lophotes*		X		
	Los Roseto (mother)	Plant				
La Playa and surrounding areas						
18	Los Tucus (mother)	*Pyrophorus noctiluca*				X
	Does not have (father)					
19	Los Liones (father)	*Panthera leo*	X			
	No data (mother)					
20	Los Chanchos (father)	*Sus scrofa domestica*	X			
	No data (mother)					
21	Los Zorrinos (father)	*Conepatus chinga*	X			
	No data (mother)					
22	Unknown (mother)					
	Unknown (father)					
23	Los Zorros (mother)	*Lycalopex gymnocercus*	X			
	No data (father)					
24	Las Iguanas (father)	Iguana			X	
	No data (mother)					
25	Los Tuquis (father)	*Empidonomus aurantioacristatus*		X		
	No data (mother)					
26	Los Zorros (father)	*Lycalopex gymnocercus*	X			
	No data (mother)					

know, I know that Colcón is a very observant creature, at night he's like listening, observing, if a fox passes by, he'll make some noise or something" (interview with M.Z.)

The Colcón (an owl, Strigidae Family) is not a bird that is part of the family's breeding spaces, but the knowledge about this animal is greater because it is the one they identify with. In contrast, the Chuña (*Cariama cristata*), despite being the name of the animal inherited from the maternal family, and being common in grasslands, are not interested in knowing about its behavior since they do not identify with this bird. This clear cultural selection regarding which species are incorporated into the mental repertoire of non-human animals also materializes when we ask about the animals they know from the environment, where, for example, the dog is not mentioned by any family, despite its fundamental role in animal herding, which implies specialized breeding.

The interaction between people and animals also shapes the cultural belief systems shared among these human communities and with their past. In this sense, domesticated animals have significant agency as people must ensure their perpetuation as a family reproduction strategy in a subsistence system. The animal thus becomes integrated into their reproduction ritual, as when they are slaughtered, part of the blood is used to draw a cross on their eye, a practice that dates back at least to the early twentieth century. In the National Folklore Survey (1921), this habit is documented in testimonies from the inhabitants of LG, where different parts of the slaughtered animals were used to maintain livestock abundance and attract good luck. Horns, entrails (librillos), and legs are integrated into households to ensure the prosperity of the livestock and in the house. However, these "parts" of the animal's body not only constitute their metonymy but also exert an influence that involves the animals in the everyday lives of the families and generates the framework of connection and bonds among them (Bastourre 2021). They thus become involved in the constant and constitutive dialogue of the familial sphere because the presence of these animals in daily practices produces and reproduces the prosperity of the livestock they raise.

The interviewed individuals also highlighted the healing functions attributed to these non-human animals, as specified by several studies conducted in the region and its vicinity (Arias Toledo and Trillo 2014; Tamburini and Cáceres 2017; Wajner 2017; Manzano-García and Martínez 2023). Many of these practices involve animals that are not part of livestock, such as the therapeutic use of puma fat and iguana, for example, for ailments related to human body joints.

3.5 The Non-human Animals in the National Folklore Survey

The National Folklore Survey from 1921 stands as a valuable document that preserves the popular memory and knowledge that are part of the idiosyncrasies of each community. It enables not only the analysis of the role of non-human animals in various contexts of beliefs and knowledge (e.g., traditional or folk medicine, and omens) but also the investigation of the different degrees of participation of these beings in the everyday lives of rural populations in Córdoba at the beginning of the twentieth century.

Sixty surveys corresponding to the departments of Ischillín, Cruz del Eje, Punilla, Minas, Pocho, San Alberto, and Colón were analyzed, representing the study areas (Fig. 3.1). To organize and classify the information, the non-human animals identified in these surveys were grouped into mammals, birds, reptiles, insects, arachnids, and amphibians.

In the narrative contexts of traditional medicine, 51 non-human animals are identified, 25 of them are classified as *bichos del monte* (wild animals, using the categories provided by the interviewed individuals). Dogs, cats, and chickens also appear as the only domesticated animals with medicinal attributes for these rural populations. To cure ailments, some parts of these animals are used; for example, the meat and/or skin of the cat are associated with the treatment of asthma and bronchitis. Nothing clarifies the reason for the healing properties of these animal parts and not others.

The most significant variability has been recorded among mammals ($N = 10$), of which the guanaco (*Lama guanicoe*) is noteworthy because it had become extinct in the high grassland region by the early twentieth century (Medina and Rivero 2007; Barri and Cufre 2014), with populations restricted to the region of the Salinas Grandes (Río and Achaval 1905), more than 100 km northern from the studied area. In this case, the bezoar stone found in the animal's kidneys is used to cure dropsy, a condition linked to the accumulation of fluids in the human body's internal tissues. Other mammals mentioned in the survey include the puma (*Puma concolor*), whose hide is used for the relief of rheumatic pains, and the skunk liver, which is used to heal pulmonary diseases. In the case of birds such as *calandrias* (*Mimus saturninus*), condors (*Vultur gryphus*), and owls (Strigiformes), parts of the animal must also be ingested to cure ailments.

The widespread use of certain animals or their anatomical parts, with magical and/or symbolic value, is included in the zooterapia dimension (Arias Toledo and Trillo 2023). Eating the brain of the *calandria*, a bird with a great ability to mimic the song of other birds and sounds with precision, helps those who have difficulty in the onset of speech. In summary, the ethological characteristics of non-human animals can be transferred to humans (Alves and Alves 2011).

Superstitions feature 26 mentions of non-human animals present in this narrative. In this case, a total of 21 species, spanning a wide range of classes except for arachnids, are recorded. Domesticated mammals ($N = 11$) and birds ($N = 7$) are the most prominent (Fig. 3.3).

In addition to the therapeutic properties and beliefs associated with non-human animals, they can also directly affect a person's life or their family by foreshadowing or predicting certain events. In this case, a total of 71 species of animals were identified, of which 19 are birds, particularly those of the wilderness, comprising 57% ($N = 12$), and insects 23% ($N = 5$). These usually herald the arrival of both negative and positive news (e.g., deaths, ill omens, misfortunes, among others) through the animal's characteristic behaviors. For example, birds express these

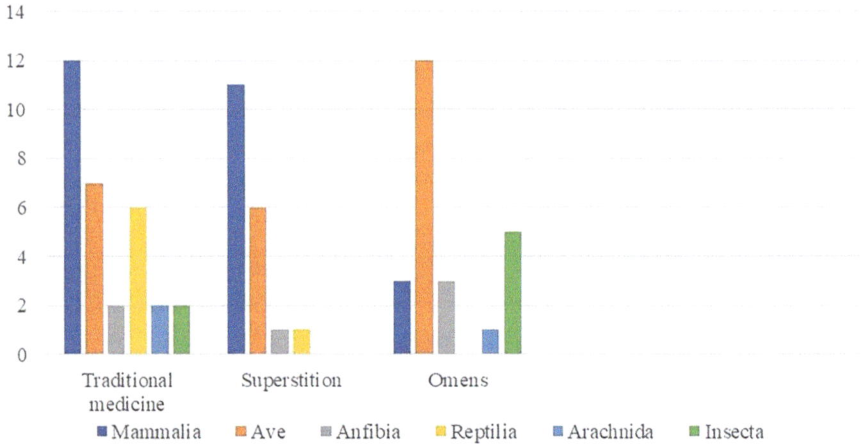

Fig. 3.3 Total number of ethnospecies identified in each of the analyzed narrative contexts from the National Folklore Survey

announcements through their songs. In the case of the few mentioned mammals ($N = 3$), they can reinforce or modify human decisions. For instance, the movement of the fox determines the success or failure of a journey depending on the direction it crosses the traveler's path.

How does animal behavior have such an effect on human behavior? Does only human interpretation of the animal's actions play a role? At this point, far from there being a unilateral relationship between humans to non-humans, it is understood that this agency of the non-human beings encapsulates mutual knowledge and dialogues forged among participants (Haber 2022) in everyday practices. This code encompasses an entire body of knowledge transmitted through multiple generations. Therefore, people do not interpret animal behavior; people engage in dialogue with the animal because they know and understand each other. This may be further strengthened when we observe the recurrence over time of certain non-human animals participating in the same social sphere as human communities.

3.6 Dialogues Between Non-human Animals Objectified in the Archaeological Evidence of the Study Areas

The archaeological record of the Sierras Centrales (current province of Córdoba, central Argentina) encompasses a temporal continuum of 12,000 years (Rivero 2010; Pastor et al. 2017). Throughout this, the human and non-human animals' relationship and dialogue were consistent. Among hunter-gatherer groups, hunting activities provide the clearest and most intense archaeological evidence of this relationship. This does not imply reducing them to a simplistic view as mere hunted animals, solely seen as providers of meat.

In this context, there exists a multitude of evidence, albeit of low intensity, that facilitates the augmentation of material studies concerning these relationships, particularly after ca. 7000 years ago. From that date onwards, artifacts have been identified in the archaeological record, suggesting the involvement of non-human animals in the dialogue between individuals and society. Mammals and gastropods have been identified in various contexts, often in the form of personal adornments. A fox tooth (*Lycalopex* sp.) was utilized as a pendant (Rivero 2010), while a necklace comprising 97 pieces of a local land snail (*Megalobulimus lorentzianus*) was identified (Pastor et al. 2017). Moreover, beads made from the same gastropod have been identified in the archaeological record since ca. 2500 BC (González 1960; Boretto et al. 2018; Gordillo 2021). Ultimately, a burial event of a female individual featured her head resting upon an armadillo (*Chaetophractus* sp.), both dated to ca. 500 BP. In summary, non-human animals are present as agents in practices that alter the body but influence the transformation of social structures (e.g., VanPool and Downs 2017), interpreted as delineating roles among hunter-gatherer groups (Pastor et al. 2017).

At the beginning of the era, archaeological evidence, depicting various social transformations that, from ca. 400 AD, indicates a process of intensification and diversification of social reproduction strategies within communities. This led to an expansion or incorporation of underutilized environments into the existing mobility circuits. One such landscape was the Guasapampa Valley, where recovered evidence suggests occupation primarily during the summer season (December–March) (Recalde 2009). Thirty-eight occupations were recorded in rock shelters and tafoni, where data (density and type of remains) indicate that they were occupied by a limited number of individuals. On the walls of these sites, panels with rock art were documented, comprising 1005 painted motifs. In the sample, non-human animals constitute 79.6% ($n = 800$), encompassing a wide variety of species, not only typical of the Sierra Chaco (i.e., *Pecari tajacu*, Ophidia) but also those whose natural habitat corresponds to high-altitude grasslands (*Lama* sp., *Ozotoceros bezoarticus*) (Fig. 3.4). A detailed analysis of the types of non-human animals represented confirms that camelids (*Lama* sp.) constitute the majority group at 67.2%, while in the universe comprising 12.3%, other species such as mara (*Dolichotis patagonum*), puma (*Puma concolor*), and *Rhea* sp. are observed (see Fig. 3.4 and Table 3.2).

Thus, through the residential mobility strategy, different landscapes were articulated, highlighting a clear process of the objectification of the role of non-human animals in the constitution of the social world. Reinforcing this process of signification of the new landscapes is the fact that in the grassland environment, which records the oldest occupation in the Sierras Centrales, only one panel with six camelids has been documented. In summary, this ancestral landscape, rooted in the social memory of the groups inhabiting the region, did not necessitate a social signification process linked to the involvement of non-human animal iconography, as it was the case of occupations in open-air sites or rock shelters in the valleys.

The iconography of non-human animals is present, albeit to a lesser extent, in other archaeological materials. Pottery and bone artifacts exhibit animal forms or representations. It is noteworthy here that, as previously mentioned, material culture

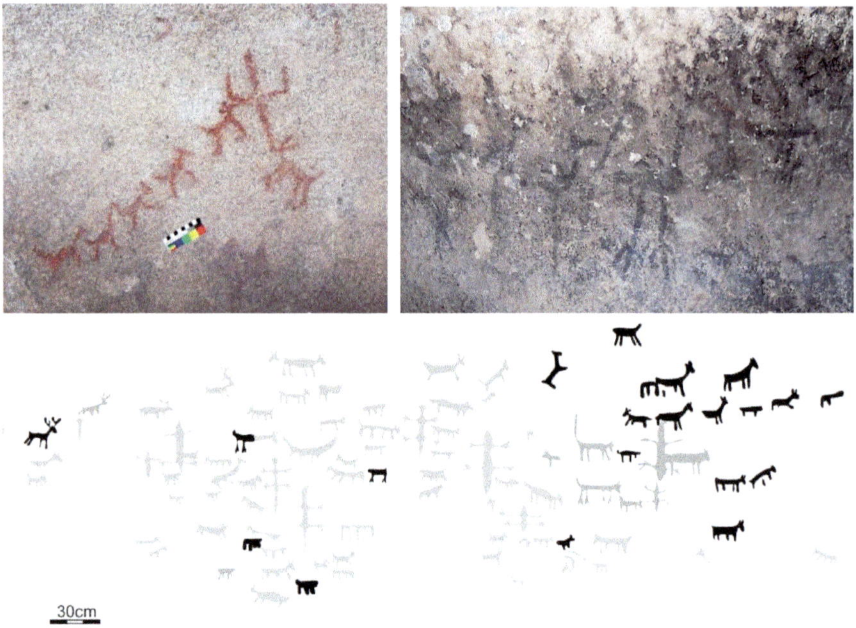

Fig. 3.4 Non-human animals' motifs in the rock art of La Playa and surrounding areas

emerges as an integral component of the material conditions that enable and regulate social action (Meskell 2008; Nanoglou 2009). Representations of non-human animals are quoting, guiding, and defining social life based on the specific characteristics of their material existence. In the cases under analysis, figurines, spatulas, and points facilitate a different type of manipulation and interaction compared to that engendered by rock art, influencing both those who produce and observe this materiality. In the cases analyzed, the figurines, spatulas, and arrowheads allow for manipulation, a coexistence different from that generated by rock art in its interaction with those who execute and observe this materiality.

As Nanoglou (2008, 2009) suggested, it is central to attend to the contexts in which they were used and the decisions regarding the type of support or object. In the central-western area of the Sierras Centrales, objects with the iconography of non-human animals and figurines have been identified. The bone points and spatulas would correspond to moments after ca. 1000 AD. The former was linked to instances of interpersonal violence (Rivero et al. 2019), while the set of spatulas would have been part of paraphernalia intended for the inhalation consumption of substances (Pastor and Moschettoni 2018). These types of artifacts have a low incidence in the archaeological record although all were documented in outdoor residential occupations or the so-called late settlements. In the central-western area of the Central Sierras, clay figurines are more abundant. Recovered from different domestic contexts associated with settlements, they present constitutive features

Table 3.2 Non-human animals depicted across various material supports in the archaeological record of central-western of Sierras Centrales from Córdoba during the Late Pre-Hispanic Period

Scientific name	Spanish Vulgar noun	Iconography			Reference
		Rock art	Bone	Ceramics	
Mammals					
Lama guanicoe	Guanaco	X	–	X	Recalde (2009); Bixio et al. (2010)
Lama glama	Llama	X	–	X	
Ozotoceros bezoarticus	Venado de las pampas	X	–	–	Recalde (2009)
Mazama gouazoubira	Corzuela	X	–	–	
Pecari tajacu	Pecarí	X	–	–	
Lycalopex sp.	Zorro	X	–	–	
Puma concolor	Puma	X	X	–	Recalde (2009); Pastor and Moschettoni (2018)
Leopardus geoffroyi	Gato montés	X	X	–	
Dolicholis patagonum	Mara	X	–	–	This paper
Birds					
Undefined	–	X	–	X	Recalde (2009): Marcellino et al. (1967)
Footprints	–	X	–	X	Recalde (2009); Marcellino et al. (1967)
Rhea americana/ pennata	Ñandú	X	–	–	Recalde (2009)
Amphibians and Reptiles					
Salvator sp.	Iguana/Lagarto	X	X	–	Recalde (2009); Pastor and Moschettoni (2018)
Ophidia	Víbora	X	X	–	Berberían (1969). This paper
Chelonoidis chilensis	Tortuga	–	X	–	Marcellino et al. (1967)
Fish					
Undefined	–	–	X	–	Pastor and Moschettoni (2018)

that allow associating the figure with quadrupeds broadly and specifically with camelids (Fig. 3.5).

However, for this study regardless of the support analyzed and the context and ritual in which it intervened, there are non human animals that unite these contexts, circulating and linking social contexts, landscapes, humans, and non-humans. Camelids, iguanas, and snakes govern social action, promoting dialogue among individuals and groups through their mediation in narratives. On the other hand, faced with the great diversity of non-human animals inhabiting both high grasslands and Sierra Chaco, it is these and not others that are selected by the groups to be represented in the different analyzed media. For example, the condor, a bird that

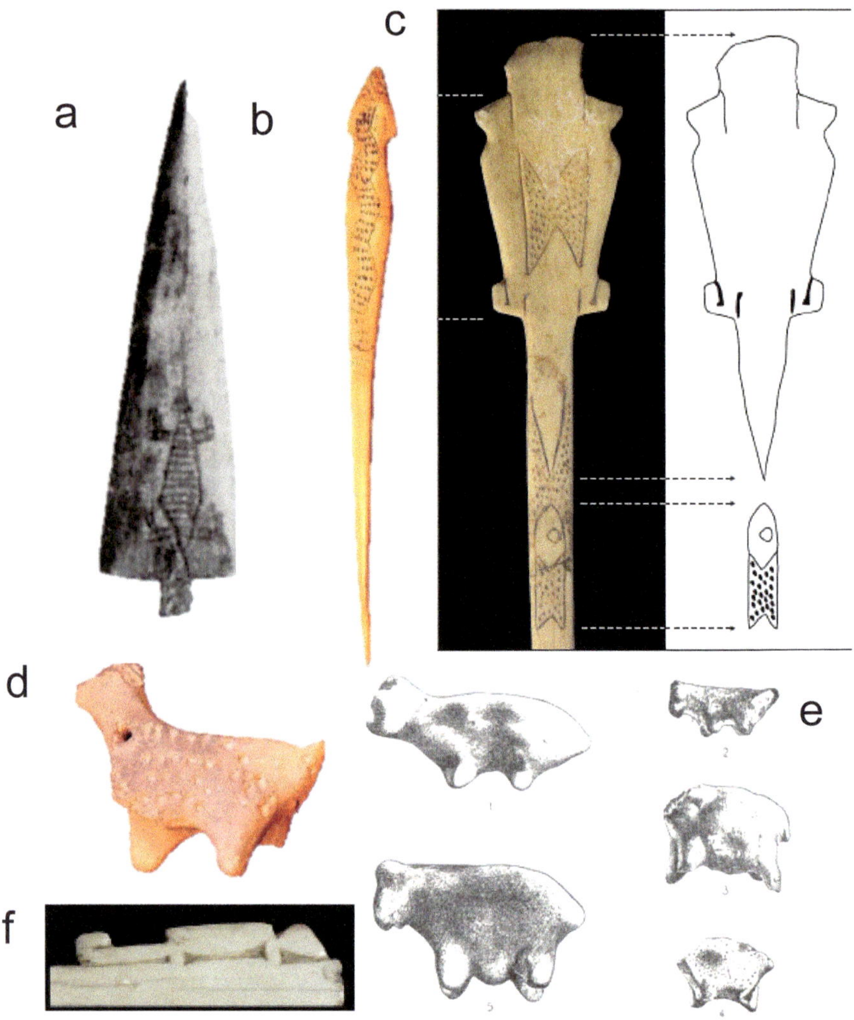

Fig. 3.5 Zoomorphic iconography on various materials supports the archaeological record of the central-western area of the Sierras Centrales, Argentina. *Salvator* sp. (**a** and **f**) (Serrano 1945; Pastor and Moschettoni 2018); Ophidia (**b**) (Bixio et al. 2010); Camelidae (**d** and **e**) (Serrano 1945; Bixio et al. 2010); Felidae and Fish (**c**) (Bixio et al. 2010)

inhabited and still inhabits the high grasslands, is not present in the repertoire. Of course, there is a bias generated by the inability to identify the real referent, as there are cases of metonymies recorded in rock art and ceramics (e.g., bird footprints) that prevent us from identifying species. But this also indicates that there are animals that were included with all their constitutive features, while for others, it was not necessary.

3.7 Integrating Perspectives and Knowledge in the Diverse Dialogues Between Humans and Non-human Animals

Interviews with the rural areas' inhabitants led to a rethinking of the categories we work with, as the traditional division between domestic and wild animals does not capture the complex relationships that occur among the various beings inhabiting La Playa and Los Gigantes landscapes. Within the concept of *bichos del monte*, lies a universe of species that become integrated into people lives' based on their behaviors and actions in the social world. In a constant dialogue with this notion is the concept of *crianza* for those animals that become integrated into the daily practices, which also presents a complex network of meanings, understanding, and reciprocity, shaping how people act (Lema 2013) and recognize themselves in the sphere of social interaction. People and animals are thus defined in their relationship with each other; it is inclusion rather than antagonism that constitutes them. The greatest expression of this complex network of meaning is embodied in the identification of families with non-human animals. These families not only acknowledge but also identify themselves through these beings, not only in their names but also in their actions, attitudes, and behaviors. They may carry the same surname, but they are families insofar as they are *Colcones*, *Zorros*, or *Calandrias*, designations that they inherit and value as an identity trait. To what extent can we trace this understanding back in time? Alternatively, is it possible to propose an ancient origin for such a conception? Is it the Western and modern perspective, as Lestel (2022) argues, that has dominated analysis, within which the animal is primarily viewed as a source of meat or a cuddly/companion creature, thus hindering our comprehension of the temporal essence of this practice?

Analyzing the data provided by the early twentieth century National Folklore Survey enables an inquiry into this essence, the root of the dialogue between humans and non-humans regarding how they mutually constitute each other. Animals herald-specific events and occurrences in people's lives, influencing them both individually and socially. As evidenced by several cases in the Survey, the fox, for instance, can sway the travelers' decisions depending on how it crosses their path. As Tola (2016) proposes, non-human beings enact actions that are experienced as real, and the relationships with these non-humans entail and induce moral values and prescribe actions that, if not adhered to, lead to disruptions in human social life. Additionally, non-human animals, through human action, also provide healing and well-being. That is to say, the combined action of the healer and the animal (either in its entirety, such as when using a toad for toothache relief, or with one of its parts, as in the use of lark brains for children who are slow to speak) is what brings about the healing or improvement of an ailment. This healing action persists today, for instance, among the rural inhabitants of Los Gigantes and neighboring areas, where G.M. mentioned that puma fat cures joint pain. These shared understandings through oral discourse constitute part of a tradition that becomes integrated into the memory of these communities (Severi 2010).

Landscapes and objects also contribute to this interaction. The landscape, since the Spanish conquest, has been shaped by a rural way of life where populations defined their practices around this changing socio-historical context (González Navarro 1999). The rearing of non-human animals was contingent upon subsistence and local demands according to historical circumstances. However, these landscapes are also constituted by social and cultural practices preexisting colonialism, which continued to be reproduced over time. Thus, the environmental knowledge shared and reproduced by pre-Hispanic communities conditioned and shaped new ways of inhabiting and living within the environment. Oral memory and the use of symbolic resources transmit this knowledge. The mutability of this memory and its transmission methods cannot be denied, as the repetition of action invokes the original meaning but under different circumstances in each period (Hastorf 2003). However, it is through the act of recollection that memory is forged, and in this sense, the expansion of knowledge is simultaneously the production of memory (Ingold 2000).

The footprints of animals like sheep, as documented in the National Folklore Survey, serve a dual role. On the one hand, they are nomothetic elements, but they also symbolize prosperity or good luck. In this sense, their mere presence in households exerts an influence that involves them in the everyday lives of families, giving rise to clear connections between them (Bastourre 2021). This leads us to consider whether the bird tracks found in the rock art of La Playa, are metonymy for the animal but also may have had agency in the interaction between birds and humans. Were these painted and engraved footprints on the rock art panels a way to obtain or expect some benefit from these animals for the reproduction of the social world?

The camelids, primarily guanacos given the absence in the local archaeological record indicating the presence of llamas, deserve particular attention, as it is a rock art and ceramic figurine most represented animals from the Late Period in the studied regions. This could have acted as a link between different landscapes, integrating not only spaces of different social interaction (the figurines in domestic contexts and the painted and engraved motifs in seasonal occupation contexts) but also different landscapes, as these occupations are recorded in valleys while this species inhabited high-altitude grasslands. The presence of these non-human animals may have also been a way to reproduce inherited worlds that reaffirm their identity and place in the landscape.

However, its absence in the surveys of the studied areas has only one exception in the practice of traditional healing, where the use of guanaco bezoar stones mixed with condor heart cures dropsy. This could be related to the drastic decrease by the late nineteenth century, which progressively impacted the interaction between people and animals. This proposal is reinforced when considering that in the Córdoba northwest (Salinas Grandes), where this species is present, people still reproduce healing practices such as the use of guanaco legs for strokes or its fiber for ear ailments (Manzano-García et al. 2019). Because of these data, the question arises as to whether, as the survey shows, for example, the sheep replaced the guanaco; or in other words, whether, faced with the extinction of this animal, so significant for domestic and social life constitution, it was replaced by another animal that was accessible and socially significant for families life.

The dialogue between humans and non-human animals can be traced over time, where landscapes and objects play a central role. It becomes evident that a perspective that separates the wild from the domestic as antagonistic pairs only limits the possibility of understanding these modes of dialogue. Only through dialogue relations between humans and non-human animals can be understood as constructions that are inherited and reproduced. Different methodologies have sought to understand how the ethnobiological present enables the elucidation of non-human animals' role in the construction of rural identities among human groups. The different ways in which people perceive nature have crystallized into conceptions about the non-human beings that coexist in the world, with whom bonds have been established that transcend time and persist in various cultural expressions regarding how to forge connections with animals.

Acknowledgments We would like to thank the editors Romina Frontini, Paula Escosteguy, and Mónica Salemme for inviting us to participate in this publication. We would also like to thank the anonymous reviewers for their contributions, which helped to improve the original version. We thank all the interviewees who generously shared their knowledge and experiences with us. Thanks to Public Education and the Argentinean Scientific System, represented by the National University of Cordoba (UNC) and the Scientific and Technical Research Council (CONICET). This work was supported by Foncyt (PICT-2021-I-A-00778) directed by one of the authors. Furthermore, the study was conducted under a Doctoral Internal Fellowship within the framework of Research Projects of Executive Units, granted by the National Scientific and Technical Research Council (CONICET).

Competing Interest Declaration The authors have no conflict of interest to declare that are relevant to the content of this chapter.

References

Alves R, Alves H (2011) The faunal drugstore: animal-based remedies used in traditional medicines in Latin America. J Ethnobiol Ethnomed 7:9–52. http://www.ethnobiomed.com/content/7/1/9

Arias Toledo B, Trillo C (2014) Animales y plantas que curan: avances sobre la farmacopea natural de los pobladores del área de Laguna Mar Chiquita. Revista de la Facultad de Ciencias Exactas, Físicas y Naturales 1(2):77–85

Arias Toledo B, Trillo C (2023) Augurios registrados en la Encuesta Nacional de Folklore de 1921: diálogos entre animales y poblaciones rurales de Córdoba que llegan hasta nuestros días. Mundo de Antes 17(1):199–213

Augé M (1998) Las formas del olvido. Editorial Gedisa

Barri FR, Cufre M (2014) Supervivencia de guanacos (*Lama guanicoe*) reintroducidos con y sin período de preadaptación en el Parque Nacional Quebrada del Condorito, Córdoba, Argentina. Mastozool Neotrop 21(1):9–16

Bastourre ML (2021) Las relaciones humanos/animales en las sociedades prehispánicas del Paraná Medio e Inferior: un giro ontológico. Rev Mus Antrop 14(2):65–80. https://doi.org/10.31048/1852.4826.v14.n2.30066

Bauman Z (2005) Identidad. Editorial Losada

Berberían E (1969) Investigaciones arqueológicas en Cosquín. Revista Ilustrada Cosquín II 8:28–29
Berkes F (1999) Sacred ecology: traditional ecological knowledge and resource management. Taylor & Francis
Bixio B, Berberián E, Pastor S (2010) Historia Prehispánica de Córdoba. Editorial Brujas
Boretto G, Gordillo S, Izeta A, Colombo F, Martinelli M, Cattáneo R (2018) Cuentas ornamentales en un contexto de cazadores-recolectores de la provincia de Córdoba: análisis mineralógico y microestructural de la concha de "Borus". Arqueología 24(1):213–223
Bugallo L, Tomasi J (2012) Crianzas mutuas. El trato a los animales desde las concepciones de los pastores puneños (Jujuy, Argentina). Rev Esp Antropol Am 42(1):205–224. https://doi.org/10.5209/rev_REAA.2012.v42.n1.38644
Cabrera M (2021) Reptiles del Centro de la Argentina. Editorial UNC
Camarena M, Necochea Gracia G (2008) Continuidad, ruptura y ciclo en la historia oral. In: Necochea Gracia G, Pozzi P (eds) Cuéntame cómo fue. Introducción a la historia oral. Imago Mundi, pp 55–62
Censo Nacional de Población de 1914 (1916) Tomo II. Población. Talleres Gráficos de LJ Rosso y Cia. República Argentina. Retrieved January 20, 2024 https://www.estadistica.ec.gba.gov.ar/dpe/Estadistica/censos/C1914-T2.pdf
Cingolani AM, Renison D, Zak M, Cabido M (2004) Mapping vegetation in a heterogeneous mountain using Landsat data: an alternative method to define and classify land-cover units. Remote Sens Environ 92:84–97
Cingolani AM, Poca M, Giorgis MA (2015) Water provisioning services in a seasonally dry subtropical mountain: identifying priority landscapes for conservation. J Hydrol 525:178–187
Colqui E, Tamburini D, Recalde A (2023) Construyendo nexos entre el presente etnográfico y el pasado prehispánico: análisis preliminar de una propuesta metodológica para un estudio local. Mundo de Antes 17(1):141–168
Connerton P (1989) How the societies remember. Cambridge University Press
Demaio P, Karlin O, Medina M (2002) Árboles nativos del centro de la Argentina. L.O.L.A. (Literature of Latin American). Colin Sharp
Descola P (2004) Las cosmologías indígenas de la Amazonía. In: García Hierro P, Surrallés A (eds) Tierra adentro. Territorio indígena y percepción del entorno. Tarea Gráfica Educativa, pp 25–36
Descola P, Tola F (2018) ¿Qué es la naturaleza?: un enfoque antropológico. Teseo
Espósito F, Di Croce EV (2013) Un archivo del folklore nacional: La Encuesta de Magisterio de 1921. Retrieved November 2023, from http://www.memoria.fahce.unlp.edu.ar/trab_eventos/ev.3849/ev.3849.pdf
González AR (1960) La estratigrafía de la gruta de Intihuasi (Provincia de San Luis, Argentina) y sus relaciones con otros sitios precerámicos de Sud América. Revista del Instituto de Antropología 1:9–296
González Navarro C (1999) Espacios coloniales. Construcción Social del Espacio en las márgenes del río Segundo. Córdoba. 1573–1650. Centro de Estudios Históricos "Profesor Carlos Segreti"
Gordillo S (2021) Registros arqueomalacológicos en el centro de Argentina: el uso de *Megalobulimus* y otros moluscos en cuentas y adornos personales. Intersecciones en Antropología 22(1):83–95. https://doi.org/10.37176/iea.22.1.2021.593
Guber R (2005) El salvaje metropolitano. Reconstrucción del conocimiento social en el trabajo de campo. Editorial Paidós
Haber A (2022) Entramado con vicuñas. In: Bugallo L, Dransart P, Pazzarelli F (eds) Animales humanos, humanos animales. Relaciones y transformaciones en mundos indígenas sudamericanas. Antropofagia, pp 243–263
Hastorf C (2003) Community with the ancestors: ceremonies and social memory in the middle formative at Chiripa, Bolivia. J Anthropol Archaeol 22:305–332
Ingold T (2000) The perception of the environment: essays on livelihood, dwelling and skill. Routledge
Ingold T (2012) Toward an ecology of materials. Annu Rev Anthropol 41:427–442

Instituto Nacional de Estadística y Censo (2010) Censo Nacional de Población, Hogares y Viviendas República Argentina. Retrieved february 15, 2024, from www.indec.gov.ar
Jones S (1998) The archaeology of ethnicity. Constructing identities in the past and present. Routledge
Jones A (2007) Memory and material culture. Cambridge
Lema V (2013) Crianza mutua: una gramática de la sociabilidad andina. X Reunión de Antropología del Mercosur, 10 al 13 de julio, Córdoba, Argentina. From https://www.researchgate.net/publication/263929196_Crianza_mutua_una_gramatica_de_la_sociabilidad_andina
Lestel D (2022) Nosotros somos los otros animales. Fondo de Cultura Económica
Lorca N (2018) Pueblos y parentelas. Una aproximación al estudio de la organización familiar en el N.O. cordobés. Ph.D. Dissertation, Universidad Nacional de Córdoba. Unpublished
Lucas G (2005) The archaeology of time. Routledge
Lucy S (2005) Ethnic and cultural identities. In: Díaz Andreu M, Sam L, Stasa B, Edwards D (eds) Archaeology of identity: approaches to gender, age, status, ethnicity and religion. Routledge, pp 86–109
Manzano-García J, Martínez G (2023) Contribución a la conservación de la fauna de interés medicinal en la región de las Salinas Grandes de Córdoba (Argentina). Ecol Austral 33:583–597. https://doi.org/10.25260/EA.23.33.2.0.2129
Manzano-García J, Costa T, Barri F, Weihmüller MP (2019) Interacciones entre el guanaco (*Lama guanicoe*) y el ser humano en el Gran Chaco: datos etnozoológicos pasados y actuales del noroeste de la provincia de Córdoba, Argentina. Rev Etnobiol 17(2):25–40
Marcellino A, Berberián E, Perez J (1967) El yacimiento arqueológico de Los Molinos (Dpto. de Calamuchita - Córdoba). Instituto de Antropología 26:1–48
Martin L, Meskell L (2012) Animal figurines from Neolithic Çatalhöyük: figural and faunal perspectives. Camb Archaeol J 22(3):401–419
Medina M, Rivero D (2007) Zooarqueología, *Lama guanicoe* y Dinámica Evolutiva del Chaco Serrano. Mundo de Antes 5:211–234
Medinaceli A, Cano E, Argueta A, Sanabria O (2018) Latin American Society of Ethnobiology's code of ethics. Ethnobiol Lett 9(1):86–89
Medrano MC (2013) Devenir-en-transformación. Debates etnozoológicos en torno a la metamorfosis animal entre los qom. In: Tola F, Medrano MC, Cardin L (eds) Gran Chaco Ontologías, poder, afectividad. Ethnographica Rumbo Sur, pp 77–101
Medrano MC (2018) Así en el agua, como en la tierra y en el cielo. El tigre entre los qom del Gran Chaco Argentino. In: Medrano MC, Vander Velden F (eds) ¿Que es un animal? Ethnographica, pp 101–122
Meskell L (2008) Memory work and material practices. In: Mills BJ, Walker W (eds) Memory work. Archaeologies of material practices. School for Advances Research Press, pp 233–246
Meskell L (2015) A society of things: animal figurines and material scales at Neolithic Çatalhöyük. World Archeol 47:6–19
Mogni P (2015) Monte es todo lo que hay acá. El naturalismo atenuado como forma de relación de los habitantes de Chancaní con su mundo circundante. Ph.D. Dissertation, Programa en Estudios Sociales Agrarios. CEA-FCA, UNC. Unpublished
Nanoglou S (2008) Representation of humans and animals in Greece and the Balkans during the earlier Neolithic. Camb Archaeol J 18(1):1–13
Nanoglou S (2009) Animal bodies and ontological discourse in the Greek Neolithic. J Archaeol Method Theory 16(3):184–204
Narosky T, Yzurieta D (2010) Aves de Argentina y Uruguay. Guía de identificación. Vázquez Mazzini Editores
National Folklore Survey (1921) Catálogo de la colección de folklore. Instituto de Literatura Argentina. Facultad de Filosofía y Letras, Universidad de Buenos Aires. From https://enf1921.cultura.gob.ar/

Nielsen A (2008) The materiality of ancestors: Chullpas and social memory in the late Prehispanic history of the South Andes. In: Mills BJ, Walker W (eds) Memory work. Archaeologies of material practices. School for Advanced Research Press, pp 207–231
Nores B, D'Andrea U (1996) Los apelativos familiares en las Sierras Centrales y el Litoral del país. Revista de la Junta Municipal de Historia 1:76–84
Pastor S, Moschettoni L (2018) Prácticas inhalatorias y redes de interacción. Análisis de espátulas óseas del centro de Argentina. Bol Mus Chil Arte Precolomb 23(1):101–115
Pastor S, Rivero D, Recalde A, Díaz I, Truyol G (2017) Procesos y paisajes sociales en las Sierras Centrales de Argentina durante el Holoceno Tardío inicial (ca. 4200-2000 años AP). Relaciones 42(2):281–303
Pen C, Romero C, Deza C, Durando P, Villar M, Barioglio C (2013) Las Prácticas Culturales de Pequeñas Productoras Caprinas de los Departamentos Cruz del Eje e Ischilín de la Provincia de Córdoba, Argentina. VII Jornadas Santiago Wallace de Investigación en Antropología Social. Sección de Antropología Social. Instituto de Ciencias Antropológicas. Facultad de Filosofía y Letras, UBA, Buenos Aires. From https://www.aacademica.org/000-063/334
Plan Estratégico Territorial de La Playa. Provincia de Córdoba, Argentina (2018) Secretaría de Planificación territorial y coordinación de obra pública
Portelli A (1991) Lo que hace diferente a la historia oral. In: Schwarzstein D (ed) La historia oral. Centro Editor de América Latina, pp 36–52
Quinlan M (2005) Considerations for collecting freelists in the field: examples from ethnobotany. Field Methods 17(3):219–234
Recalde A (2009) Representaciones rupestres y sitios de ocupación transitoria en el período agroalfarero tardío: los casos de Charquina 2 y Cerco de la Cueva 3 en el Valle de Guasapampa (Córdoba). In: Sepúlveda M, Briones C, Chacama J (eds) Crónica sobre la Piedra. Arte rupestre de las Américas. Universidad de Tarapacá. Chile, pp 281–192
Río M, Achaval L (1905) Geografía de La Provincia de Córdoba, vol I. Compañía Sudamericana de Billetes de Banco
Rivero DE (2010) La transición Pleistoceno-Holoceno (11.000–9.000 AP) en las Sierras de Córdoba (Rep. Argentina). Rev Arqueol 16:175–189
Rivero D, Cornero S, Truyol G, Neves W (2019) Human cranium of Candonga cave site and its implications for the initial peopling of South America. Homo. Journal of Comparative Human Biology 70:283–295. https://doi.org/10.1127/homo/2019/1118
Rodríguez Becerra S (2021) A vueltas con el concepto de superstición. Religión y religiosidad en los primeros folcloristas. Demófilo 53:67–91
Rosalía P (2018) La cultura ancestral cordobesa desde la Encuesta Nacional de Folklore de 1921. Asociación Cultural Relatos del Viento
Ryan S (2004) Pobreza y desarrollo rural: la implementación del Programa Social Agropecuario en la región de Pampa de Achala. I Jornadas de Estudios Agrarios y Agroindustriales de la Región NOA. Salta. 24 y 25 de noviembre
Serrano A (1945) Los Comechingones. Editorial Instituto de Arqueología, Lingüística y Folklore de la Universidad Nacional de Córdoba
Severi C (2010) El Sendero y la Voz. Una Antropología de la Memoria. Editorial SB
Sutrop U (2001) List task and a cognitive salience index. Field Methods 13(3):263–276
Tamburini D, Cáceres DM (2017) Estrategias de Uso de la Fauna Silvestre por las Comunidades Campesinas de Argentina Central. Etnobiología 15(3):5–23
Thompson P (1988) "La evidencia", en La voz del Pasado. Alfons el Magnànim. Institució Valenciana D'Estudis I Investigació 26: 117–169
Thompson P (2004) Historia oral y contemporaneidad. Anuario UNR 20:15–34
Tola F (2016) Giro ontológico, múltiples mundos y animismo. Reflexiones desde el Gran Chaco. Apuntes de Investigación del CECYP 27: 128–139
Torres R, Tamburini D (2019) Mamíferos de Córdoba y su estado de conservación. Editorial UNC

Vander Velden F (2021) Sobre fracasos y éxitos relativos Técnica, política y ontología en los proyectos de crianza animal en aldeas indígenas de Amazonia. Runa 42(1):391–407. https://doi.org/10.34096/runa.v42i1.8287

VanPool CL, Downs L (2017) Dressing the person: clothing and identity in the Casas Grandes world. Am Antiq 82(2):262–287. https://www.jstor.org/stable/26340756

Wajner M (2017) Conocimiento Ecológico Local sobre animales silvestres y sus redes de interacción etnobiológicas por parte de pobladores rurales de los alrededores de la Cuchilla Nevada, Sierras Grandes-Córdoba. Undergraduate Thesis. Facultad de Ciencias Exactas, Físicas y Naturales, Universidad Nacional de Córdoba. Unpublished

Chapter 4
Fleece and Leather Analysis in Human and Animal Archaeological Assemblages. Southern Puna of Argentina (ca. 50 BC to AD 1800)

Andrés S. Romano, Sara M. L. López Campeny, Sabrina Donadio, and Florencia Figuerero Minetti

4.1 Introduction and Conceptual Framework

The present study aims to generate a space for reflection, from a post-humanist perspective (Barad 2003), on the mutual relationships between animals and humans over time, considering the multiplicity of aspects that may have been involved (Overton and Hamilakis 2013). In line with this, we adopt a perspective that does not presuppose unbalanced or hierarchical relationships between humans and other than-humans (Latour 2005; Witmore 2007; Olsen et al. 2012; Harris 2018). This does not mean that the world is conceived as uniform, but rather that a reality is recognized in which there is heterogeneity and differentiation, but not necessarily in terms of oppositions, dualities, or negativities. This view, in turn, aligns with an animist and relational ontology such as that accounted for in Andean ethnography (Arnold et al. 1998; Arnold 2016; Bugallo et al. 2022; Bolton and Loovers 2023).

To address this matter, we directed our attention to a portion of the faunal record that has been partially "forgotten", namely, that composed of fleeces and hides with fibres preserved. Through our research, we have observed that this materiality is rarely considered in faunal studies, and we include ourselves self-critically in this

A. S. Romano (✉) · S. M. L. López Campeny (✉)
Facultad de Ciencias Naturales e Instituto Miguel Lillo (FCN and IML), Universidad Nacional de Tucumán (UNT), San Miguel de Tucumán, Argentina

Instituto de Arqueología y Museo (IAM), FCN and IML, UNT, and Instituto Superior de Estudios Sociales (ISES), CONICET-UNT, San Miguel de Tucumán, Argentina
e-mail: asromano@csnat.unt.edu.ar; lopezcampeny@csnat.unt.edu.ar

S. Donadio · F. Figuerero Minetti
Facultad de Ciencias Naturales e Instituto Miguel Lillo (FCN and IML), Universidad Nacional de Tucumán (UNT), San Miguel de Tucumán, Argentina
e-mail: sdonadio23@alumnos.csnat.unt.edu.ar; fafiguerero21@alumnos.csnat.unt.edu.ar

R. Frontini et al. (eds.), *Non-human Animals and People: Interactions, Agents and Materialities*, The Latin American Studies Book Series,
https://doi.org/10.1007/978-3-031-97495-3_4

observation. When they are addressed, it is generally from a utilitarian perspective that emphasizes asymmetrical relationships of human domination and manipulation (Reigadas 2014). Although the lack of treatment might be justified by its lower frequency of conservation, it is at least striking (or suspicious) that in zooarchaeological analysis, the presence of meat in the past is assumed since this is the main focus of much of the research, even in its absence. In contrast, the fibres and hides that covered the meat and bones do not seem to receive the same importance in formulating hypotheses about their significance and relevance in the relationships between animals and people in the past. Based on the analysis of animal-based fabric and yarns, we have proposed a shift in focus from the central theme of many archaeofaunal studies: nutritional utility (López Campeny and Romano 2021), in line with other precedents for the southern Andean region (Dransart 2002, pp. 12–14).

We aim to apply this approach, which acknowledges the multiplicity of relationships, also to the methodological realm, in order to draw attention to the excessive fragmentation among disciplinary specializations and fields of study in our archaeological practice. Instead, we seek to promote paths of greater convergence, unity, and assemblages of perspectives (Haber 1999; Deleuze and Guattari 2004; González Ruibal 2007), as an underlying principle in the constitution of the archaeological record itself. For this reason, both from a methodological and interpretative perspective, we have integrated these results with information from other associated materialities at the same sites, aiming to achieve a more interconnected interpretation of the archaeological record. To do so, we have relied on previously obtained results from the analysis of various textile technologies elaborated with animal fibres (López Campeny 2000, 2014; López Campeny and Romano 2021), as well as the zooarchaeological record of both sites (López Campeny et al. 2005; Urquiza et al. 2013).

Finally, this characteristic of multiplicity also relates to the perception of time, as it is understood that in some way the past continues to be active and mediating in the lives of present communities (Bugallo et al. 2022). The aim is to rescue the points of proximity and connections that exist between different "non-past" pasts, from a notion of multi-temporality embedded in the materialities themselves. This notion implies that objects like fleece and leather do not merely exist as remnants of a single past moment but encapsulate on going layered relationships with time (Witmore 2007). We considered that these materials continue to mediate between the past and present, influencing contemporary practices and beliefs.

4.2 Study Area and Archaeological Sites

The term "Puna" originates from Quechua and means "high and cold land". The Argentine Puna is a natural extension of the high mountain plateaus in Perú and Bolivia, forming part of the Atacama Puna. It consists of two sectors: the Dry and the Salty Puna (Cabrera 1968). The study area is located in the latter, which is drier and characterized by extensive salt flats due to low humidity and high evaporation

(Grau et al. 2018). These local climatic conditions allow an exceptional degree of conservation of various organic archaeological remains.

The micro-region of Antofagasta de la Sierra is situated in the northwest corner of the Catamarca province, Argentina. Despite its biogeography corresponding to that of a highland desert, characterized by an arid and cold climate, intense solar radiation, a wide day/night thermal range, scarce precipitation, highly unpredictable rainfall patterns with long periods of drought, and occurrences of frosts and strong winds, it forms a series of "oases" within this Puna environment. This is based on an endorheic basin and a network of permanent narrow watercourses (Aschero et al. 2012). The Las Pitas River stands out among these tributaries, and the two archaeological sites we will briefly examine are situated near its banks in the middle of the river (Fig. 4.1).

4.2.1 Punta de la Peña 9 Site

Punta de la Peña 9 (PP9) represents a multi-component open-air settlement situated at an elevation of 3600 m.a.s.l. Its successive re-occupations unveil a complex formation history, dating from approximately ca. 100 BC to AD 1600 (López Campeny 2000; Somonte and Cohen 2006). The site exhibits three distinguishable sectors. The first sector (I) comprises a collection of dispersed sub-circular stones structures

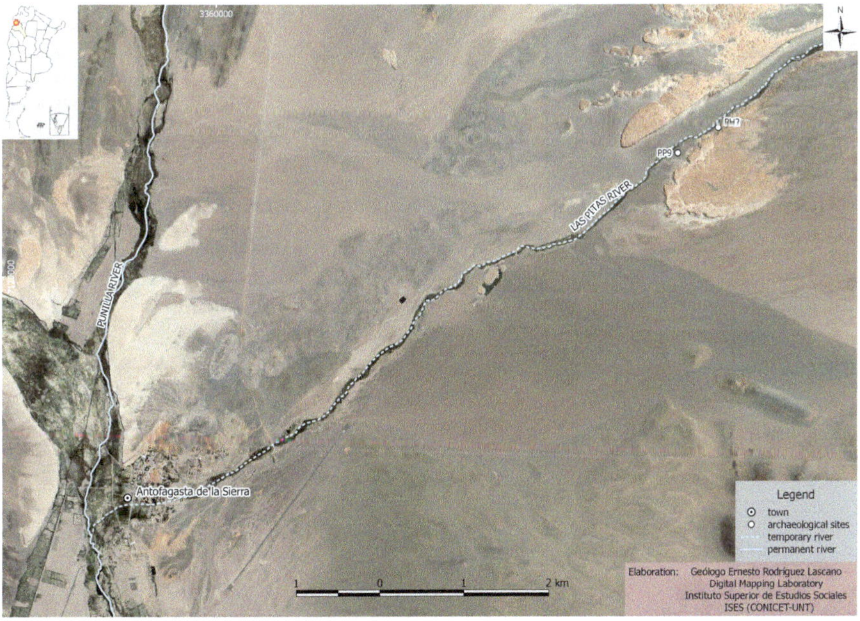

Fig. 4.1 Map of the study area with location of archaeological sites

situated on the high terrace of the Las Pitas River. These structures reflect varied patterns of spatial utilization during a period spanning approximately from ca. AD 530 to, likely, AD 900 (Babot et al. 2006; López Campeny 2009). The second sector (II) comprises large rocky blocks featuring multiple grinding mortars and various examples of rock carving situated within the debris at the base of the slope. The execution of these features is estimated to have occurred between the beginning of the Common Era and AD 500 (Aschero et al. 2009). The third sector (III), also situated at the base of this slope encompasses a cluster of residential, ritual, and productive structures of sub-circular form. These structures are positioned in proximity to an ignimbrite outcrop that delineates the eastern boundary of the site (López Campeny 2000; Somonte and Cohen 2006).

The analysed materials come from a residential structure (E2) located within the aforementioned third sector (III). The boundaries of this structure are defined by large rock blocks and a stone wall. Indications of a partially roofed space include a series of holes in the blocks, coupled with traces of posts found in the stratigraphy (Fig. 4.2). The covered space, enclosed by the rock blocks and the stone wall, spans an area of approximately 12 m^2 (A sector). A second, smaller sector (B) with a surface area of about 4 m^2, lacking evidence of roofing, has been interpreted as an access corridor to the enclosure (López Campeny 2000). The integration of stratigraphic, taphonomic, ecofactual, and artefactual analyses led to the conclusion that levels 1 (A and B) and 4 (A) are associated with periods of enclosure vacancy, while levels 2, 3 (A and B), and 5 and 6 (A) correspond to phases of effective occupation. Four radiocarbon dates (Table 4.1) provide evidence of a persistent and variable use of this sector spanning just over a millennium and a half, from ca. 50 BC to AD 1500. This period encompassed various domestic activities such as habitation, processing, consumption, production, discarding, as well as the creation of rock engravings, pastoral activities, and funerary practices (López Campeny 2010). The funerary context (B sector) involves a limited assemblage of bone remains[1] attributed to a male infant individual (age: 5–7 years old), accompanied by a collection of yarns and woven garments, two ceramic bowls, mineral and malacological-based beads (mollusc from the Pacific Ocean), and an assortment of seeds and fruits (mostly of non-local origin), among other elements with diverse functions (López Campeny 2014). The integrated analysis of architectural, stratigraphic, associational, and distributional features has led to the proposition that, subsequent to the original burial but prior to the formation of level 2, the tomb was reopened. This process involved the removal and transport—to another location—of a significant portion of the bodily remains, the reorganization of the remaining materials, and subsequent re-burial (López Campeny 2000).

[1] A fibula, a scapula, a fragment of the coxal bone, a clavicle, and unidentified rib fragments (López Campeny 2000).

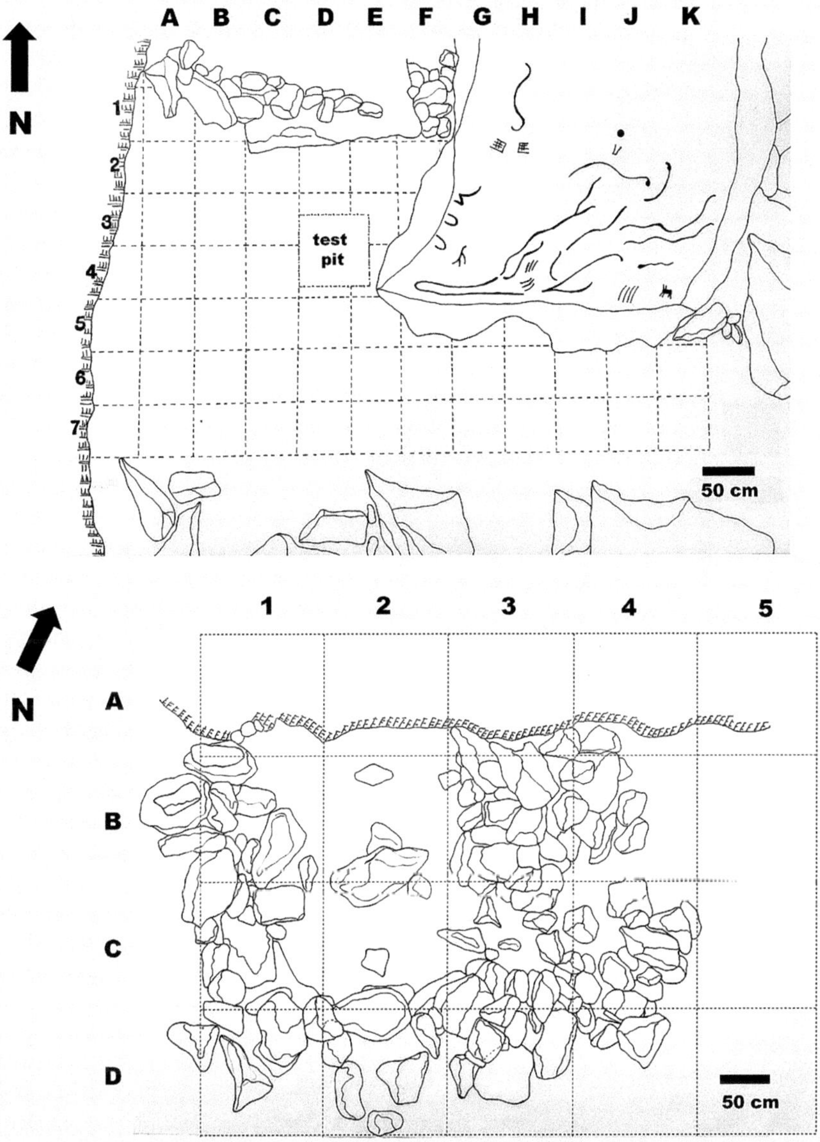

Fig. 4.2 Top: plan view of E2, site PP9. Bottom: plan view of E1, site PH2

Table 4.1 Compilation of archaeological dates from the PP9 and PH2 sites

Site and unit	Level and sector	Conventional and calibrated date		Dated material and associated context	References
		^{14}C YBP	cal. BC/AD[a]		
PP9 (III) E2	N6 (A)	1970 ± 50	[50 cal BC to 205 cal AD] $p = 95.4$	Camelid faeces Stockyard	López Campeny (2000, 2010)
	N3 (B)	1480 ± 40	[542–565 cal AD] $p = 6.1$ [570–669 cal AD] $p = 89.4$	Seeds Funerary deposit of infantile individual	
	N5 (A)	1150 ± 150	[658–1214 cal AD] $p = 95.4$	Plant carbon remains Hearth area	
	N2 (B)	530 ± 50	[1389–1496 cal AD] $p = 90.9$	Plant carbon remains Rubbish area	
PH2 E1	N4	690 ± 60	[1270–1416 cal AD] $p = 95.4$	Plant charcoal remains Ritual structure monolith (*huanca*)	López Campeny (2009) Urquiza et al. (2013)
	N3	580 ± 60	[1299–1365 cal AD] $p = 31.1$ [1377–1456 cal AD] $p = 64.4$	Plant charcoal remains Hearth area	
	N1	220 ± 30	[1649–1705 cal AD] $p = 23.7$ [1720–1814 cal AD] p = 65.9	Grasses Deposit between stones of the wall	

[a] Calibrated with v4.4 (Bronk Ramsey 2021). Atmospheric data from Hogg et al. (2020)

4.2.2 *Piedra Horadada 2 Site*

The Piedra Horadada 2 site (PH2) is located 600 m upstream from PP9, near the southern bank of the Las Pitas River, at an elevation of 3640 m.a.s.l. Like PP9, PH2 is an open-air site characterized by stone structures of varying dimensions with a sub-circular plan, dispersed among blocks from landslides and making use of the natural slope of the terrain. Different sectors have been interpreted as corrals, residential units, processing and consumption areas, zones for lithic production, and sites for various rituals. The occupation sequence begins around ca. AD 50 and extends up to recent historical periods (ca. AD 1800) although asserting continuous occupation throughout the entire period is not definitive (López Campeny 2009).

The study sample comes from an enclosure (E1) with a chronology ranging from ca. AD 1270 to AD 1815 (Table 4.1). The final configuration of this enclosure results from an architectural history that can be summarized in four construction phases (Fig. 4.2): (1) Associated with the oldest stratigraphic unit (level 5), the shelter of a large collapsed block (west) and a rocky outcrop (north) is utilized; (2) A semi-circular stone perimeter wall is added (south), along with a distinctive feature—a massive and heavy rock block (a monolith)—supported at its base by a pile of rocks forming an ad hoc foundation (level 4); (3) A second wall (east), of lesser depth, is constructed, resting on the previous one (south) and confining the space around the monolith. A circular structure is identified for maintaining a hearth, filled with ashes and coals, along with a seat for pots (level 3); and (4) Adjacent to the west block, a mound of rocks (levels 2 and 1), reminiscent of the so-called *apachetas* or *mojones* (in Spanish), similar accumulations of stones ritually acknowledged in the Andes to this day.[2] Finally, a deposit or intentional arrangement of objects—yarns, camelid fleeces, and fragments of equine mandibles (*Equus* sp.) from a neonate individual—was documented in the interstices or niche of a wall that was part of the enclosure (level 1). The presence of equine remains confirms the continuity in the use of the archaeological structure during the Hispanic period, according to the radiocarbon dates (Urquiza et al. 2013).

By integrating evidence from artefacts and ecofacts, as well as architectural, stratigraphic, and contextual features, it is proposed that the E1 enclosure was associated with pastoralist propitiatory practices and/or rituals linked to caravan activity, centred on monoliths (Quechua *huanca* or Sp. *mojón*). These rituals were practised by indigenous herding populations until sometime after Spanish contact, with textiles playing a notable role in the associated symbolism (López Campeny 2009, 2014).

4.3 Study Sample and Methodology

As previously mentioned, the analysis focused on a collection of fleeces (N: 62) and hides with fibres (N: 8) from two archaeological sites. Within the second subgroup, a collection of artefacts (n: 5) crafted from this material can be discerned (Table 4.2 and Fig. 4.3). It consists of a total of four "bundles" or "wrappings" of untreated leather (fleece-covered skin), three of which present fillings of herbaceous (Fig. 4.3a, c) and woody plant elements (Fig. 4.3d). One of these bundles also features a cord of vegetable fibre on the exterior (Fig. 4.3a), and another one of them presents a yarn of animal fibre (Fig. 4.3d), included in this analysis. Two of the leather "bundles" feature perforations on the surface of the skin, resembling eyelets (Fig. 4.3a, c). A fifth specimen is included, comprising a small leather fragment, folded in a manner that creates a fold at the base resembling a small bag, within which a fleece or bundle of fibres was found (Fig. 4.3e). It is noteworthy that the skin's surface exhibits a treatment involving fleece removal (*depiling*). Consequently, fibre

[2] The *mojón* (Sp.) alludes to a large stone or conical rocky hummock that represents the reproduction of the family's herd animals. Each year during the marking ritual (Sp. *señalada*), stones are added to this rocky hummock, so that this *mojón* "grows" with the increase of the herds (García and Rolandi de Perrot 2000). This lithic feature has been noted in relation to the ritual sites frequented by llama caravans too (Berenguer 2004).

Table 4.2 Description of artefacts crafted from hides with fibres

#ID	Maximum dimensions (cm)	Interior (filling)	Exterior (modification or addition)	Other analyses
154	6.7 × 2.15 Figure 4.3e	Fleece or bundle of fibres[a] (Tables 4.3 and 4.4)	Fleece removal (*depiling*) and folded or bent	–
161	22.5 × 11.4 Figure 4.3c	Bundles of straw or grasses *Deyeuxia deserticola.* Poaceae[b] *Festuca orthophylla.* Poaceae	Folded or bent Eyelet (perforation) with a fleece inserted	Identification of pollen grains preserved[a] Stable isotope analysis[c] (C–N–S)
190	50.2 × 14.28 Figure 4.3a	Bundles of straw or grasses *Stipa vaginata.* Poaceae[b]	Folded or bent Vegetal fibres yarn (2zS) inserted through the eyelet (perforation) *Deyeuxia eminens.* Poaceae *Deyeuxia curvula.* Poaceae[b]	–
191	10.5 × 5.75 Figure 4.3d	A bundle of fractured stems at extremity *Atriplex imbricata* Chenopodiaceae[b]	Yarn (2zS) of animal fibre wrapped around the stems[a] (Tables 4.3 and 4.4)	–
202	33.5 × 13.1 Figure 4.3b	–	Folded or bent	–

[a] Analysed in the present work
[b] López Campeny (2010)
[c] Araníbar et al. (2007)

recovery from it for analysis has not been feasible. Nevertheless, a sample corresponding to the fleece found inside was obtained.

The taxonomic sample analysed is distributed as follows. The most diverse is the one from PP9, as it includes fleeces (n: 27), small fragments of fleece-covered skin (n: 4), leather artefacts (n: 5), and the yarn (n: 1) recovered in association with the leather artefact. In the case of site PH2, the sample analysed here is exclusively composed of fleeces (n: 35). In the results section, their spatial distribution is detailed and analysed. In regard to the sample considered for the examination of pollen content, it was more limited. For site PP9, one of the bundles of fleece-covered skin, or leather artefact filled with grasses from the funerary context (No. 161) was selected. In the case of PH2, a set of four fleeces (n: 4) from different stratigraphic levels and sectors of the E1 was sampled for palynological analysis. Details about their spatial provenance are presented in the results section, while also being integrated into the taxonomic analysis.

The materials enumerated were examined at a microscopic observation scale to: (a) taxonomically identify cohabiting species, (b) characterize the fleece transformation processes, and (c) recover and identify pollen grains preserved in the fleece.

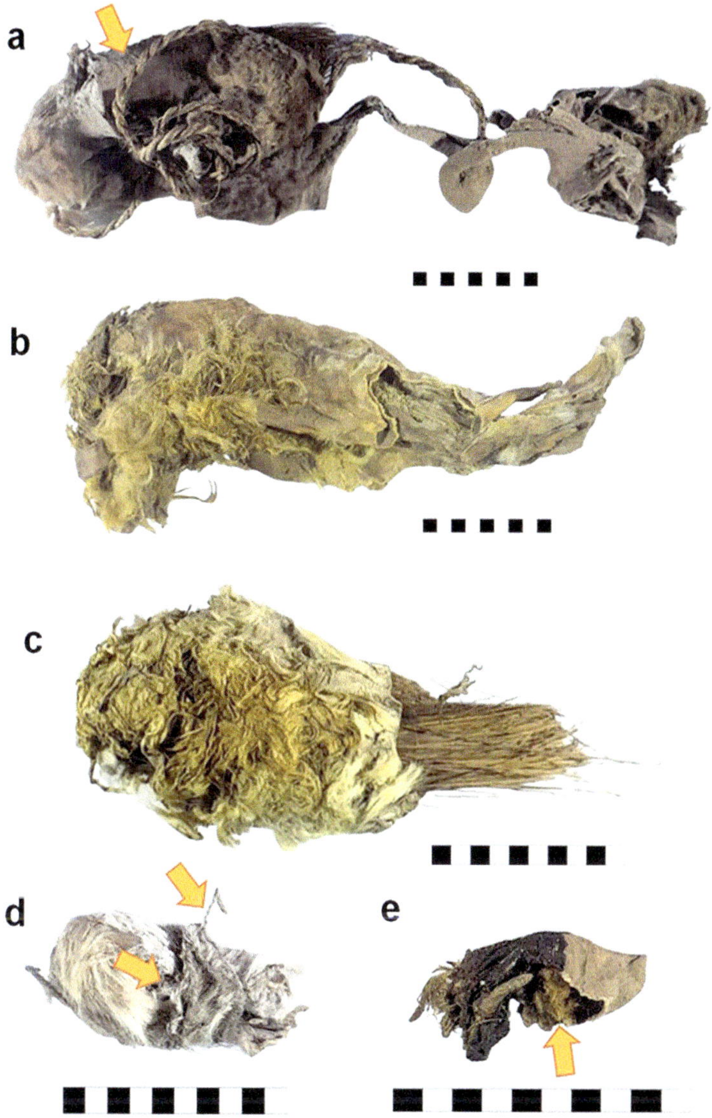

Fig. 4.3 Collection of artefacts crafted from leather: The arrows indicate the external vegetal fibre (**a**) and animal fibre (**d**) yarns, as well as the bundle of inner filling fibres (**e**), of the leather artefacts. (**a**) #ID 190; (**b**) #ID 202; (**c**) #ID 161; (**d**) #ID 191, and (**e**) #ID 154

The observation, recognition, and recording of variables pertinent to these three objectives were conducted using a trinocular biological microscope (Olympus CX31). Observation scales ranged from 40× to 400× for fibres and 1000× with oil immersion for pollen. Measurements were captured using a digital camera installed in the third eyepiece of the microscope, and the obtained images were analysed using the public domain software ImageJ/Fiji 1.53.

With regard to the first two objectives, 40 individual fibres were sampled from each fleece and hide. Emphasis was placed on examining the three primary components of each individual fibre: the cuticle, the cortex, and the medulla. These components are arranged concentrically, from the surface (cuticle) to the central canal (medulla). For the medulla and cortex, the following intrinsic variables (Reigadas 1995) were observed and recorded under the optical microscope: (1) total diameter (μ), (2) type, distribution, and thickness (μ) of the medulla, and (3) medullation index (MI). The medullation index (MI) corresponds to the ratio between the thickness of the medulla and the total diameter of the fibre. It serves as an indicator of the degree, or proportion, of the fibre that is affected by the medulla in its continuous form. Similarly, the percentage of total medullation (PTM) was calculated, representing the proportion of fibres, per sample, that exhibit medulla in any of its forms. Additionally, the type of medulla refers to the observed pattern in its morphology, serving as a distinctive feature for various genera and families of mammals. This facilitates the identification of different taxa among the fibres using reference collections and literature. As a complementary approach, the observation of the cuticle or outer layer of the fibres contributes to the taxonomic characterization of animal-origin samples (Juárez et al. 2010). For microscopic observation, negative casts were previously created from the external surface of the fibres, on a thin layer of colourless enamel. These casts were instrumental in recording and describing the cuticular pattern of the outer layer of the fibres. With regard to extrinsic variables (Reigadas 1992), the following were recorded: (4) the type of fibre, determined based on the range of thickness or total diameter, and (5) the colour, defined through direct observation. To define the fibre type in the archaeological samples, the criteria proposed by Reigadas (1995) for camelids were followed. This involves determining a thickness range of 5–31 μ for thin fibres (*lanilla*); a range between 31 and 66 μ for intermediate fibres and measurements exceeding 66 μ for thick hairs (*pelo*) (Figs. 4.4 and 4.5).

The methodology for analysing pollen preserved in fleeces and hides comprises three main stages, as outlined by Romano (2019): (1) Field sample collection, (2) Extraction of palynomorphs, and (3) Identification and pollen counting.

1. The materials analysed were collected on-site at the time of discovery. The samples were placed in sterile bags and immediately sealed to prevent contamination from contemporary pollen. Traceability of the sampling and subsequent handling conditions was ensured throughout the process.
2. In the laboratory, the samples were manipulated and processed under a fume hood. A standardized protocol (Faegri and Iversen 1989) was followed, which included a physical-chemical treatment applying the acetolysis procedure as proposed by Erdtman (1960). The sample mounting was also conducted within the laboratory hood. Using a micropipette, 1–2 drops of the sample were applied

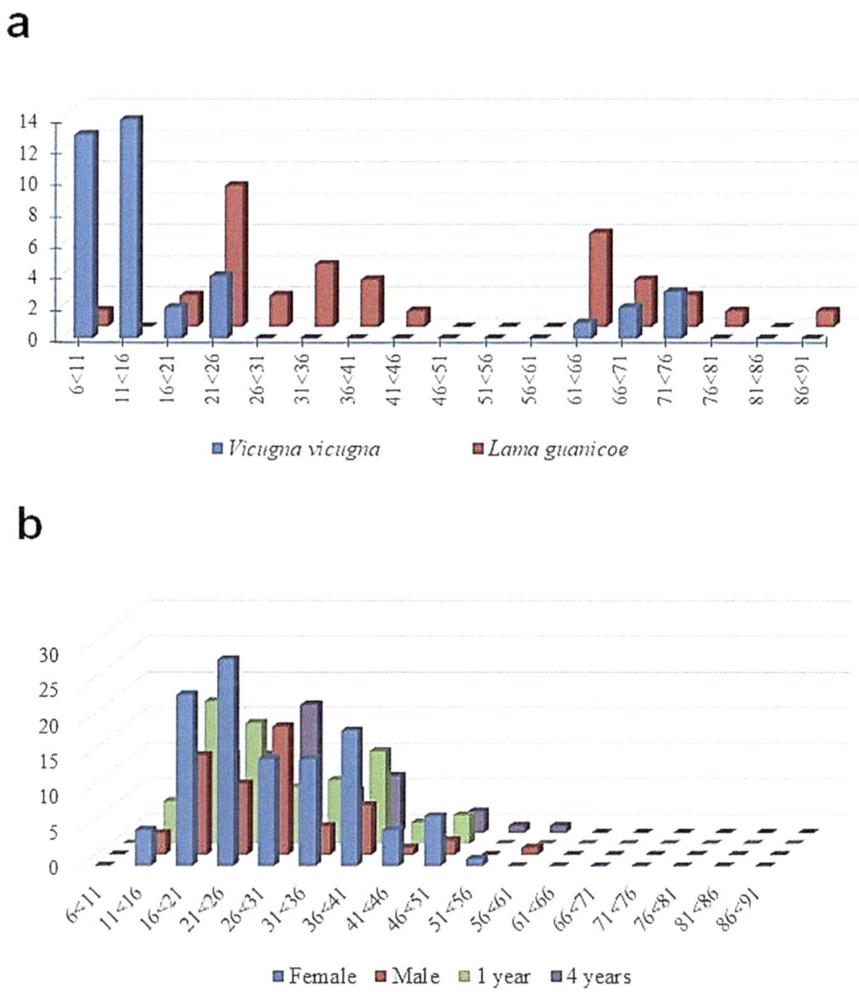

Fig. 4.4 (**a**) Distribution of fibre thickness by type for *Vicugna vicugna* (*vicuña*) and *Lama guanicoe* (andean *guanaco*); (**b**) distribution of fibre thickness by type for *Lama glama* (*llama*)

to a slide and stained with safranin diluted in glycerine. The coverslip was then affixed using paraffin. This mounting technique allows for the movement of palynomorphs, facilitating observation under the microscope—an advantage not afforded when using balsam or gelatine. Samples were deemed valid if, following five longitudinal scans at 400× magnification: (a) 250 or more pollen grains were counted, (b) at least 12 different taxa were identified, and (c) the percentage of indeterminable pollens did not surpass 50% of the total pollen sum.

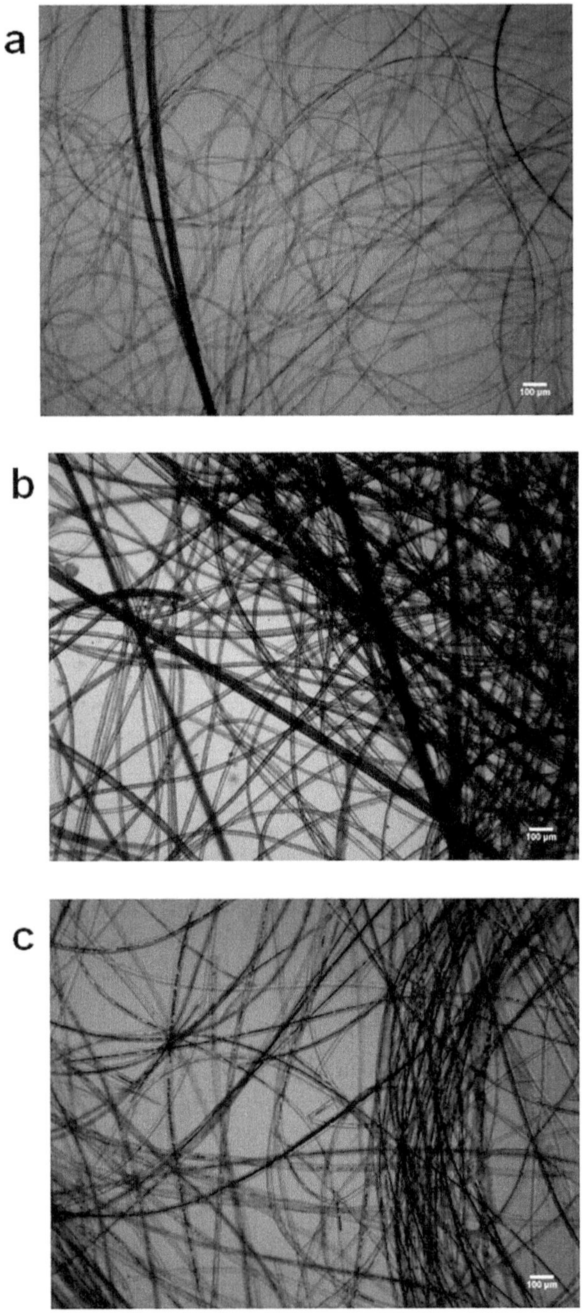

Fig. 4.5 Examples of microphotographs of: (**a**) *Vicugna vicugna* (40×); (**b**) *Lama guanicoe* (40×); (**c**) *Lama glama* (40×). Scale Bar = 100 µ

3. The identifications were carried out by comparison with our own reference palynotheca, palynological atlases (Heusser 1971; Markgraf and D'Antoni 1978) and web databases.[3] For the comparison, only acetolysed samples were taken into account, considering the following variables and attributes: grain shape; range (AMB); size in microns; apertures: type, distribution, quantity; surface type, and ornamentation. Pollen diagrams, on the other hand, were created using TILIA.GRAPH software (TGView 2.0.4; Grimm 2011).

4.4 Results

4.4.1 Punta de la Peña 9 Site

Regarding the provenance of the fleeces, findings are documented for all levels, except for level 4, interpreted as a moment of vacancy. This inference is consistent with the observed differential weathering state in the faunal bone material (López Campeny et al. 2005). The highest density (67%) comes from levels 2 and 3, especially 2(B) and 3(A) sectors. Concerning their spatial distribution, a similar pattern is observed across all levels. The findings correspond with marginal areas of the enclosure, specifically adjacent to the collapsed blocks (N and S), and the stone wall (W), that constitutes the architectural boundaries. In the lower levels (5 and 6), the fleeces were concentrated in the NW corner of the enclosure, where the layer of excrement associated with the stockyard was preserved. In regard to the leather artefacts, all have been recovered in association with the funerary context (B sector, level 3).

The results of the taxonomic identifications are presented below (Tables 4.3 and 4.4, Figs. 4.6a and 4.7). From a distributional perspective, the highest specific diversity corresponds to levels 2 and 3 (with 5 and 4 identified taxa, respectively), aligning with the levels where the highest density of fleece findings was recorded. Conversely, the highest taxonomic specificity corresponds to level 6 (sector A), with the exclusive identification of llama fleece. It can be observed that, considering the total analysed assemblage, the absolute number of identified taxa (N: 38) shows a clear predominance of Camelidae (81%). Nearly half of the total (47%) corresponds to *Lama glama* (llama), followed by a relatively high proportion (29%) of specimens of *Vicugna vicugna* (vicuña) and, in third place (5%), *Lama guanicoe* (guanaco). The sample is completed with the presence of rodents (10%), which includes the determination of *Lagidium* sp. (viscacha) in half of the cases. The remaining minor proportions (3%) correspond to human hair, fibres from *Ovis* sp. (sheep), and Cervidae. Regarding this latter Family, fragments of antlers corresponding to

[3] PalDat—a palynological database (2000 onwards, www.paldat.org); http://lillo.org.ar/atlaspolinicodelnoa/; http://www.europeanpollendatabase.net/index.php (Bush and Weng 2006).

Table 4.3 Results of taxonomic identifications and characterization from PP9 (III) site: (A) for fleeces and hides with fibres (−)

Provenance		#ID	Thickness (μ) and percentage (%) Thin 5–31 μ	Intermed. 32–66 μ	Hairs > 66 μ	PTM	$\bar{X}$	δ	Results
N1 (A)	A2	1	36	61	3	–	38.5	16.1	Hair cuticular scale patterning: *Ovis* sp. full coat
N1 (B)	F6	66	–	100	–	80	44.4	7.4	Patterning: *Lama glama.* Modified coat
	G6	102	100	–	–	70	15.3	2.9	Patterning: *Vicugna vicugna.* Modified coat for “desmoteado” o dehairing
	J6	98	23	77	–	43	40.9	12.5	Patterning: *Lama glama.* Full coat from the legs
N2 (A)	A3	25	38	55	7	100	39.9	14	Patterning: *Lama glama.* Full coat from the legs
	A4	39	45	45	10	–	37.1	18.6	Patterning: *Lama glama.* Full coat from the legs. With rodent medullary patterned fibres
	E2	11	77	23	–	20	24.7	7.3	Patterning: *Lama glama.* Full coat from the back
N2 (B)	F6	61	55	45	–	80	30.9	9.7	Patterning: *Lama glama.* Fleece without hairs, possibly from the legs
	I6	100	70	30	–	32.5	28.2	8.3	Patterning: *Lama glama.* Fleece without hairs, possibly from the back.
		100-2	–	–	–	–	–	–	Fibres with uni-, bi-, and multi-seriate medulla. Patterning: Rodent. *Lagidium* sp. with cervidae medullary patterned fibres
	J6	101	90	7	3	20	20.5	10.4	Patterning: *Vicugna vicugna.* Full coat
	H6	134(−)	66	33	–	70	28.3	9.2	Patterning: *Lama glama.* Full coat from the back with leather
		134-2	82	18	–	52	20.5	7.9	Patterning: *Vicugna vicugna.* Modified coat for “desmoteado” o dehairing
		99	97	3	–	22	16.4	7.5	Patterning: *Vicugna vicugna.* Modified coat for “desmoteado” o dehairing
		99-2	12	18	70	84	48	20.2	Patterning: *Lama guanicoe.* Full coat.

(continued)

Table 4.3 (continued)

Provenance		#ID	Thickness (μ) and percentage (%)			PTM	$\bar{X}$	δ	Results
			Thin 5–31 μ	Intermed. 32–66 μ	Hairs > 66 μ				
N3 (A)	D2	19	90	7	3	55	21.2	9.9	Patterning: *Lama glama*. Modified coat for "desmoteado" o dehairing
		21(−)	10	90	–	95	38	9.3	Patterning: *Lama glama*. Full coat from the legs with leather
		21-2(−)	96	–	4	40	17.3	10.9	Patterning: *Vicugna vicugna*. Full coat with leather
	E2	22	–	–	–	–	–	–	Fibres with uni-, bi-, and multi-seriate medulla patterning: Rodent. *Lagidium* sp.
		34	75	20	5	62	26.5	15.6	Patterning: *Lama glama*. Full coat.
		20	–	–	–	–	–	–	Fibres with uni-, bi-, and multi-seriate medulla. Patterning: rodent
		20-2	100	–	–	42	16	8	Patterning: *Vicugna vicugna*. Modified coat for "desmoteado" o dehairing
N3 (B)	I6	117	17	80	3	70	34.8	10.4	Patterning: *Lama glama*. Full coat
		253	90	10	–	40	16.7	8.6	Patterning: *Vicugna vicugna*. Modified coat for "desmoteado" o dehairing
N5 (A)	A4	52	–	–	–	.	–	–	Patterning: Human hair
	E2	45	82	18	–	67	22.5	10.1	Patterning: *Lama glama*. Modified coat for "desmoteado" o dehairing
		45-2	100	–	–	25	13.4	3.4	Patterning: *Vicugna vicugna*. Modified coat for "desmoteado" o dehairing
N6 (A)	E2	47(−)	85	15	–	40	21.4	10.6	Patterning: *Lama glama*. Modified coat for "desmoteado" o dehairing with leather
		92	7	90	3	100	34.4	10.3	Patterning: *Lama glama*. Full coat possibly from the legs
		57	10	68	22	100	40.9	12.9	Patterning: *Lama glama*. Full coat possibly from the legs

Table 4.4 Results of taxonomic identifications and characterization from PP9 (III) site: (B) for leather artefacts and associated materials

Provenance		#ID	Thickness (μ) and percentage (%) Thin 5–30 μ	Intermed. 31–66 μ	Hairs >66 μ	PTM	$\bar{X}$	δ	Results
N3 (B)	H7	154[a]	30	70	–	80	35.2	11.7	Fleece filling: Patterning: *Lama glama* Modified coat for "desmoteado" o dehairing With feathered barbules
	H6	161	80	5	15	70	20.5	4.7	Patterning: *Vicugna vicugna.* Full coat
	H6-H7	190	67	23	10	33	22.9	10.9	Patterning: *Lama guanicoe.* Full coat
	H7	191	87	8	5	70	20.3	8.7	Patterning: *Vicugna vicugna.* Full coat
	H7	191[a]	–	52	48	65	32.4	6.4	Yarn: Patterning: *Lama glama.* Modified coat for "desmoteado" o dehairing
	H6	202	82	12	6	65	19.8	6.4	Patterning: *Vicugna vicugna.* Full coat

[a] "Full coat" and "modified coat" refer to the natural fleece or fur of camelids

Hippocamelus antisensis (Andean deer or Sp. *taruca*) were identified in the bone record (López Campeny et al. 2005).

In the case of hide fragments with fibre, all have been assigned to camelids, with llama being the most common (75%), and the remaining percentage corresponding to one vicuña specimen. However, when considering leather artefacts from the funerary context, the preponderance is of wild specimens (vicuña: 60% and guanaco: 20%). This dominance becomes absolute if we take into account that the only identification of llama (20%) does not correspond to the fibre of the leather itself—which could not be identified as it was completely "skinned" or "depilated"—but to the fleece recovered as filling, which also includes barbs of feathers. The remaining identification of llama, associated with this assemblage, corresponds to the yarn wrapped around the artefact filled with *Atriplex* stems.

In regard to the second aspect analysed, we can conclude that in all levels, both complete fleeces and processed fibres are recorded. Focusing on the Camelids subgroup (n: 25), the total set includes: 40% complete llama fleeces and 24% that show treatment (dehairing or *desmoteado*). In the case of vicuña, the proportion of modified fleece (24%) is equal to the percentage of *llama* showing manipulation, while 8% present the complete fleece. Finally, the remaining 4% corresponds to guanaco fibres exhibiting the complete fleece. In the case of small leather fragments with fibres, only one of them (*llama*) shows evidence of having been modified. The three bundles or artefacts made with *vicuña* skin and the one made with *guanaco* present complete fleeces, while the yarn (*llama*) associated with one of them (*vicuña*),

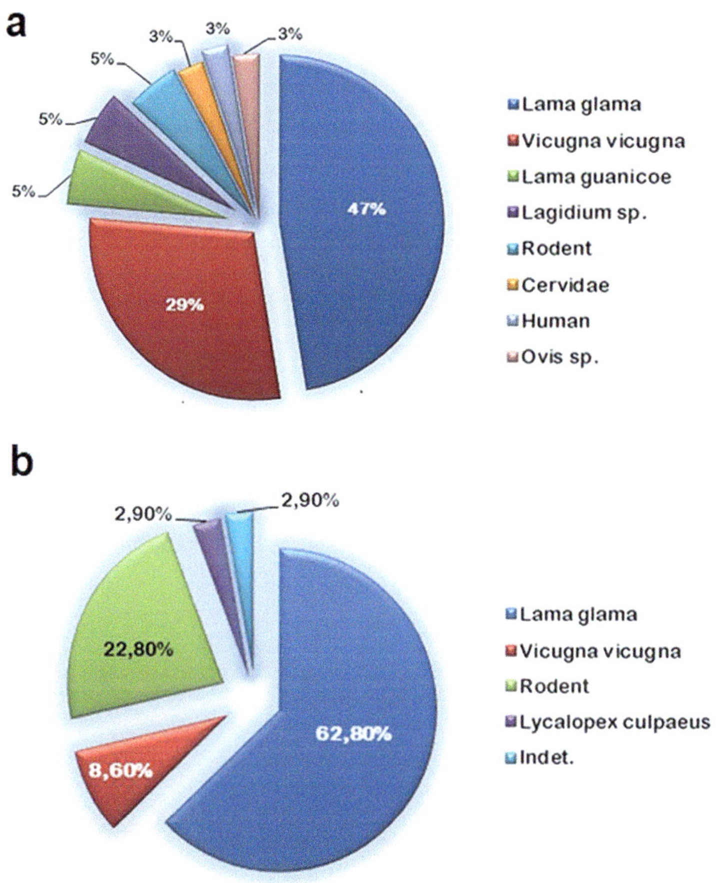

Fig. 4.6 Results of the taxonomic identifications of fibres: (**a**) PP9 (III) site, (**b**) PH2 site

shows evidence of dehairing, as would be expected. Finally, in the case of the fleece recovered from inside the small leather container, it also shows evidence of processing (dehairing or *desmoteado*). The microscopic observation of the colour variable led to the conclusion that there are no signs of dyeing or artificial pigmentation processes in the fleeces, but rather the fibres exhibit their natural hues.

Regarding the results of the pollen analysis, it is important to highlight firstly that in all the samples analysed, the pollen count was consistently higher than 200 grains, sometimes exceeding 250 specimens. Additionally, a high proportion of identification was observed, which may be related to excellent preservation conditions. As anticipated, in the case of the PP9 site, the sample was obtained from the artefact corresponding to one of the bundles made with *vicuña* leather. Pollen grains were identified mainly from plants with a botanical habit of herbs (54%), followed by shrubs (45.9%). No tree species were recorded. The botanical families are local, with Poaceae (36%) and Asteraceae (Composites 18%) being dominant, followed

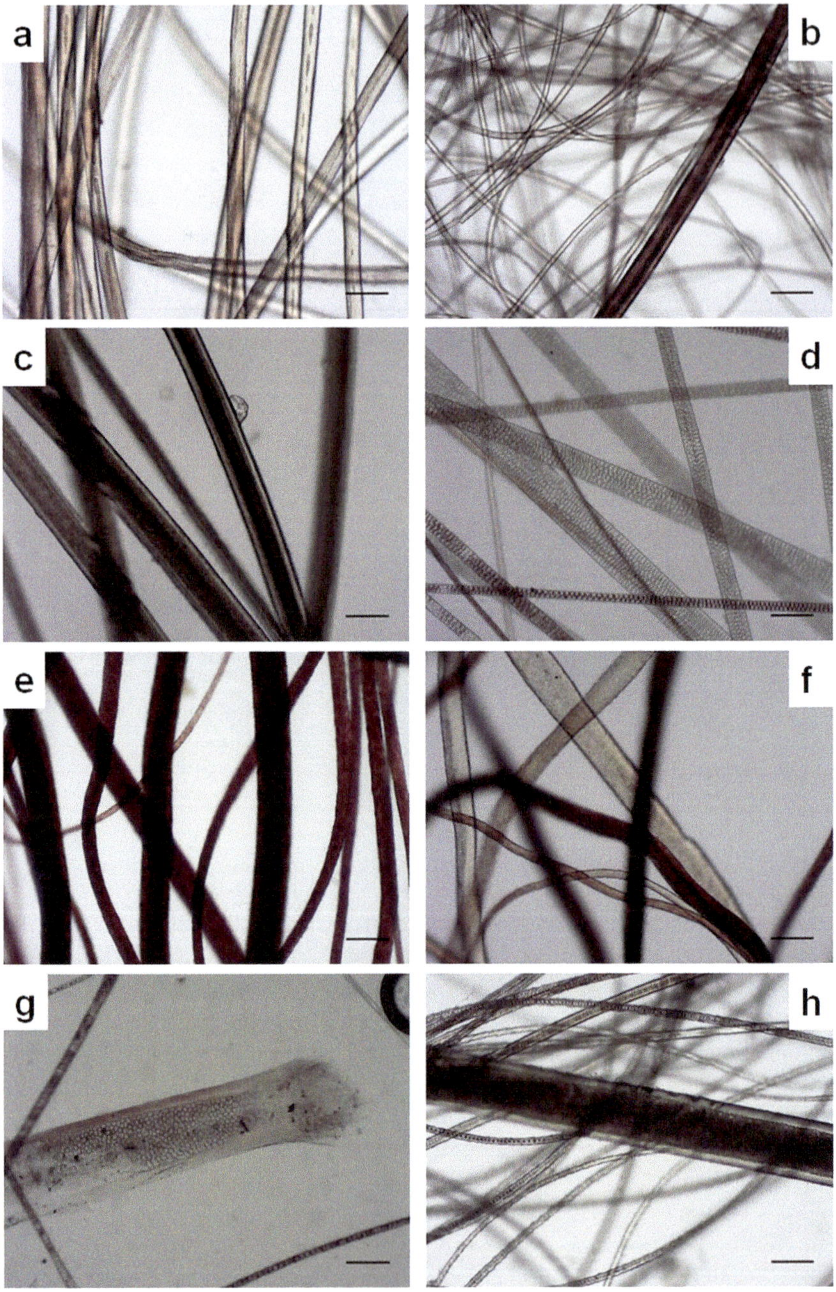

Fig. 4.7 Examples of microphotographs of identified fibres: (**a**) *Lama glama*; (**b**) *Vicugna vicugna*; (**c**) *Lama guanicoe*; (**d**) Rodentia: *Lagidium* sp.; (**e**) Human hair; (**f**) *Ovis* sp.; (**g**) Cervidae; (**h**) Canidae: fox. Scale Bar = 100 μ

Table 4.5 Percentage of pollen grains recovered in the PH2 and PP9 samples. Source: Tilia 3.0.3 software

Code	Name	Group	PH2 B1 Wall	PH2 B2N2	PH2 B1N3	PH2 A4N4	PP9 E2
	Poaceae	Herbs	32.3	9.2	30.3	6.9	38.5
	Verbenaceae	Shrubs	17.5	32.6	33.0	61.0	10.8
	Asteraceae	Shrubs	13.1	17.0	8.3	11.8	19.5
	Chenopodeaceae	Shrubs	7.0	7.8	11.0	6.5	12.1
	Ephedraceae	Shrubs	0.0	0.0	0.0	0.8	3.5
	Rosaceae	Herbs	5.2	5.5	2.7	0.8	8.2
	Cyperaceae	Herbs	3.5	1.8	0.0	0.0	4.8
	Malvaceae	Herbs	0.0	2.8	0.0	0.8	0.0
	Hydrophillaceae	Herbs	7.0	10.6	9.8	0.0	2.6
	Amarantaceae	Shrubs	0.0	0.0	0.4	0.0	0.0
	Juncaginaceae	Herbs	0.0	0.0	0.0	0.0	0.0
	T. Polemoniaceae	Herbs	0.0	0.0	4.5	0.0	0.0
	Plantaginaceae	Herbs	0.0	0.0	0.0	0.0	0.0
	Geraniaceae	Herbs	0.0	0.0	0.0	0.0	0.0
	Alnus sp.	Tree	8.3	7.3	0.0	6.1	0.0
	Celtis sp.	Tree	6.1	5.5	0.0	5.3	0.0
SUM (herbs)			48.0	29.8	47.3	8.5	54.1
SUM (shrubs)			37.6	57.3	52.7	80.1	45.9
SUM (tree)			14.4	12.8	0.0	11.4	0.0
SUM (excluded)			14.2	13.8	7.7	9.6	12.8
SSUM (s)	Sum pollen	Herbs, shrubs, tree	229	218	264	246	231
SSUM (t)	Sum totals	Herbs, shrubs, tree, excluded	267	253	286	272	265

by Verbenaceae and Chenopodiaceae with a representation of 10% and 11%, respectively. Lastly, Rosaceae was recorded at 8%, and the remainder at percentages less than 5%. Among the herbs, there is a high percentage of species that grow in meadow areas or *vegas* (Cyperaceae and Rosaceae). It can be concluded that in this sample, no pollen of non-local botanical species have been recovered (Table 4.5, Fig. 4.8).

4.4.2 *Piedra Horadada 2 Site*

From a distributional perspective, the majority of the fleece sample (64%) is concentrated in levels 1 and 3. The primary provenance areas in the two upper levels (1 and 2) are associated with the northwest rocky mound (Sp. *apacheta*) and the

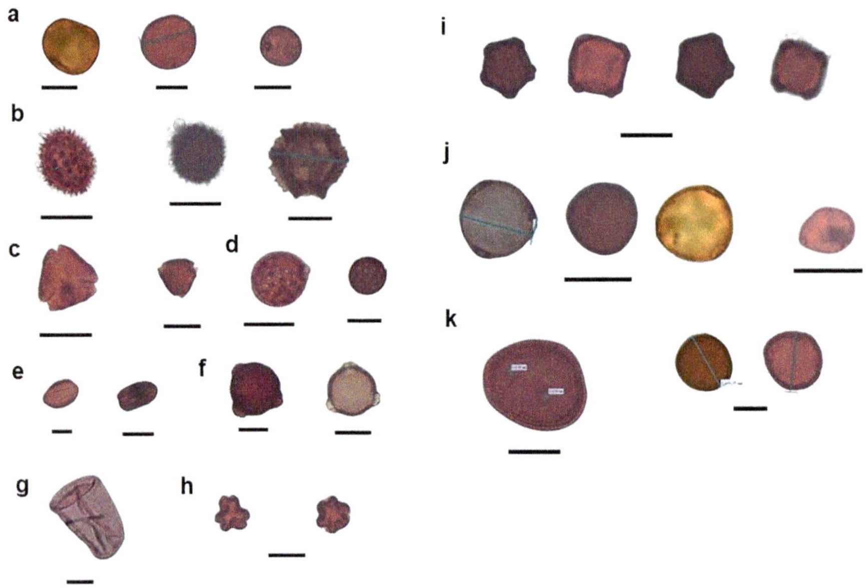

Fig. 4.8 Local pollen Families: (**a**) Poaceae; (**b**) Asteraceae (Compositae); (**c**) Verbenaceae; (**d**) Chenopodiaceae; (**e**) Ephedraceae; (**f**) Rosaceae; (**g**) Cyperaceae; (**h**) Hydrophyllaceae. Non-local pollen Families and species: (**i**) *Alnus* sp.; (**j**) *Celtis* sp.; (**k**) Type Polemoniaceae. Scale Bar = 25 μ

vicinity of the southern wall (intentional arrangement). The distribution changes in the subsequent two levels (3 and 4), and the findings were concentrated in the central space of the enclosure, in the proximity of the monolith, and slightly less in association with the southern wall.

In terms of the results of the taxonomic identifications (Fig. 4.6b), according to the previous exposition in Urquiza et al. (2013), the fleece samples from PH2 exhibit a clear predominance of Camelidae specimens (71.4%). Within this group, 62.8% correspond to *Lama glama*, and 8.6% to *V. vicugna*. Continuing with the taxonomic representation of the sample, the fleeces show a high frequency (22.8%) of rodent fibres. These identifications were made based on the uniseriate, biseriate, triseriate, and multiseriate shape that the medulla acquires for this taxon. It is also noteworthy that all fibres are medullated, which is characteristic of fibres from these mammals. Regarding the colours recorded for these specimens, various shades of grey and yellowish tones were observed in all cases. Hypothetically, we can infer the presence of fibres from *Ctenomys* sp. (tuco-tuco). This inference is supported by the mentioned characteristics and their comparison with the reference collection of fibres by one of the authors (A. S. Romano), and by the bone record recovered at the site (Urquiza et al. 2013). The remaining 5.8% of the fleece sample is distributed between a possible specimen of fox and undetermined specimens. The presence of foxes would also be supported by the identification of coprolites in the archaeological record (Urquiza et al. 2013).

The results of the fleece characterization analysis show a different panorama compared to what was observed at the PP9 site. This is because, out of the total set of fibres corresponding to camelids, only two fleeces (5.5%) show evidence of having been processed or manipulated. These samples come from the same unit and excavation level (micro sector B1, level 3) and were identified as originating from *vicuña*. It is notable that one of them is composed of a set of thin fibres (Sp. *lanilllas*) while the other includes the portion of coarser hairs (Sp. *pelo*). The observation of similarities in colour attribute, along with the coincidence in their spatial distribution, suggests that both sets of fibres originally belonged to the same fleece. Additionally, during the microscopic observation of the fibres corresponding to the hairs, the presence of hair bulbs at the end of the stems could be identified.

As anticipated, the pollen content analysis for this site was conducted using four fibre samples. In all cases, the fleeces have been identified as corresponding to *llama* specimens. The results are as follows. In the fleece recovered in association with the East wall (B1), pollen grains corresponding to plants, according to their botanical habitat, are distributed as follows: herbs (48%), followed by shrubs (37.6%) and trees (14.4%). The local botanical families are dominated by Poaceae (32%), followed by Verbenaceae (17.5%) and Asteraceae (Composites 13.1%). Non-local tree species are represented by 8.3% of pollen grains identified as *Alnus* sp. (Alder tree) and 6.1% of *Celtis* sp. (Hackberry tree). The next sample comes from a fleece from level 2 (B2). Pollen grains have been identified in the following distribution according to their botanical habitat: shrubs (57.3%), followed by herbs (29.8%) and trees (12.8%). Local botanical families are dominated by Verbenaceae (32.6%), followed by Asteraceae (Composites 17%), while Hydrophyllaceae accounts for 10.6% and Chenopodiaceae for 7.8%. Non-local tree species are represented by pollen grains identified as *Alnus* sp. (7.3%) and *Celtis* sp. (5.5%). For the micro-region, Hydrophyllaceae specimens were only recorded in high-altitude ravines (Quebrada Seca and Real Grande, ca. 4000 m a.s.l.) or on sandy soils over ignimbrite cliffs (Cuello 2006). The results obtained for the fleece from level 3 (B1) are distributed as follows: shrubs (52.7%), followed by herbs (47.3%). Local botanical families are dominated by Verbenaceae (33%) and Poaceae (30.3%), while Chenopodiaceae (11%), Hydrophyllaceae (9.8%), and Asteraceae (Composites 8.3%) are recorded in smaller proportions. Although no non-local species were surveyed, the presence of Polemoniaceae pollen (4.5%) can tentatively be included in this group. This is because the only specimens belonging to this botanical family were collected on the way to Antofagasta de la Sierra in the grasslands of Peña Fría, at an approximate distance of 50 miles from Piedra Horadada 2 (Cuello 2006). Finally, for the fleece from level 4 (A4), pollen grains from plants with botanical habitat are distributed as follows: shrubs (80.1%), followed by trees (11.4%), and a low percentage of herbs (8.5%). Local botanical families are dominated by Verbenaceae (61%) and Asteraceae (Composites 11.8%), complemented by Chenopodiaceae (6.5%) and only 6.9% of Poaceae. Non-local tree species are again represented by pollen grains identified as *Alnus* sp. (6.1%) and *Celtis* sp. (5.3%) (Table 4.5, Fig. 4.8).

4.5 Discussion

The results of the distributional analysis reveal some differences among the samples studied at each site.

In the case of PP9, a recurrence in the concentration of fleeces and small leather fragments with camelid fibres is observed in peripheral areas, specifically in sectors adjacent to the collapsed blocks and wall. The signs of manipulation (depiling) suggest the possibility that their distribution in the sector with evidence of roofing (A) responded to temporary storage, with the intention of being used in the subsequent production of yarns. These characteristics are, in turn, consistent with the results of the textile ensemble characterization (López Campeny and Romano 2021). In this sense, the walls and rocky shelters may have favoured the protection of this material, sheltering it from spaces of greater circulation or intensive use. In other sectors of the structure (level 2, sector B), coinciding with the record of the highest specific diversity, the dispersion of fleeces is associated with the accumulation of a highly variable set of secondary character residues. In these cases, we propose that their location could respond to discard situations (space maintenance), in which the effect of the wind may also have played a relevant role, as it is a sector without evidence of roofing. At level 6 (A), all the fibres recovered in association with the layer of *guano* correspond to *llama*, which would be consistent with the interpretation of the space as a pen or stockyard, coupled with the fact that most of the fleeces do not show signs of processing. At the same time, it is relevant to highlight that the general spatial arrangement of the fleeces differs from that of the faunal bone remains, which instead concentrate in central areas, distant from the walls and associated with combustion features (hearths) and the dispersal of lithic flakes. This bone assemblage is characterized by its small size, high fragmentation, and signs of thermal alteration, with a very low identification percentage, while also showing a low incidence of taphonomic factors. These factors indicate that these are primary residues resulting from consumption activities, which were preserved in a sandy matrix that favoured rapid burial, in an area with frequent trampling and periodic cleaning. This interpretation is reinforced by the fact that only the bone sample exceeding 5 cm in length (1%) has been recovered in areas near the walls (López Campeny et al. 2005).

In the case of PH2, the first difference observed compared to PP9 is that the distribution of the fleeces does not show a recurrent pattern across all levels. In the upper levels (1 and 2), the fleeces are located near to two architectural elements: the accumulation of rocks (Sp. *apacheta*) and the south wall. The latter coincides, in turn, with the feature interpreted as an intentional deposit or offering. However, this distribution changes in the subsequent levels (3 and 4), where the highest recovery of fleeces is associated with another architectural feature of E1: the monolith (Qch. *huanca*), located in a central position. Furthermore, similar to what was recorded in PP9 site; there is no spatial coincidence between the fleeces and the faunal remains. The highest proportion of faunal remains with anthropogenic cut marks and evidence of exposure to fire was recovered associated with the combustion pit of level

3. Additionally, the increase in bones with taphonomic marks (rodent and carnivore action, mineral precipitations) and an increase in weathering stages in levels 2, 4, and 5 suggests recurrent moments of abandonment or prolonged disuse of the structure (Urquiza et al. 2013).

Regarding the leather artefacts collection from PP9, as previously mentioned, all of them have been recovered within the funerary context, and crafted from hides of wild species. However, they feature external (yarn) and internal (fleece) elements identified as *llama.* This assemblage thus brings together a variety of beings that, from our modern perspective, belong to different domains (domestic and wild). Regarding the vegetal filling, identifications indicate that they are species with local distribution (Carilla et al. 2018), which are also part of the diet of these camelids (see discussion on the pollen content analysis below). This collection exclusively originates from the funerary context, prompting us to consider its role in this ritual event. For archaeological contexts of the Late Archaic period in northern Chile (Tulán Site), Dransart (2002, pp. 226–228) reports the discovery of items crafted from hide and animal skin. The author notes that some yarns were pierced through, stitched and/or tied to pieces of rodent skin and camelid hide. Particularly interesting is a piece of camelid hide stitched using two different yarns (item N°538, TU 54, Fig. 4.8, p. 228), which Dransart interprets as possible clothing. It is highlighted that the piece from TU 54 did not have the fleece removed, which is similar to the cases found in PP9. She then expresses that: "Garments made from animal hides are rarely mentioned in archaeological literature on the area, although bodies in coastal graves are frequently mentioned as having been wrapped in animal or bird skins". (Dransart 2002, p. 227). However, the presence of vegetable fillings is not mentioned in any case. In summary, we are unable to provide any definitive hypothesis regarding the functionality of this artefact assemblage from PP9, as no references to similar findings have been found. Additionally, the reopening of the tomb and the removal of the body prevent us from discerning the original arrangement and associations among the accompanying elements.

From a comparative taxonomic perspective, significant differences can be observed between both sites. On the one hand, both assemblages exhibit a predominance of camelids (80%–70%), with a higher percentage found in PP9. This dominant trend aligns with findings from pastoral sites in the area based on osteo-archaeological analyses (Olivera 1997). However, the main difference lies in the distribution of camelid species. While *llama* identification is predominant in both sites, it does not reach 50% in PP9, followed by nearly 30% of *vicuña* and 5% of *guanaco*. Remarkably, there is a high percentage of *vicuña* fibres and artefacts in Structure 2 of PP9, where a considerable number (n: 10) of projectile points were recovered during occupation levels (López Campeny 2010). In contrast, in PH2, the proportion of *llama* is nearly absolute, with *vicuña* representing less than 10%. Regarding *guanaco*, its absence in PH2 aligns with the observed trend in the Antofagasta de la Sierra micro-region post-2000 BP, as documented by López Campeny et al. (2023).

Another notable difference between both sites, in terms of sample composition, lies in the proportion of rodents, with their representation in PH2 being more than double (22%) that in PP9. Regarding this, the abundance of pellets regurgitated by birds (*egagrópilas*) could also account for their prevalence in PH2's E1. The identification of fibres and skeletal remains of small-sized rodents with burrowing habits, such as *Ctenomys* and *Phyllotis*, suggests their presence as bioturbating agents (Díaz and Bárquez 2003). On the other hand, in the case of *Cavia* sp., the presence of anthropic marks indicates their consumption. A similar situation may be inferred for the significant presence of *Lagidium* sp. among the rodent fibres identified in PP9 although further research is needed to confirm this hypothesis.

Finally, regarding the minor species, in the case of *Ovis* sp., it is only recorded in the most recent occupation level of PP9, which is consistent with the chronology, but it is not present in PH2, which does show dated and bone evidence of other introduced species, such as *Equus* sp., during colonial times (Urquiza et al. 2013). The presence of fox in PH2, identified both from fibres (attributes of medulla and cuticle) and excrement, supports the existence of significant periods of abandonment and vacancy of E1. On the other hand, in PP9, the Cervidae fibres suggest access to non-local resources and the possible introduction of portions of the post-cranial skeleton, providing additional information to the previous limited identification of Andean deer antlers, exclusively related to their technological use (López Campeny et al. 2005).

When comparing the fleece assemblage with the textile record, in the case of PP9 the presence of all three species of camelids was recorded in fabrics and cordage, in similar proportions to those of the fibre assemblage (López Campeny and Romano 2021). This suggests that fibre processing activities in the structure 2 may be related to textile manufacturing. This hypothesis is supported by the presence of an important proportion of fleece in the process of manipulation and transformation, along with unaltered material (possibly used for storage). Furthermore, it is interesting to note that, in terms of processing, a high proportion of vicuña fibres with modifications were found (in equal proportion to llama), suggesting that this species was exploited not only (or predominantly) for its meat. In the case of the PH2 sample, however, the record of processed fibres is exceptional, corresponding to the only case (two samples) of a *vicuña* fleece that we interpret as having been processed in situ. This latter inference is supported by the finding of different parts of the coat (thin fibres and hairy or guard fibres) in the same micro sector and archaeological level. Regarding the presence of hair bulbs at the ends of the fleece, it suggests that these were obtained by being pulled from the skin, rather than by shearing or cutting (*esquila*). This, in turn, leads us to infer that the fleece was obtained from the skin of a deceased animal (Dransart 2002, p. 110). Furthermore, we note a lack of correspondence between the origin of this fibre and the materiality that forms the yarns and cords recovered in the E1 of PH2 site (*llama* and human hair). In other terms, there is no evidence to suggest on-site textile production.

Regarding the results obtained through the application of the sampling and pollen analysis protocol, the *vicuña* leather artefact (PP9 III, E2) has contributed pollen grains from local plant species of Puna and High Andes eco-regions (Carilla et al. 2018). It is relevant to highlight that the *vicuña* diet reported in the literature (Koford 1957; Cajal 1989; Borgia et al. 2010, among others), and inferred from micro histological studies of faeces (Mosca Torres 2010), consists of plant species whose pollen grains coincide with those recovered. In this sense, the different sources show a preference of these camelids for consuming grasses, especially Poaceae, with a lower incidence of Cyperaceae. Additionally, some shrubs (Verbenaceae, Chenopodiaceae, and Asteraceae) and certain malvaceous plants are included as dietary supplements. Importantly, the PP9 sample did not show the presence of pollen from tree species, nor was pollen from non-local species documented. On the other hand, samples from fleece obtained from site PH2 identified as llama, contributed pollen grains from the aforementioned families (local), along with pollen from *Plantago* sp., Type Polemoniaceae, and tree species from the *Yungas*, such as *Alnus* sp. and *Celtis* sp. In this context, we believe that the presence of non-local taxa in these pollen records is explained as a result of the movement of these animals through different eco-regions (Morello et al. 2012). These findings would constitute evidence of the extensive mobility of these camelids outside the Antofagasta de la Sierra micro-region, which includes areas near sources of tree species dispersion, such as certain Fabaceae, Betulaceae (*Alnus* sp.), and Ulmaceae (*Celtis* sp.).

4.6 Conclusions

From a spatial distribution perspective, certain architectural elements can metaphorically be considered as "attractors", as they appear to influence the arrangement of fleeces and hides within the analysed spaces. Based on specific contextual analysis for each site, we interpret this to be attributable to various reasons. Firstly, it relates to the dynamics inherent in the organization of intramural space, encompassing domestic and productive uses, such as preferences for activity areas, differentiation in space usage, and maintenance and cleaning tasks. Secondly, it is associated with aspects linked to ritual practices and symbolic meanings of space, such as burials, the presence of the monolith or *huanca*, the *apacheta* or *mojón,* and offering deposits. Considerations should also be given to the effects associated with taphonomic factors and the activity of bioturbation agents. Additionally, it is important to note that no correlation in spatial distribution between fleeces and faunal bone remains was observed in the cases studied. This suggests that fibres exhibit differential treatments likely associated with different modes of interaction between humans and camelids, emphasizing the relevance of their discriminated analysis, but complementarity with the bone assemblage.

From a taxonomic perspective, the analysis of fibres, complemented by the examination of bone assemblages and textile artefacts, has facilitated the reinforcement of previously posited hypotheses in certain instances. In others, it has enabled the addition of novel data concerning increased diversity of coexisting species, variations in modes of procurement and utilization, and fresh inferences regarding the body parts introduced into intramural spaces, among other factors. An integral and specialized use of animal fibre emerges, stemming from the multiple utilization of locally available camelids and the distinctive qualities of their coats. Such an approach assumes particular significance in instances where bone assemblages exhibit a limited degree of identification, attributable to processing activities and/or conservation conditions. The concept of fluidity (Ingold 2007) is crucial for understanding the fleece and hides studied here in their fluid state, and how they evolve in their roles as functional, symbolic, and ritual objects. These materials are in continual development, engaged in sustained dialogue through networks of mediation and exchange with humans and other non-human entities. This concept refers to the idea that materials are not fixed entities but constantly transform in response to environmental, social, and temporal factors (Ingold 2007; Witmore 2007).

The application of the pollen analysis protocol on the samples of fleeces and leathers with fibres yielded results with very promising prospects for application. It is possible to argue that camelid fleeces are excellent collectors of environmental pollen and promote good preservation of its grains. This is evidenced by the high number of grains recovered and the high proportion of identifications. Thus, when these supports or "soft containers" are compared with others traditionally approached by palynology (e.g., sediments), we can conclude that they constitute "mobile pollen traps". One of their main advantages is that they have a temporal resolution limited to the animal's lifespan (between 15 and 20 years), which is further reduced (2 or 3 years), if regular shearing events are considered (Dransart 2002). Additionally, it was noted that the pollen grains exhibit an excellent state of preservation (Romano and Baied 2012; Romano 2019). These pollen evidences allow us to propose a close relationship between humans and domesticated camelids, which includes, among other daily aspects, the interregional mobility of certain animals from the herd with their human owners. These results are consistent with those obtained from other lines of contextual analysis in the site. Therefore, we believe that samples of fibres and leathers with camelid fleeces have a high potential to assess archaeological topics, such as the mobility of camelids and their relationship with people and the "things" in transit.

The analysis of fleece and leather assemblages from these archaeological sites of the Southern Puna of Argentina provides significant insights into human-animal relationships beyond their material utility. Although much of the archaeological record focuses on the nutritional or economic role of camelids, the findings presented here offer an opportunity to reflect on more fluid and interconnected interactions between humans and non-human animals. This requires a shift from a utilitarian perspective towards understanding these relationships as part of a broader ontological framework, where animals and humans co-exist in interdependent, yet asymmetrical relationships.

Anthropological studies, such as those by Dransart (2002), have emphasized the importance of moving beyond hierarchical views of human-animal interactions. In Andean contexts, animals are not simply domesticated resources but are part of mutual care relationships (*uywa*), particularly between camelid herders and their animals (Haber 1999, 2007). This suggests that the presence of fleece and leather in ritual and funerary contexts, such as those analysed here, points towards practices of reciprocal communication and care, rather than solely exploitation of resources. The integration of these animals into rituals, including the burial contexts discussed, highlights their symbolic significance, a theme explored in Arnold (2016) and Bugallo et al. (2022).

Moreover, the co-existence of different species within these assemblages, particularly *llama* and *vicuña*, reflects not only the practical use of these animals but also their interconnected roles in Andean cosmologies. The findings challenge strict boundaries between wild and domestic animals, suggesting a fluid continuum where human-animal relationships are embedded in the environment and cosmological beliefs. These insights align with Bolton and Loovers (2023), who argue for a deeper understanding of these entanglements as part of a sentient world where both humans and animals actively shape social and material landscapes.

A particularly revealing aspect of this hybrid world is the evidence of shared journeys between humans and camelids. The identification of exotic pollen in the fleece samples suggests that these animals travelled with their human counterparts across different ecological zones. These journeys not only reflect the mobility of the animals but also their participation in complex networks of exchange, ritual, and subsistence. The pollen evidence highlights the interwoven relationships between people, animals, and landscapes, as well as the vital role camelids played in connecting diverse regions. This echoes Ingold's (2007) notion of material fluidity, where both humans and animals are active participants in processes that blur the lines between nature and culture, wild and domestic, ritual and everyday life.

In conclusion, while material evidence provides valuable insights into the economic aspects of human-animal interactions, it is crucial to incorporate a more anthropological perspective that considers the symbolic, ritual, and relational dimensions of these interactions. This shift in focus allows for a more nuanced interpretation of how humans and non-human animals have co-existed, not only in terms of survival but also through shared rituals, mutual care, and ongoing communication.

Acknowledgements An unpublished version of this work was presented under the title *A flor de piel… las fibras como ensambles animales y humanos* [Skin Deep… Fibres as Animal and Human Assemblages] at the symposium *Los animales y las personas. Interacciones, agentes y materialidades* [Animals and People. Interactions, Agents, and Materialities], held during the XXI National Congress of Argentine Archaeology in Corrientes, Argentina, in July 2023. The symposium was coordinated by Romina Frontini and Paula Escosteguy, and chaired by Mónica Salemme, to whom we are grateful for the opportunity to participate and for fostering a fruitful environment for peer discussion. We also extend our thanks to them for the opportunity to contribute to this thematic volume. We further wish to thank two anonymous reviewers for their valuable comments and suggestions, which helped improve the final version of this work.

Competing Interest Declaration The authors have no conflict of interest to declare that are relevant to the content of this chapter.

References

Araníbar J, López Campeny SML, Colaneri MG, Romano AS, Macko SA, Aschero C (2007) Dieta y sociedades agropastoriles: Análisis de isótopos estables de un sitio de la Puna Meridional Argentina (Antofagasta de la Sierra, Catamarca). Comechingonia Rev Arqueol 10(1):29–48. https://doi.org/10.37603/2250.7728.v10.n1

Arnold D (2016) Territorios animados. Los ritos al Señor de los animales como una base ética para el desarrollo productivo en los Andes. In: Román-López Dollinger E, Galarza Mendoza H (eds) Símbolos, desarrollo y espiritualidades. El papel de las subjetividades en la transformación social. Instituto Técnico Ecuménico Andino de Teología, La Paz, pp 111–159

Arnold D, Jiménez A, Yapita J (1998) Hacia un orden andino de las cosas: tres pistas de los Andes meridionales. Hisbol, La Paz

Aschero C, Martel AR, López Campeny SML (2009) El sonido del agua… Arte rupestre y actividades productivas. El caso de Antofagasta de la Sierra, Noroeste Argentino. In: Sepúlveda M, Briones C, Chacama J (eds) Crónicas sobre la piedra. Arte Rupestre de Las Américas. Ediciones Universidad de Tarapacá, Tarapacá, pp 257–270

Aschero C, Izeta A, Hocsman S (2012) New data on south American bone size changes during middle-late Holocene: osteometry at Peñas Chicas 1.5 (Antofagasta de la sierra, Argentinian Puna). Int J Osteoarchaeol 24(4):492–504. https://doi.org/10.1002/oa.2235

Babot MP, Aschero CA, Hocsman S, Haros C, González Baroni L, Urquiza S (2006) Ocupaciones agropastoriles en los sectores intermedios de Antofagasta de la Sierra (Catamarca): Un análisis desde punta de la peña 9.1. Comechingonia 9:57–75. https://revistas.unc.edu.ar/index.php/comechingonia/article/view/27615/29080

Barad K (2003) Posthumanist performativity: how matter comes to matter. Signs J Women Cult Soc 28(3):801–831

Berenguer J (2004) Caravanas, Interacción y Cambio en el Desierto de Atacama. Sirawi Ediciones, Santiago de Chile

Bolton M, Loovers JP (eds) (2023) Sentient entanglements and ruptures in the Americas; human-animal relations in the Amazon, Andes, and Arctic. Series: human-animal studies, vol 27. Brill, Leiden

Borgia M, Vilá BL, Cassini MH (2010) Foraging ecology of vicuña, *Vicugna vicugna*, in dry Puna of Argentina. Small Rumin Res 88:44–53. https://doi.org/10.1016/j.smallrumres.2009.11.009

Bugallo L, Dransart P, Pazzarelli F (eds) (2022) Animales humanos, humanos animales: relaciones y transformaciones en mundos indígenas sudamericanos. Antropofagia, Buenos Aires

Bush M, Weng C (2006) Introducing a new (freeware) tool for palynology. J Biogeogr 34:377–380. https://doi.org/10.1111/j.1365-2699.2006.01645.x

Cabrera A (1968) Geo-ecología vegetal de las regiones montañosas de las Américas tropicales. Colloquium Geogr 9:91–116

Cajal JL (1989) Uso de hábitat por vicuñas y guanacos en la Reserva San Guillermo, Argentina. Vida Silvestre Neotrop 2:21–31

Carilla J, Grau A, Cuello S (2018) Vegetación de la Puna Argentina. In: Grau HR, Babot MP, Izquierdo A, Grau A (eds) La Puna Argentina. Naturaleza y Cultura, Serie Conservación de la Naturaleza, 24. Fundación Miguel Lillo, Tucumán, pp 143–156. https://www.lillo.org.ar/revis/cnaturaleza/2018-scn-v24.pdf

Cuello AS (2006) Guía Ilustrada de la flora de Antofagasta de la Sierra, Catamarca (Puna Meridional Argentina). Curso de Entrenamiento para la obtención del Grado de Licenciatura en Ciencias Biológicas - Orientación Botánica. Facultad de Ciencias Naturales e Instituto Miguel Lillo. Unpublished manuscript

Deleuze G, Guattari F (2004) A thousand plateaus. Continuum, London

Díaz MM, Bárquez RM (2003) Los Mamíferos de Jujuy. Editorial LOLA, Buenos Aires

Dransart P (2002) Earth, water, fleece and fabric: an ethnography and archaeology of Andean camelid herding. London, Routledge

Erdtman G (1960) The acetolysis method. A revised description. Svensk Botanisk Tidskrift 54:561–564

Faegri K, Iversen J (1989) Textbook of pollen analysis. Wiley, Hoboken

García S, Rolandi de Perrot D (2000) Relatos y ritual referidos a la Pachamama en Antofagasta de la Sierra, Puna Meridional Argentina. Relac Soc Argentina Antropol 25:7–25

González Ruibal A (2007) Arqueología Simétrica: Un giro teórico sin revolución paradigmática. Complutum 18:283–285. https://revistas.ucm.es/index.php/CMPL/article/view/CMPL0707110283A/29416

Grau MJ, Babot MP, Izquierdo A, Grau A (eds) (2018) La Puna Argentina. Naturaleza y Cultura. Serie Conservación de la Naturaleza. Fundación Miguel Lillo, Tucumán. https://www.lillo.org.ar/revis/cnaturaleza/2018-scn-v24.pdf

Grimm E (2011) TILIA software version 1.7.16. Illinois State Museum, Research and Collection Center, Springfield. http://intra.museum.state.il.us/pub/grimm/tilia/

Haber A (1999) Uywaña, the house and its indoor landscape: oblique approaches to, and beyond, domestication. In: Gosden C, Hather J (eds) The prehistory of food. Appetites for change. Routledge, London, pp 57–82. https://doi.org/10.4324/9780203203385

Haber A (2007) Arqueología de uywaña: un ensayo rizomático. In: Nielsen AE (ed) Producción y circulación prehispánicas de bienes en el sur andino. Brujas, Córdoba, pp 13–34

Harris OJ (2018) More than representation: multi-scalar assemblages and the Deleuzian challenge to archaeology. Hist Hum Sci 31(3):83–104. https://doi.org/10.1177/0952695117752016

Heusser C (1971) Pollen and spores of Chile. The University of Arizona Press, Tucson

Hogg A, Heaton TJ, Hua Q, Palmer JG, Turney C, Southon J, Bayliss A, Blackwell PG, Boswijk G, Bronk Ramsey C, Pearson C, Petchey F, Reimer P, Reimer R, Wacker L (2020) SHCal20 southern hemisphere calibration, 0–55,000 years cal BP. Radiocarbon 62(4):759–778. https://doi.org/10.1017/RDC.2020.59

Ingold T (2007) Material against materiality. Archaeol Dialog 14(1):1–16. https://doi.org/10.1017/S1380203807002127

Juárez D, Estrada C, Bustamante M, Quintana Y, Moreira J, López J (2010) Guía ilustrada de pelos para la identificación de mamíferos mayores y medianos de Guatemala, 2nd edn. Universidad de San Carlos de Guatemala, Guatemala

Koford CB (1957) The vicuña and the Puna. Ecol Monogr 27:153–219. https://doi.org/10.2307/1948574

Latour B (2005) Reassembling the social: an introduction to actor-network theory. Oxford University Press, Oxford

López Campeny S (2000) Tecnología, iconografía y ritual funerario. Tres dimensiones de análisis de los textiles formativos del Sitio Punta de la Peña 9 (Antofagasta de la Sierra, Argentina). Estud Atacameños 20:29–65. https://doi.org/10.22199/S07181043.2000.0020.00003

López Campeny S (2009) *Asentamiento, Redes Sociales, Memoria e Identidad. Primer milenio de la era. Antofagasta de la Sierra, Catamarca*. PhD. Dissertation. Facultad de Ciencias Naturales y Museo, Universidad Nacional de la Plata. https://doi.org/10.35537/10915/4341

López Campeny S (2010) De un hogar en la Puna… Relatos de idas y vueltas. In: Albeck ME, Scattolín C, Korstanje MA (eds) El hábitat prehispánico. Arqueología de la arquitectura y de la construcción del espacio organizado. Universidad Nacional de Jujuy, Jujuy, pp 215–242

López Campeny S (2014) The agency of textile technology in some archaeological ritual contexts of Northwest Argentina. J Anthropol Archaeol 2(2):39–75. https://doi.org/10.15640/jaa.v2n2a3

López Campeny S, Romano A (2021) Siguiendo el hilo: ensamble camélidos-humanos-textiles. Una mirada multiescalar y multitemporal. Relac Soc Argentina Antropol 46(1):77–111. https://doi.org/10.24215/18521479e003

López Campeny S, Olivera D, Fernández Varela V, Peña J (2005) Procesos tafonómicos, subsistencia y uso del espacio. Análisis de la arqueofauna de un sitio agropastoril de la Puna Meridional Argentina (Punta de La Peña 9, Antofagasta de la Sierra, Catamarca). Intersec Antropol 6:11–28. http://www.scielo.org.ar/scielo.php?script=sci_arttext&pid=S1850-373X2005000100003&lng=es&tlng=en

López Campeny S, Martínez MS, Romano AS (2023) Persistencias y transformaciones en la producción de hilados. Puna de Antofagasta de la Sierra (siglos VI a fines del XIX). Relac Soc Argentina Antropol 48(2):e079. https://doi.org/10.24215/18521479e079

Markgraf V, D'Antoni H (1978) Pollen flora of Argentina: modern spore and pollen types of Pteridophyta, Gymnospermae, and Angiospermae. The University of Arizona Press, Tucson

Morello J, Matteucci SD, Rodriguez AF, Silva M (2012) Ecorregiones y complejos ecosistémicos argentinos. Editorial Orientación Gráfica Argentina, Buenos Aires

Mosca Torres ME (2010) Selección de alimento y hábitat por vicuñas silvestres de la reserva provincial Los Andes (Salta, Argentina). Ph.D. Dissertation, Universidad de Buenos Aires, Facultad de Ciencias Exactas y Naturales. https://hdl.handle.net/20.500.12110/tesis_n4902_MoscaTorres

Olivera D (1997) La importancia del recurso Camelidae en la Puna de Atacama entre los 10.000 y 500 años AP. Estud Atacameños 14:29–41. http://www.jstor.org/stable/25674670

Olsen B, Shanks M, Webmoor T, Witmore C (2012) Archaeology: the discipline of things. University of California Press, Berkeley

Overton NJ, Hamilakis Y (2013) A manifesto for a social zooarchaeology. Swans and other beings in the Mesolithic. Archaeol Dialog 20(2):111–136. https://doi.org/10.1017/S1380203813000159

Reigadas MC (1992) La punta del ovillo: Determinación de domesticación y pastoreo a partir del análisis microscópico de fibras y folículos pilosos de camélidos. Arqueología 2:9–52. http://repositorio.filo.uba.ar/handle/filodigital/6941

Reigadas MC (1995) Criterios metodológicos para la detección del proceso de domesticación de camélidos en el NOA a partir del análisis microscópico de fibras y cueros arqueológicos. Cuad Inst Nac Antropol Pensamiento Latinoamericano 5:151–168

Reigadas MC (2014) Explotación y aprovechamiento de la fauna en el noroeste argentino: ¿qué informan las fibras y pieles arqueológicas? Etnobiología 12(2):64–80. https://revistaetnobiologia.mx/index.php/etno/article/view/267/268

Romano A (2019) Vamos al grano… Análisis palinológico de restos camélidos y sus implicancias para la arqueología de pastores puneños de Antofagasta de la Sierra (Catamarca, Argentina). Undergraduate thesis. Facultad de Ciencias Naturales e IML, Universidad Nacional de Tucumán. Unpublished

Romano A, Baied C (2012) Análisis palinológico de restos camélidos y sus implicancias para el estudio de la movilidad de pastores puneños. In Pensando la multiplicidad y la unidad en los Andes. Resúmenes Expandidos de las Jornadas de Estudios Andinos, Tilcara, Jujuy. Editorial de la Facultad de Filosofía y Letras, Universidad de Buenos Aires, pp 40–42

Somonte C, Cohen ML (2006) Caracterización de los conjuntos líticos del sitio agropastoril Punta de La Peña 9-III: Un aporte a la dinámica ocupacional puneña (Antofagasta de la Sierra, Catamarca, Argentina). Werkén 9:135–158. http://hdl.handle.net/11336/148060

Urquiza S, López Campeny S, Romano A (2013) Historia ocupacional y prácticas sociales: Un análisis arqueofaunístico contextual. Sitio Piedra Horadada 2, Antofagasta de la Sierra, Catamarca, Argentina. In: Izeta A, Mengoni Goñalons G (eds) De la Puna a las Sierras: Avances y Perspectivas en Zooarqueología Andina. Archaeopress, Oxford, pp 121–144. https://ri.conicet.gov.ar/handle/11336/233074

Witmore C (2007) Arqueol Simétrica 18:305–313

Chapter 5
Living in Quebrada de Los Corrales: Camelids, Humans, and Landscape within the First Village Settlements in El Infiernillo (Tucumán, Argentina)

María Eugenia Naharro, Jorge Gabriel Martínez, Andrés Darío Izeta, and Nurit Oliszewski

5.1 Introduction

The archaeology developed in recent decades in the Argentine Northwest and the South-Central Andes has enabled the interpretation of the social and economic transformations experienced by societies over the past 7000 years (e.g., Núñez et al. 2006; Hocsman 2010; Marsh 2015; Correa et al. 2018; Núñez and Perlès 2018; Oliszewski et al. 2018). As a result, the processes that led these communities to gradually replace, though not entirely, predatory relationships between hunters and wild prey (vicuña and guanaco) with relationships involving the management of domesticated species (llama and alpaca) began to be understood. This ultimately would have implied a transition from a hunter-gatherer economy to a productive way of life, whether as communities with mixed economies, agro-pastoral, or pastoral economies (e.g., Olivera 1997; Dransart 1999; Mengoni Goñalons and Yacobaccio 2006; Cartajena et al. 2007; Izeta 2004, 2007; Grant Lett-Brown 2008, Martínez et al. 2013).

In Northern Chile, the transitional process was evidenced in the Tulán and Puripica ravines (Fig. 5.1), around 5000–3800 years BP, within a framework of

M. E. Naharro (✉) · J. G. Martínez · N. Oliszewski
Consejo Nacional de Investigaciones Científicas y Técnicas (CONICET) - Grupo de Investigación en Arqueología Andina (ARQAND), Facultad de Ciencias Naturales e Instituto Miguel Lillo, Universidad Nacional de Tucumán, Tucumán, Argentina
e-mail: mariaeugenianaharro@csnat.unt.edu.ar; jormartin@csnat.unt.edu.ar; nuritoli@csnat.unt.edu.ar

A. D. Izeta
IDACOR-CONICET, Instituto de Antropología de Córdoba, Facultad de Filosofía y Humanidades, Universidad Nacional de Córdoba, Córdoba, Argentina
e-mail: aizeta@ffyh.unc.edu.ar

R. Frontini et al. (eds.), *Non-human Animals and People: Interactions, Agents and Materialities*, The Latin American Studies Book Series,
https://doi.org/10.1007/978-3-031-97495-3_5

Fig. 5.1 Location of the Quebrada de Los Corrales archaeological site (marked in red) along with other sites and localities mentioned in this study (marked in blue)

paleoclimatic changes. This process is associated with dynamic sedentary mobility, characterized by the formation of villages, the adoption of pottery, the domestication of camelids, and the emergence of pastoral practices (ca. 3100–2400 years BP), transhumant practices, and the intensification of resource exploitation (Cartajena et al. 2007). These developments created the conditions for the transition within the same territories toward the emergence of social complexity (Núñez et al. 2006). Additionally, the presence of a complex structure, Tu-54, characterized by its specific architecture, scale, and interior organization required considerable energy expenditure. Among the recorded activities were the preparation and consumption of food and beverages *in situ*, the burial of 28 infants accompanied by numerous offerings inside, combustion activities with remains of meals and other waste, along with secondary redeposited discards, and adult burials outside the structure. This suggests that the structure likely functioned as a ceremonial center (temple), enabling gatherings in this sacred space (Núñez and Perlès 2018).

For the Argentine Northwest, and particularly for the Puna of the provinces of Jujuy and Catamarca, this transition process has been proposed to have begun sometime between 6000 and 3000 years before the present (e.g., Hocsman and Babot 2018; Yacobaccio 2001; Aschero et al. 2012). For the mid-altitude valleys, however, this situation is different, as the identification of sites assignable to hunter-gatherer communities is still very scarce (e.g., Martínez et al. 2013; Somonte and Baied

2013; Carbonelli et al. 2021). Nevertheless, for both environmentally diverse regions, it is interpreted that in the last 3000 years, the economy shifted from an exclusively hunter-gatherer mode to a range of new forms. This includes a mixed mode (extractive+productive economy), an agro-pastoral economy, and an exclusively pastoral economy. All of this was complemented by an extensive network for the circulation of people and goods, allowing the acquisition of resources from various areas (e.g., Scattolin and Lazzari 1997; Yacobaccio et al. 2004; Aschero and Hocsman 2011; Martínez and Funes Coronel 2020).

Based on this, we can agree that the last 3000 years of history for both regions include the adoption of a different way of using space and natural resources. This was, because production and habitation areas became spaces with deliberately constructed infrastructure, moving away from the use of natural spaces such as caves and rock shelters. In fact, the spatial analysis of the houses suggests that these (now private) spaces not only functioned as areas for domestic use but also as places where social, economic, and religious structures were reproduced and reinforced (e.g., Scattolin et al. 2009; Scattolin 2010). This pattern is comparable to what has been observed in other regions of the Argentine Northwest and other areas of the South-Central and Southern Andes, as previous studies have proposed (e.g., Tarragó 1999; Oliszewski 2017; Oliszewski and Di Lullo 2020). Additionally, the layout of the houses and the distances between them indicate a clear delineation of private and public spaces, with possible areas of exchange and negotiation identified in circulation spaces and productive lands (Oliszewski and Di Lullo 2020).

This new vision and structuring of space and resources undoubtedly had an impact on the possible and chosen subsistence strategies of these human groups. Therefore, the subsistence models of pastoral and/or agro-pastoral societies in Puna environments, dated between 4000 and 3000 years BP, were characterized by implementing strategies linked to factors such as changes in environmental conditions—increased aridity and low resource availability—and social factors—population growth and greater use of various spaces (e.g., Madero 1992; Elkin 1996; Yacobaccio et al. 1997; Olivera 1998; Haber 1999; López Campeny 1999). Olivera (1997, 1998) proposed a model of dynamic sedentarism associated with pastoralism for the southern Argentine Puna, characterized by the presence of residential dwellings and corrals where herders and their families lived for extended periods (permanent or semi-permanent) in ravines or basin bottoms. They had access to pastures and water resources. There are also temporary settlements (seasonal stays) where part of the family group moved, known as high-altitude hunting and/or herding posts, to access temporary pastures near tolares and grasses, and to exchange resources with other areas (Raffino 1977; Olivera 1997, 1998; Yacobaccio et al. 1997).

However, the quantity and intensity of research focused on early village societies with mixed agro-pastoral economic systems in the valleys of Northwest Argentina is relatively lower compared to the Puna environments (Izeta 2008). Among the pioneering studies in this environmental region are the investigations conducted in the Lerma Valley (Quebrada del Toro, Raffino et al. 1977) and in the northern Calchaquí Valley (Tarragó 1994). Later, studies were carried out in the southern Calchaquí Valleys by Belotti López de Medina (2011), Izeta (2004, 2007), Scattolin et al. (2001), and Scattolin (2019), among others. Through zooarchaeological analyses, Izeta (2004) established that during the first millennium AD in the southern Calchaquí Valleys, there was a tendency to prioritize the preservation of domestic livestock (camelids) until adulthood, possibly for wool production. This would imply a relatively higher presence of other species (non-camelids) in archaeological contexts. However, López and de Medina (2011) identified a different pattern of camelid uses at the Soria 2 site, located at the bottom of the Santa María Valley (Fig. 5.1). There, the age pattern of camelids shows a higher frequency of subadult individuals, resembling more closely the pattern identified at many Puna sites. This adds a greater, expected diversity to the ways camelids were utilized by mixed-economy societies.

In this context, this work presents a new case that aims to expand our knowledge of camelid resource use in the mid-altitude valleys of Northwest Argentina. It focuses particularly on the assemblages of faunal bone remains recovered from domestic contexts of the Puesto Viejo village. This site is part of the Quebrada de Los Corrales (QDLC), an archaeological locality that has been investigated for over 20 years. As a result, evidence from various periods, ranging from the early Middle Holocene to the Late Holocene, has been identified. Many of them leading to the interpretation of archaeological contexts that reflect a way of life transitioning from hunting and gathering to sedentary societies with an agro-pastoralist mode of production in a lapse spanning through ca. 7820–630 years BP (Martínez et al. 2013, 2020; Oliszewski 2017; Oliszewski et al. 2018).

With this, we are seeking to contribute further data on the possible ways this resource was utilized in the context of the expansion of an innovative economic form for the area and period. Therefore, the aim of this work is to define and understand the strategies employed by the village societies of QDLC in developing pastoral practices, through the zooarchaeological analysis of materials from a residential unit in Puesto Viejo (PV) village. This analysis will also take into account isotopic evidence and the productive structures identified in the study area.

5.2 Quebrada de Los Corrales

5.2.1 *Location and Environmental Characteristics*

Quebrada de Los Corrales is located in the area known as Abra del Infiernillo, at an altitude of approximately 3100 m.a.s.l., in the western of Tucumán province, within the northern sector of the Aconquija mountain range. It belongs to the Yungas ecoregion, specifically within the montane forests and grasslands complex (Morello et al. 2012). The area is bordered by a humid valley to the east (Tafí del Valle, Tucumán) and a semi-arid valley to the west (Valle de Santa María, Catamarca) (Fig. 5.1).

Currently, the study area is characterized by a dry and warm climate, with exclusively summer rainfall ranging between 150 and 200 mm annually. Cabrera (1976)

Table 5.1 Flora of Abra del Infiernillo and its photosynthetic pathway

Taxon	C_3	C_4	CAM
Cereus sp.			X
Echinocactus sp.			X
Digitaria sp.	X		
Stipa sp.	X		
Agrostis sp.	X		
Abromeitiella brevifolia	X		
Dickia velascana	X		
Dickia chaguar	X		
Deuterocohnia longifolia	X		
Deuterocohnia haumani	X		
Hypseocharis pimpinellifolia	X		

described the flora as xerophytic (Table 5.1), with a predominance of plants employing C_3 and CAM photosynthetic pathways. A more recent study by Martín et al. (2002) recorded the plant diversity along a transect from Tafí del Valle to Amaicha del Valle. They identified some species present in El Infiernillo, such as *Festuca hieronymi* Hack. and *Stipa ichu* (Ruiz & Pav.) Kunth, which make up the high-altitude grasslands.

5.2.2 Archaeological Background of the Study Area

As previously mentioned, various archaeological evidence corresponding to different periods within the timeframe of approximately 7820–630 years BP has been identified in QDLC. This body of evidence corresponds to multiple interconnected spatial areas along the entire Los Corrales river (Fig. 5.2). In the lower basin, there is a site under shelter, Cueva de Los Corrales 1 (CC1), which features sporadic and short-term occupations dated between 3040 ± 30 ^{14}C year BP (1163–1631 d.C. -cal a.C./d.C. 2 σ) y 630 ± 140 ^{14}C year BP (1387–1114 a.C.-cal a.C./d.C. 2 σ). This cave is characterized by containing a total of 14 rock-carved mortars within a context of multiple activities. Inside, activities related to the consumption/discard of faunal remains (Camelidae, Dasypodidae) and plant materials (*Geoffroea decorticans*, *Neltuma* sp., *Cucurbita* sp., *Phaseolus vulgaris*, *Zea mays* y *Chenopodium quinoa*) would have been particularly developed, as well as the maintenance of lithic and bone tools. The mortars were filled with materials similar to those found in the associated stratigraphic layers (Oliszewski et al. 2008, 2022).

In the middle/upper basin, the agro-pastoral productive areas are located, which are separated from the residential area. These areas cover approximately 500 hectares of slopes modified for agricultural use (terraces and platforms) within a dryland irrigation system. Microfossil analyses identified the presence of phytoliths from Poaceae with affinities to *Zea mays* (Gómez Augier et al. 2008; Caria et al. 2010; Di Lullo 2010; Oliszewski et al. 2015, 2018).

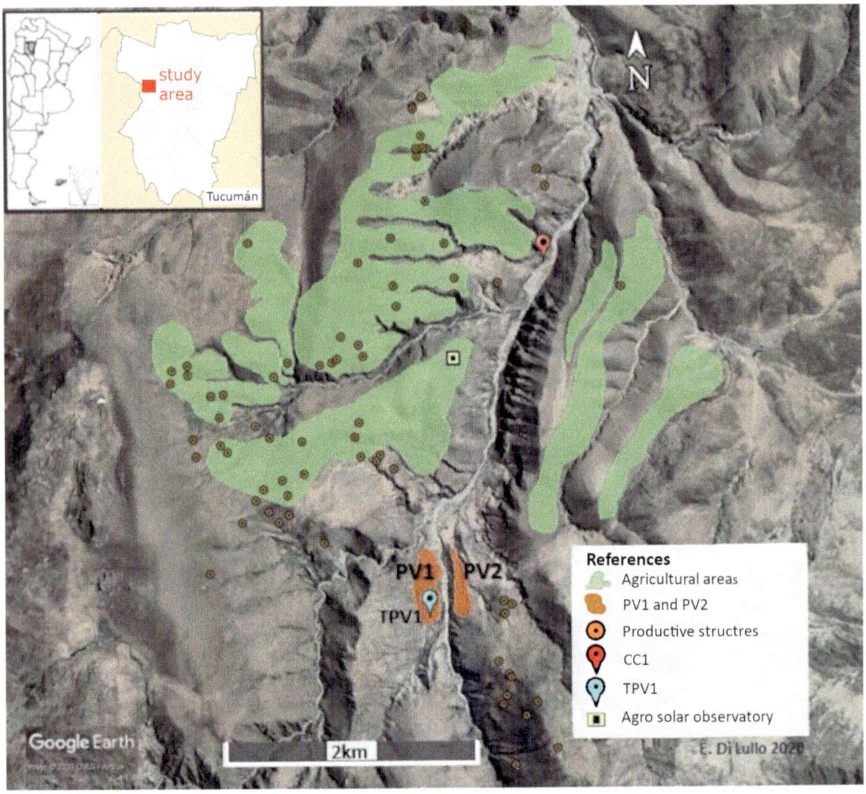

Fig. 5.2 Location of archaeological sites within the study area. Key sites include: Cueva de Los Corrales 1 (CC1), Puesto Viejo 1 (PV1), Puesto Viejo 2 (PV2), and Taller de Puesto Viejo 1 (TPV1) (adapted from Oliszewski and Di Lullo 2020, p. 327)

In the upper basin is the village complex known as Puesto Viejo (Oliszewski and Di Lullo 2020), which consists of nearly a hundred residential units assignable to the "Tafí pattern" (*sensu* Berberián and Nielsen 1988). Only 47 are in good condition and have been mapped on both sides of the Los Corrales river. This residential complex was divided as Puesto Viejo 1 (PV1, to the west) and Puesto Viejo 2 (PV2, to the east), and shows occupations dated between 1850 and 1560 years BP. The diameter of the central domestic structures ranges from 7 to 15 m, and the number of attached lateral structures can vary from 2 to 5 m. The different units are separated by distances ranging from 5 to 20 m (Di Lullo 2010). Excavations carried out in six residential units confirmed that these structures were domestic in nature, where various activities such as preparation, consumption, and discard of faunal (*Lama glama, Lama guanicoe* and *Hippocamelus antisensis*) and plant resources—*Neltuma* sp., *Geoffroea decorticans*, and *Zea mays*—took place. It also carried out tasks related to the cooking of ceramic artifacts, the maintenance of lithic artifacts and a simple burial of an adult male in a cist dated to 1560 years BP (Muntaner

Fig. 5.3 Bone artifact for textile use made from animal fiber recovered in TPV1 (from Martínez et al. 2020, p. 305)

2012; Martínez et al. 2013; Mercuri and Mauri 2013; Oliszewski et al. 2015; Mercuri 2017; Oliszewski and Di Lullo 2020).

In addition, to the south of PV1 is the site Taller de Puesto Viejo 1 (TPV1), where various human occupation evidence in stratigraphy was recovered, dated between 7800 and 1750 years BP. The transitional process from hunter-gatherer to agro-pastoral societies occurred between 4000 and 2000 years BP, based on the presence of lithic points associated with the bow and arrow system, mortars and manos (grinding stones), coarse pottery fragments and camelid remains (*Lama guanicoe*). Additionally, a cremation context dated between 3800 and 3500 years BP was identified, among which a bone artifact associated with textile activities was recovered (Oliszewski et al. 2018; Martínez et al. 2020) (Fig. 5.3). Furthermore, a simple secondary burial of an adult male individual, dated to 3330 years BP, was documented (Naharro 2019, 2020).

5.2.3 Other Lines of Evidence Related to Camelid Herding: Productive Structures and Isotopic Analyses

The survey through pedestrian prospecting combined with drone use allowed for the identification of structures assignable as corrals, located within terraced areas and also isolated to the south of PV2. Regarding architectural design, Oliszewski and Di Lullo (2020) state that these are circular enclosures that are attached in different ways (linearly or radially; Fig. 5.4). They are characterized by having a circular structure of small dimensions, from which larger circular structures are attached. It is suggested that they "show a high investment of labor in their construction: first, their dimensions undoubtedly involved a huge movement of rocks; second, all cases present their enclosures cleared of stones and some even internally leveled or terraced" (Oliszewski and Di Lullo 2020, p. 331). The authors propose that some of these may have had a mixed use related to the cultivation of plant species, according to soil analyses conducted (Di Lullo pers. comm., 2023), while others would have functioned purely as corrals.

Oliszewski et al. (2020) conducted stable isotope analyses on human and faunal bone remains, but here we will focus only on the latter. These authors found that wild species (*Lama guanicoe* and *Hippocamelus antisensis*) yielded δ^{13}Ccol‰

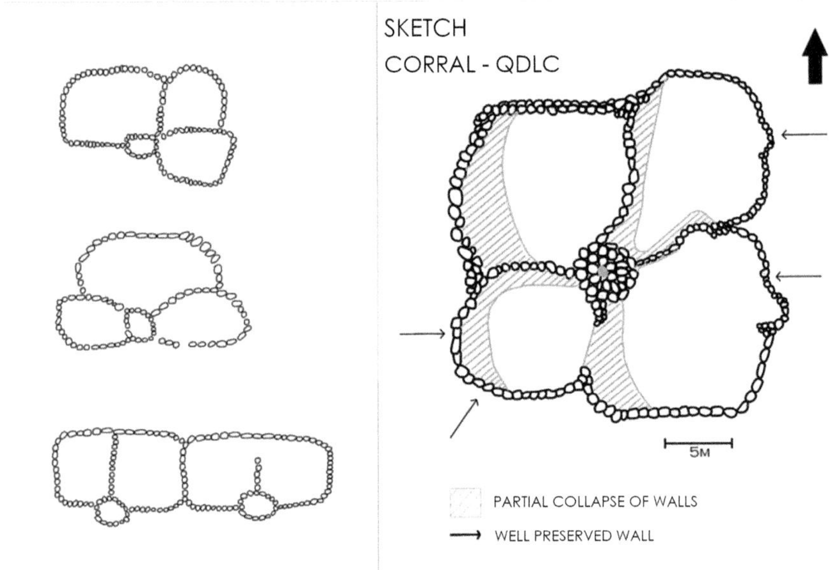

Fig. 5.4 Types of productive structures (left) and a sketch of one of the productive structures (right) (adapted from Oliszewski and Di Lullo 2020, p. 331)

values ranging from −18.9 to −19.2, while *Lama glama* values ranged from −19.1 to −19.5 (Oliszewski et al. 2020; Tabla 4), indicating a diet based on C_3 photosynthetic pathway plants. They establish that these values cluster with those from the Valle del Cajón; however, in the case of El Infiernillo, the variability in sample values is lower, likely because almost all of its vegetation cover consists of species with a C_3 photosynthetic pathway (Tabla 1).

5.2.4 Andean Pastoral Practices as Assemblages of Humans and Camelids

The concept of landscape has been defined from multiple perspectives, among which we agree with Tilley (1994). This author defines the landscape as a space to be considered as a "medium" in which various connections are generated: the physical/natural space, perceived and represented space, the interaction, and interrelation between agents and their surrounding environment. From a subjective dimension, landscapes should be understood as socially constructed spaces, penetrated with meanings, experiences, and actions. Therefore, as Ingold (1993) states, the landscape is the result of ways of dwelling, that is, as a lasting record of testimonies of past lives, which, by dwelling, left something of themselves.

The concept of assemblage, proposed by Deleuze and Guattari (1987), emerges as a theoretical alternative to the concept of organic totality for analyzing societies.

An assemblage is constituted by different components or entities (both material and immaterial) that interrelate with each other, possess certain autonomy, and, if placed in another assemblage, would generate different interactions. The relationships established create processes that allow for the stability of the identity of that assemblage, such as territorialization (DeLanda 2006). Territorialization processes are those that define or delineate the spatial boundaries of real territories (DeLanda 2006) and can materialize in affective/sensory, mnemonic/temporal, and political mechanisms. Affective/sensory mechanisms refer to productive force, which causes bodies to act and establish connections with material and immaterial elements. Mnemonic/temporal mechanisms establish the coexistence of all pasts condensed in the present. Political mechanisms involve the capacity and power of a social agent, which is subject to rules governing productive forces.

Pastoral practice was a way of life in Andean societies, both in the past and in some present ones, in which various entities such as herders, domestic animals, and a specific landscape interrelate, generating particular relationships. Therefore, we consider various lines of analysis (zooarchaeology, stable isotopes, productive structures, etc.) that allows a better understanding of the assemblage in QDLC between herders, camelids, and the highland *quebrada* landscape, which produced distinctive forms of inhabiting space within the agro-pastoral societies of northwest of Argentina.

5.3 Materials and Methods

The set of faunal remains analyzed consists of bone materials recovered through stratigraphic excavations in 2019 from the residential structure named as Unit 3 of PV2 (Fig. 5.5). Unit 3 (U3) is a typical dwelling of the previously mentioned Tafí pattern. The zooarchaeological materials analyzed come from the excavation of the central room (RC, western half), Room 1 (R1), and Room 2 (R2). All the analyzed bone assemblage is associated with a single occupation floor of U3, which was dated to approximately 1650 years BP based on a bone sample from the RC (Molar et al. 2024). Architectural and stratigraphic evidence suggests that this represents a synchronous context although R2 would correspond to a later, yet quasi-synchronous moment, as the general dates for PV2 fall within the range of 1767–1600 years BP (Oliszewski and Di Lullo 2020).

Taxonomic identification of each element in the zooarchaeological assemblage was carried out (Izeta 2007), representation of parts indices (MNE by diagnostic zones using GIS software, mxMNI, and MAU%), and economic indices (degree of fragmentation vs. complete elements, basic skeletal parts -PEB-, and processing marks) were calculated (Binford 1981; Mengoni Goñalons 1988, 1999; Yacobaccio et al. 1997; De Nigris 1999; Outram 2001; Izeta 2007). Additionally, bone tool production was analyzed (Izeta et al. 2013). A scale adapted to the fauna of northwestern Argentina for body size (BDSZ, Table 5.2) was used, based on the classification proposed by Brain (1981). The impact of taphonomic factors, such as weathering

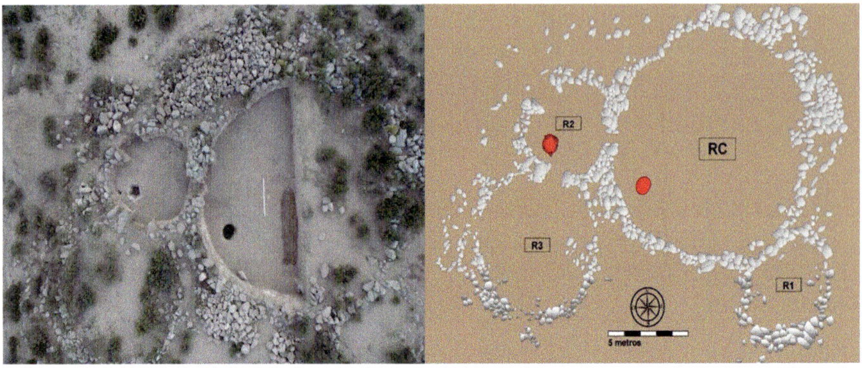

Fig. 5.5 Drone image (left) and plan of U3 of PV2 (prepared by Lic. Juan Montegú)

Table 5.2 Body sizes of the most common animals in the region under study

Body sizes	Taxon	Approximate weight (kg)
1	*Ctenomys*	0.50
1	Rodentia	>0.50
2	*Chaetophractus vellerosus*	0.900
3	*Lycalopex culpaeus*	11
4	*Lama guanicoe*	70
4	*Lama glama*	120
4	*Equus ferus caballus*	380
9	Non-identifiable	Unknown

(Todd 1987), the action of rodents and carnivores on the assemblage and anthropogenic modifications (Binford 1981; Mengoni Goñalons 1999), was also studied to understand the preservation state of the sample.

On Camelidae remains, osteometric analyses were conducted following the protocol commonly used in this type of study (see Menegaz et al. 1989; Cardich and Izeta 1999-2000; L'Heureux 2005; Mengoni Goñalons and Yacobaccio 2006; Cartajena 2009; Izeta et al. 2009, 2012; Aschero et al. 2012; Gasco et al. 2014; Le Neün et al. 2023; among others). Measurements were taken using a digital caliper (e.g., Kent 1982; Izeta 2004) and were compared with the following collections: (a) modern llama from Laguna Blanca (Catamarca, Argentina; L150 and L151) housed in the Museo de Antropología of Córdoba, and an adult female specimen from La Quebrada (Cajón Valley, Argentina; L148; Izeta 2004); (b) adult guanaco from the Sierra del Aconquija (Argentina; G149); (c) adult vicuña from Laguna Blanca (Catamarca, Argentina; VLB) from the Elkin's Collection (Elkin 1996); and (d) adult alpaca from the Museo de Historia Natural (Mendoza, Argentina; code VP-HMN). The results were grouped into tables and analyzed statistically using (a) UPGMA (Unweighted Pair Group using Arithmetic Averages) cluster analysis and (b) Principal Component Analysis.

In this study, we used some osteometric variables systematized by Izeta et al. (2012). Although this guide develops multiple variables for analysis, we selected those variables that are compatible with existing databases. Thus, the measurements used in this work are for the first phalanx (FP1V1, FP1V2, FP1V3, FP1V4, FP1V5; Kent 1982), second phalanx (P2V6, P2V7, P2V8, P2V9, P2V10; Kent 1982); third phalanx (P3V11, P3V12, P3V13, P3V14; Kent 1982); radial carpal (P, A, L; Izeta 2004); third carpal (P, A, L; Izeta 2004); fourth carpal (P, A, L; Izeta 2004); pelvis (HAm, HII, HIS, BA, BFvc, SDA, LA; Kaufmann and L'Heureux 2009); tibia (87, 88, 91, 92, 93, 94, 95, 96,114; Kent 1982); astragalus (LM, LI, Ad, Ap, Pm, PI; Izeta 2004); calcanei (L, Ap, La, Pd, Ad, Pp, Lt; Izeta 2004).

Additionally, mortality profiles were constructed considering epiphyseal fusion: early (young), ranging from neonate to 12 months; intermediate (juvenile), from 12 to 36 months; and late (adult), older than 36 months. Dental eruption and wear patterns were also considered (Kent 1982; Wheeler 1982; Puig 1988; Kaufmann 2004).

5.4 Results

The zooarchaeological analyses established an NSP = 1163 (Table 5.3), of which 54.51% (N = 634/1163) of the specimens were identified, allowing for species-level identification in some cases. Concerning the preservation state of the sample, it can be established that 94.58% (N = 1100/1163) is at a weathering grade 2 according to Todd (1987), indicating relatively low weathering, which has similarly affected all body sizes, suggesting that the remains were deposited contemporaneously. Concerning modification by natural agents, 98.97% (N = 1151/1163) shows no marks from rodents or carnivores.

5.4.1 Interspecific Identification of the Camelidae Sample

From the camelid sample, 11 postcranial elements were selected that met the necessary conditions for analysis (Table 5.4). The results allowed for progress in the interspecific identification within the Camelidae remains, contributing to a better understanding of the sample's composition and characteristics. These data were complemented with statistical analyses to ensure the accuracy of the interpretations. These results allowed to identify two specimens as *Lama guanicoe*, two to *Lama glama* and one to possible *Vicugna pacos* (Figs. 5.6 and 5.7).

Based on the morphological analysis and the degree of dental wear of two mandible fragments, we interpret that one of the specimens could correspond to a juvenile *Lama guanicoe* (24–30 months), while the other could belong to an adult *Lama glama* (4–5 years). These assignments are based on the observed characteristics although further studies would be necessary to confirm these interpretations and rule out other possibilities.

Table 5.3 NISP (number of identified specimens), indet, and body size of the *taxa* recorded in each room of Unit 3 (U3) at PV2

Taxon	Body size	NISP			Total
		RC	R1	R2	
Amphibia	1	2	0	0	2
Aves	2	0	1	0	1
Rodentia	1	28	14	11	53
Ctenomyidae	1	18	3	4	25
Mammalia	2	4	2	0	6
Dasypodidae	2	3	0	0	3
Chaetophractus vellerosus	2	3	0	8	11
Dicotyles tajacu	2	0	0	3	3
Lycalopex sp.	2	1	0	0	1
Mammalia	3	0	0	1	1
Mazama gouazoubira	3	1	0	0	1
Mammal	4	1	0	1	2
Ungulate	4	9	0	1	10
Lama sp.	4	182	227	89	498
Lama glama	4	6	3	2	11
Lama guanicoe	4	8	0	5	13
Vicugna pacos	4	3	0	1	4
Hippocamelus antisensis	4	0	0	5	5
Subtotal identified		258	249	127	634
Indet	3	5	0	0	5
Indet	4	137	61	174	372
Indet	9	41	38	73	152
Total		441	348	374	1163

Table 5.4 Measurements (in mm) of U3 Archaeological Camelids. Osteometric variables are named according to Izeta et al. (2012)

Samples	Element	Taxon	Measures (in mm)
PV2–15	First phalanx (fore limb)	*Lama guanicoe*	FPV1V1 = 69.23; FPV1V2 = 20.77 FPV1V3 = 17.52; FPV1V4 = 17.15 FPV1V5 = 16.09
PV2–17	First phalanx (fore limb)	*Lama guanicoe*	FPV1V1 = 69.64; FPV1V2 = 21.06 FPV1V3 = 17.58; FPV1V4 = 17.06 FPV1V5 = 16.11
PV2–20	First phalanx (rear limb)	*Vicugna pacos*	FPV1V1 = 58.43; FPV1V2 = 18.21 FPV1V3 = 16.09; FPV1V4 = 16.42 FPV1V5 = 13.78
PV2–21	First phalanx (rear limb)	*Vicugna pacos*	FPV1V1 = 59.34; FPV1V2 = 18.32 FPV1V3 = 16.63; FPV1V4 = 16.50 FPV1V5 = 13.65
PV2–22	First phalanx (fore limb)	*Lama guanicoe*	FPV1V1 = 69.69; FPV1V2 = 19.57 FPV1V3 = 17.62; FPV1V4 = 17.17 FPV1V5 = 15.80

(continued)

Table 5.4 (continued)

Samples	Element	Taxon	Measures (in mm)
PV2–24	First phalanx (fore limb)	*Lama guanicoe*	FPV1V1 = 69.45; FPV1V2 = 21.59 FPV1V3 = 18.12; FPV1V4 = 17.50 FPV1V5 = 15.97
PV2–285	First phalanx (fore limb)	*Lama glama*	FPV1V1 = 69,61; FPV1V2 = 20,92 FPV1V3 = 18,42; FPV1V5 = 13,75
PV2–18	Second phalanx	*Lama guanicoe*	P2V6 = 32.40; P2V7 = 17.72 P2V8 = 12.71; P2V9 = 15.89 P2V10 = 13.91
PV2–23	Second phalanx	*Lama guanicoe*	P2V6 = 33.24; P2V7 = 16.73 P2V8 = 12.93; P2V9 = 15.67 P2V10 = 13.21
PV2–83	Second phalanx	*Lama guanicoe*	P2V6 = 29.83; P2V7 = 17.32 P2V8 = 12.35; P2V9 = 15.29 P2V10 = 13.78
PV2–84	Second phalanx	*Lama guanicoe*	P2V6 = 30.45; P2V7 = 16.96 P2V8 = 12.75; P2V9 = 15.18 P2V10 = 13.52
PV2–34	Third phalanx	*Vicugna pacos*	P3V11 = 10.78; P3V12 = 12.74 P3V13 = 11.74; P3V14 = 20.66
PV2–202	Third phalanx	*Lama glama*	P3V11 = 12.63; P3V12 = 14.46 P3V13 = 14.55; P3V14 = 22.39
PV2–121	Radial carpal	*Lama glama*	P = 25.05; A = 15.95 L = 17.52
PV2–109	Third carpal	*Lama guanicoe*	P = 20.47; A = 18.55; L = 12.41
PV2–406	Third carpal	*Lama* sp.	A = 15.66; L = 9.81
PV2–261	Fourth carpal	*Lama guanicoe*	P = 25.83; A = 19.10; L = 12.01
Pv2–52	Pelvis	*Vicugna pacos*	HAm = 18.00; HII = 27.10; HIS = 31.10 BA = 30.83; BFvc = 23.00; SDA = 9.6 LA = 35.00
PV2–223	Tibia	*Lama glama*	87 – 03.1; 88 = 08.29; 91 = 24.00 92 = 26.8; 93 = 35.25; 94 = 27.25 95 = 14.40; 96 = 10.00; 114 = 8.00
PV2–490	Astragalus	*Lama glama*	LM = 39.5; LI = 44.5; Ad = 27.4 Ap = 26.6; Pm = 23.8; PI = 23.00
PV2–287	Calcaneus	*Lama glama*	L = 98.5; Ap = 21.58; La = 29.00 Pd = 38.5; Ad = 31,21; Lt = 24
PV2–289	Calcaneus	*Lama glama*	L = 93.00; La = 23.00; Lt = 27.10

5.4.2 *Camelid mxMNI, MNE, MAU %*

Based on the epiphyseal fusion, dental eruption and wear, and the MNE, an mxMNI of 13 was estimated: two young of *Lama* sp. (11 months of gestation to 12 months; Table 5.5); four juveniles, three of *Lama* sp., and one of *Lama glama* (12–30 months; Table 5.6); and seven adults, two of *Lama* sp., two of *Lama guanicoe*, two of *Lama glama*, and one of possible *Vicugna pacos* (36–72 months; Table 5.7).

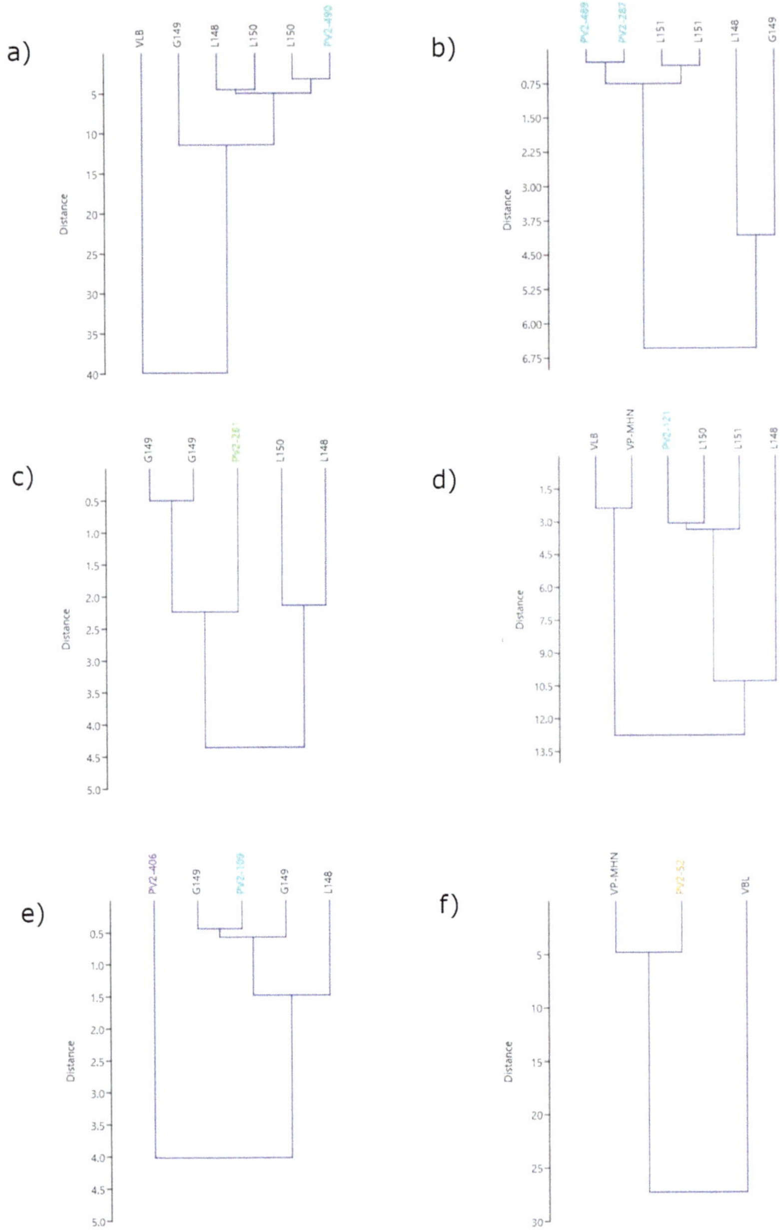

Fig. 5.6 Clustering of postcranial elements from the Camelidae in U3. The X-axis shows how the elements are grouped relative to the comparative samples of the Camelidae, while the Y-axis displays the distance of dissimilarity measures among all analyzed samples. Identified elements include: (**a**) astragalus of *Lama glama* (PV2–490); (**b**) calcaneus of *Lama glama* (PV2–287 and PV2–489); (**c**) fourth carpal of *Lama guanicoe* (PV2–261); (**d**) radial carpal of *Lama glama* (PV2–121); (**e**) third carpal of *Lama glama* (PV2–109); and (**f**) pelvis of possible *Vicugna pacos* (PV2–52). References: G149 (guanaco); L148/L150/L151 (llama); VLB (vicuña); VP-MHN (alpaca)

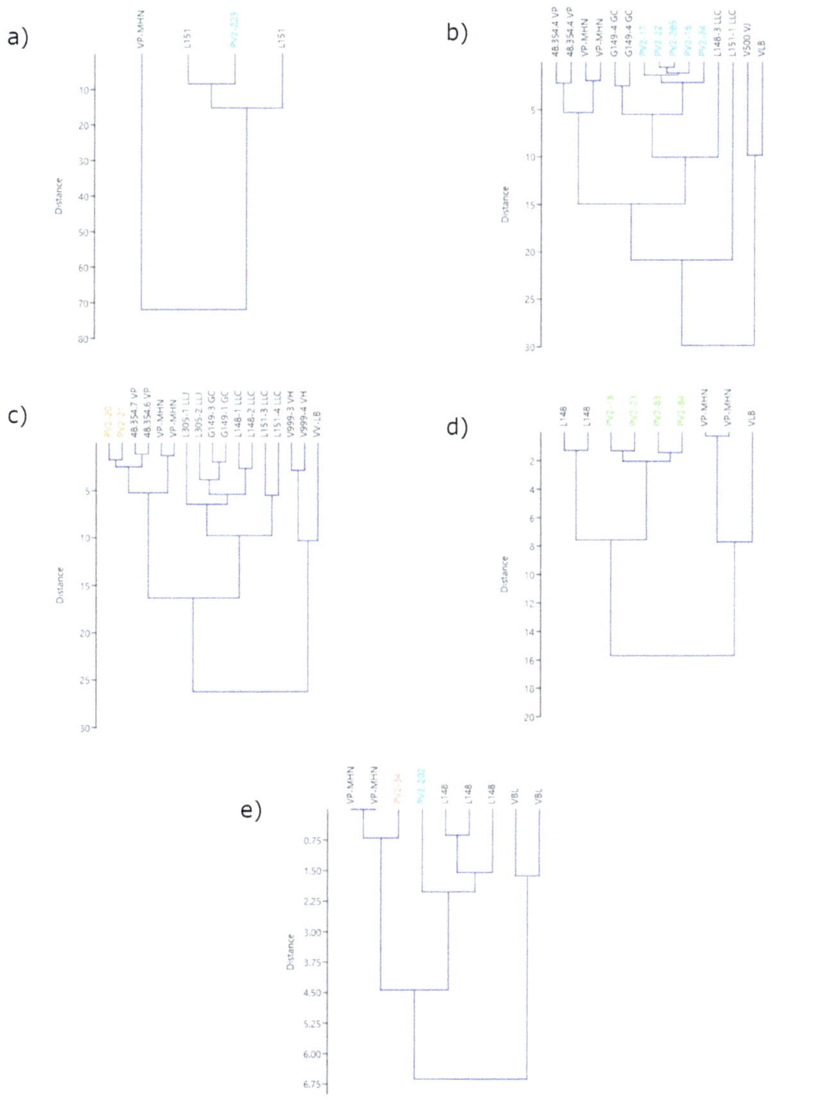

Fig. 5.7 Clustering of postcranial elements from the Camelidae in U3. The X-axis shows how the elements are grouped relative to the comparative samples of the Camelidae, while the Y-axis displays the distance of dissimilarity measures among all analyzed samples. Identified elements include: (**a**) tibia of *Lama glama* (PV2–223); (**b**) first front phalanges of *Lama guanicoe* (PV2–15, 17, 22, 24, and 285); (**c**) first rear phalanges of possible *Vicugna pacos* (PV2–20 and 21); (**d**) second phalanx of *Lama guanicoe* (PV2–18, 23, 83, and 84); and (**e**) third phalanx of possible *Vicugna pacos* (PV2–34) and *Lama glama* (PV2–202). References: G149 (guanaco); L148/L150/L151 (llama); VLB (vicuña); VP-MHN (alpaca)

Table 5.5 NISP, MNE, MNI, MAU, and MAU% of young individuals. References: *R* (right), *L* (left); *S/D* (laterality not determined)

Element	NISP	*R*	*L*	*S/D*	MNE	MNI	MAU	MAU%
Cranium	66	–	–	1	1	1	1.00	60.24
Mandible	3	–	–	1	1	1	1.00	60.24
Incisor	6	–	–	6	6	1	0.75	45.18
Third premolar	2	1	1	–	2	1	0.50	30.18
Fourth premolar	1	–	–	1	1	1	0.25	15.06
Ribs	168	11	10	7	28	2	1.66	100
Scapula	5	1	2	–	3	2	1.50	90.36
Humerus	4	1	1	–	2	1	1.00	60.24
Ulna	1	–	1	–	1	1	0.5	30.12
Radioulna	1	1	–	–	1	1	0.5	30.12
Pelvis	3	1	1	–	2	1	1.00	60.24
Metapodial	8			2	2	1	0.25	15.06
TOTAL	268					2		

Table 5.6 NISP, MNE, MNI, MAU, and MAU% of juvenile individuals. References: *R* (right), *L* (left); *S/D* (laterality not determined)

Element	NISP	*R*	*L*	S/D	MNE	MNI	MAU	MAU%
Cranium	4	–	–	1	1	1	1.00	40
Mandible	1	–	–	1	1	1	0.50	20
Maxilla	2	–	–	1	1	1	0.50	20
Canine tooth	1	–	–	1	1	1	0.25	10
First molar	1	–	–	1	1	1	0.25	10
Second molar	1	–	–	1	1	1	0.25	10
Cervical vertebrae (3–7)	1	–	–	1	1	1	0.20	8
Lumbar vertebrae	11	–	–	5	5	1	0.71	28.4
Ribs	28	–	–	20	20	1	0.83	33.2
Sternum	1	–	–	1	1	1	0.17	6.8
Scapula	1	1	–	–	1	1	0.50	20
Radioulna	6	1	2	–	3	2	1.50	60
Ulnar carpal	2	1	1	–	2	1	1.00	40
Fourth carpal	1	–	1	–	1	1	0.50	20
Metacarpal	2	–	1	–	1	1	0.50	20
First front phalanx	5	–	–	4	4	1	1.00	40
Femur	2	2	–	–	2	1	1.00	40
Tibia	3	1	2	–	3	2	1.50	60
Pelvis	4	1	1	–	2	1	1.00	40
Astragalus	4	1	3	–	3	3	1.00	40
Metatarsal	4	2	2	–	4	2	2.00	100
First rear phalanx	3	–	–	2	2	1	0.50	20
First phalanx	8	–	–	7	7	1	0.25	10
Second phalanx	11	–	–	5	5	1	0.12	4.8
Third phalanx	4	–	–	4	4	1	0.5	20
Sesamoid	3	–	–	3	3	1	0.19	7.6
Metapodial	11	–	–	7	7	2	1.75	87.5
Vertebrae	2	–	–	1	1	1	0.3	1.2
Total	127					4		

Table 5.7 NISP, MNE, MNI, MAU, and MAU% of adult individuals. References: *R* (right), *L* (left); *S*/*D* (laterality not determined)

Element	NISP	*R*	*L*	S/D	MNE	MNI	MAU	MAU%
Cranium	1	–	–	1	1	1	0.50	25
Mandible	3	–	–	2	2	2	2.00	100
Maxilla	1	–	–	1	1	1	1.00	50
Maxilla molar	1	–	–	1	1	1	0.16	8
Incisor	6	–	–	6	6	1	0.75	37.5
First molar	1	–	–	1	1	1	0.25	12.5
Second molar	1	–	–	1	1	1	0.25	12.5
Third molar	1	–	–	1	1	1	0.25	12.5
Axis	1	–	–	1	1	1	1.00	50
Cervical vertebrae (3–7)	6	–	–	6	6	1	1.20	60
Thoracic vertebrae	3	–	–	2	2	1	0.16	8
Lumbar vertebrae	1	–	–	1	1	1	0.14	7
Caudal vertebrae	1	–	–	1	1	1	0.25	12.5
Ribs	10	–	–	10	10	1	0.42	21
Sternum	3	–	–	3	3	1	0.50	25
Scapula	1	1	–	–	1	1	0.5	25
Humerus	1	–	1	–	1	1	0.50	25
Radioulna	1	1	–	–	1	1	0.5	25
Intermediate carpal	2	–	2	–	2	2	1.00	50
Third carpal	2	–	2	–	2	2	1.00	50
Fourth carpal	1	1	–	–	1	1	0.50	25
Radial carpal	1	1	–	–	1	1	0.50	25
Ulnar carpal	1	–	1	–	1	1	0.5	25
First front phalanx	5	–	–	5	5	2	1.25	62.5
Astragalus	1	–	1	–	1	1	0.50	25
Pelvis	1	–	–	1	1	1	0.50	25
Femur	1	1	–	–	1	1	0.50	25
Tibia	2	1	1	–	2	2	1.00	50
Calcaneus	2		2	–	2	2	1.00	50
Fourth tarsal	1	1	–	–	1	0.5	0.5	25
First rear phalanx	3	–	–	3	3	2	0.75	37.5
Second phalanx	5	–	–	5	5	2	1.25	62.5
Third phalanx	2	–	–	2	2	2	0.5	25
Sesamoid	3	–	–	3	3	1	0.37	18.5
Metapodial	6	–	–	6	6	2	1.50	75
TOTAL	82					7		

The MAU% was grouped into Basic Skeletal Parts (PEB) (Yacobaccio et al. 1997), which revealed a predominance of skeletal parts with high meat content (ribs) in the young individuals' category, followed by parts with high marrow content and medium/low meat content (scapula, humerus, skull). In juveniles, there was a predominance in the selection of parts with high marrow content (metatarsals, metapodials), followed by parts with high/medium meat content (radio ulna, tibia, femur, pelvis). In adults, a predominance in the selection of parts with high marrow and meat content (metapodials, tibia, cervical vertebrae) was recorded, followed by parts with high marrow content and low meat content (radio ulna, metatarsal, phalanges).

5.4.3 The Central Room (RC)

About 30.65% ($N = 61/199$) of the Camelidae subset consists of fragments measuring 0–20 mm, followed by 19.10% ($N = 38/199$) of complete elements (e.g., phalanges, carpals, scapula). Cut marks related to primary processing were recorded in 17.5% ($N = 7/40$) of cases, located on the proximal area of the dorsal surface of some ribs and on the distal articular area of metapodials from young individuals. Regarding secondary processing, 38.89% ($N = 35/90$) of saw marks were observed on ribs, along with impact marks and numerous fresh fractures on the diaphysis of metapodials, metacarpals, metatarsals, radioulna, and tibia (Fig. 5.8) from juvenile and adult individuals. Additionally, a fragment of a bone tool made from a *Lama* sp. metapodial was recovered, featuring a polished blunt tip, transverse striations, and a fresh transverse fracture resulting from use (Fig. 5.9a, b). A metapodial from an adult individual, which was transversely and perpendicularly sawed, was also identified, possibly a preform for a tool. Furthermore, several scutes of *Chaetophractus vellerosus* were recorded.

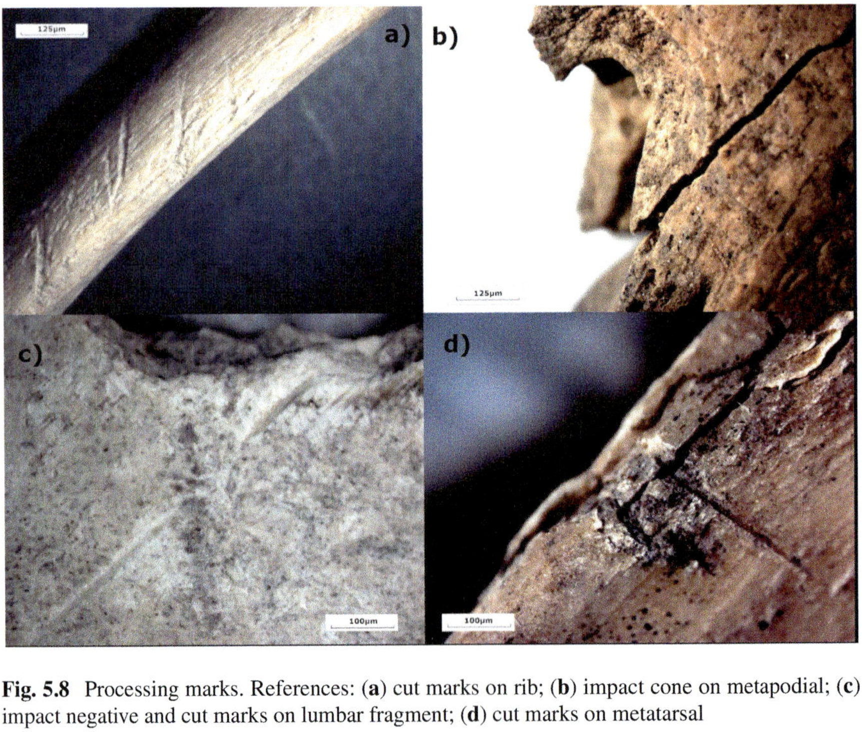

Fig. 5.8 Processing marks. References: (**a**) cut marks on rib; (**b**) impact cone on metapodial; (**c**) impact negative and cut marks on lumbar fragment; (**d**) cut marks on metatarsal

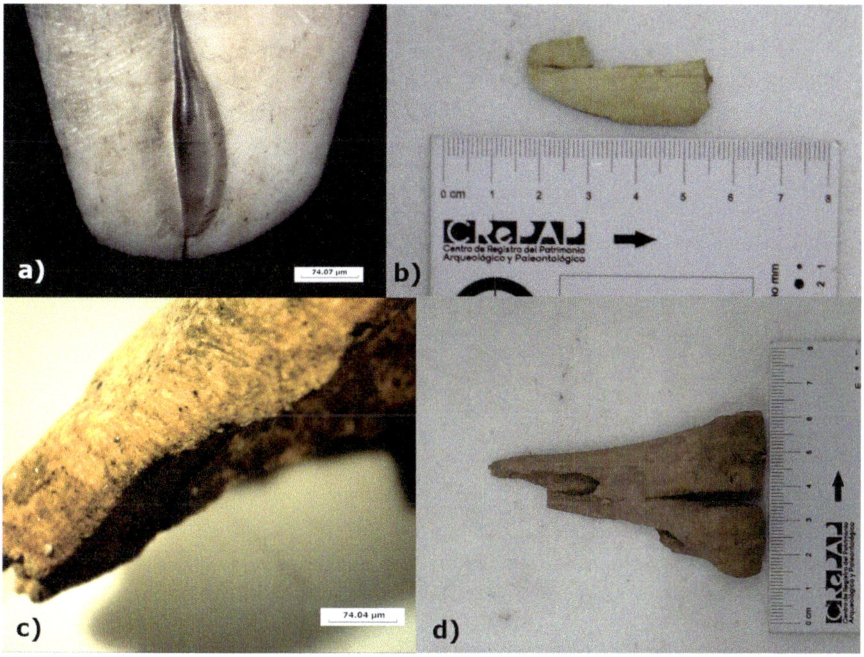

Fig. 5.9 Bone artifacts: (**a**) magnified image of a polished blunt point with transverse use marks on a metapodial; (**b**) modified metapodial; (**c**) ventral view showing polishing marks and transverse use marks on a metapodial; (**d**) modified metapodial

5.4.4 Room 1

Approximately 64.91% (*N* = 148/228) of the Camelidae subset consists of fragments measuring 0–20 mm, followed by 10.53% (*N* = 24/228) of complete elements. Cut marks related to primary processing were recorded on two ribs from young individuals and on the proximal epiphysis of a tibia from a juvenile. Additionally, 11.11% (*N* = 5/45) of the sample showed fresh fractures on the diaphysis of metatarsals, ribs, and long bones from juveniles, while 15.38% (*N* = 4/26) exhibited fresh fractures on radioulna and long bones from adults.

5.4.5 Room 2

Approximately 13.27% (*N* = 13/98) of the Camelidae subset consists of fragments measuring 0–20 mm, while 30.61% (*N* = 36/98) are represented by complete elements. Primary cut marks were observed on the angular region of a juvenile mandible, and secondary processing marks, such as cut marks, impact marks, and scrapings, were recorded on the transverse process of a lumbar vertebra from an adult. Cut marks related to meat extraction were also identified on a juvenile metatarsal and on long bones from an adult. Additionally, 44.26% (*N* = 27/61) of the sample exhibited fractures and impacts on the diaphysis of long bones, metapodials, radioulna, and tibia from both juveniles and adults. Furthermore, another bone tool, made from the distal end of a metapodial of an adult *Lama* sp., was recovered. On the ventral view, polished marks and transverse use marks were recorded, along with an impact cone showing characteristics of a dry fracture, possibly resulting from reworking (Fig. 5.9c). Moreover, a cut mark was documented on the mandible of an adult *Dicotyles tajacu*, and a helical fracture was observed on a humerus of *Hippocamelus antisensis*, as well as postcranial elements (ribs, pelvis fragments, radioulna) of *Chaetophractus vellerosus*.

5.5 Discussion

The taphonomic analyses of the zooarchaeological assemblage from Unit 3 (U3) at PV2 have shown low weathering across all *taxa* due to the high sedimentation rate in the area, with minimal evidence of rodent and carnivore activity. Additionally, considering the small number of identified elements, (NSP = 1163), we can conclude that the preservation of camelid remains is the result of anthropogenic actions (e.g., cleaning of the dwelling before the abandonment).

Regarding the consumption of species, such as *Hippocamelus antisensis*, *Lama glama*, and *Lama guanicoe*, it can be established that the inhabitants of U3 shared similar dietary habits with other residential structures of the QDLC. Concerning the recorded osteological elements of *Hippocamelus antisensis* (a humerus with cut marks and phalanges), part of the left forelimb would have been brought in, suggesting a probable distribution of the hunted specimen among other households. It is worth noting that the presence of *Chaetophractus vellerosus* had not been recorded in other residential structures, except for CC1. Considering the presence of diverse elements (osteoderms, ribs, part of the pelvis, a radioulna, and a cervical vertebra) and the fact that this is a species inhabiting the study area, we can infer that it likely entered the site whole for consumption. It also indicates that both domesticated and wild species were consumed, reaffirming that hunting continued as a key practice within the agro-pastoral lifestyle.

It is noteworthy that, consistent with previous findings from other sites in the region (e.g., Izeta 2004, 2008), our results also reveal a predominant presence of bone remains from wild animals (*N* = 33/48). Furthermore, the diversity of species—including *Lama guanicoe*, *Hippocamelus antisensis*, *Mazama gouazoubira*, *Chaetophractus vellerosus*, and *Dicotyles tajacu*—reflects a wide range of body sizes and ecological habitats within this wild faunal assemblage. For instance, the peccary is found in lower areas with "vegetation cover consisting of open or closed layers of woody plants up to 15 m in height" (Sistema de Información de Biodiversidad de la Administración de Parques Nacionales de Argentina n.d.). Indeed, current occurrences and sightings are exclusively recorded in such environments, like the eastern edge of the Yungas. Therefore, it is important to note that the incorporation of domestic animals may not have greatly altered the exploitation of wild resources, which were likely used on a similar scale as in previous times of hunting and gathering only (e.g., Izeta 2008; Belloti López de Medina 2011; Miyano 2018). While the adoption of domestic animals and agricultural practices certainly impacted other aspects of ancient societies, this is reflected in the diverse material culture and archaeological records found throughout the NWA region.

The identification of elements assignable to possible *Vicugna pacos* represents a significant archaeological finding for two main reasons. First, this species typically inhabits high-altitude regions between 4000 and 5000 m.a.s.l. Second, the majority of archaeological findings related to this species have been documented in Perú, where it was domesticated around 6000 years BP, and in Chile, such as the remains identified at Tulán 54 (2380–3080 BP; Cartajena et al. 2007). Therefore, the presence of alpaca bone remains in archaeological sites in Argentina is rare. Though, in the literature, there are two mentions on the presence of alpaca remains in archaeological contexts for Northwestern Argentina. The first one identified two alpaca incisors from an occupation level, dated at ca.3390 years BP, from the Tomayoc site (Jujuy, Argentina; Lavallée et al. 1997, pp. 14–15); the second one assign the presence of alpaca bone elements in sites from the "Aguada culture" with chronologies of 600–900 AD in the province of La Rioja, Argentina (Callegari 1988 in González 1998, p. 74). These identifications have been subject to debate, and many colleagues do not agree with what was published at the time.

However, some bone elements allowed us to identify using osteometry a morphotype very similar to modern *Vicugna pacos* samples. The elements preliminarily assigned to this taxon at QDLC consist of several hind limb elements, including the pelvis, astragalus, and hind phalanges. These do not show anthropogenic marks related to processing. As this appears to be a unique animal for the site, the possible origin of this individual should be evaluated. Thus, it makes us question whether the animal entered alive or just those meat-rich parts. For this purpose, stable isotope analyses are currently being conducted on these bone elements to gain further insights into the introduction of this material into the residential structure U3.

Based on studies on the use of neonate camelid carcasses, a predominance of parts with medium meat content (e.g., ribs, humerus, scapula) was recorded, followed by contributions with little meat and marrow content (skull); this could be interpreted as a strategy for optimizing available resources (intentional slaughter of premature animals and/or those that died from environmental stress), a practice recorded in northwest Argentine pastoralists (Izeta 2004). Given the recorded elements (skull, ribs, pelvis, metapodials, ulna) and processing marks (cut marks on the distal portion of the metapodials and proximal ends of the ribs), we can infer that one of the juvenile individuals was likely transported whole to the site.

In contrast, juveniles and adults show a higher frequency of whole elements (phalanges, metapodials, carpals, and tarsals) which could be assumed as a discard pattern as these bones are less suitable for marrow extraction or as blanks for making bone instruments. Additionally, fresh fractures and impact cones in metapodials, radioulnar, and diaphyses of adult long bones ($N = 10/79$) are observed, which, this time, can be more accurately related to marrow extraction. Regarding cut marks, these were recorded on the mandible, intermediate carpal, rib, lumbar, and two fragments of long bone, and are associated with primary processing. Based on the presence of axial skeletal elements (skull, mandible, lumbar vertebrae, ribs, and sternum) and appendicular elements (scapula, radioulna, carpals, metacarpals, femur, tibia, pelvis, sesamoids, and phalanges), we can infer that a juvenile individual was likely transported complete into the site. As for the adults, the presence of axial elements (skull, mandible, cervical/thoracic/lumbar vertebrae, ribs, and sternum) and appendicular elements (scapula, humerus, radioulna, carpals, tarsals, femur, tibia, tarsals, sesamoids, and phalanges) suggests that they, too, were introduced complete.

These analyses in RC established that cut marks on ribs and metapodials of young individuals (neonates and juveniles) are related to dismemberment and meat acquisition. The impacts and fresh fractures observed in metapodials, metacarpals, metatarsals, radioulnar, and tibias of juveniles and adults are the result of marrow extraction. There is evidence of a mix of primary processing waste (adult phalanges, carpals, tarsals, scapulas, ribs), secondary processing waste (limb parts), as well as some fragmented bone artifacts. Based on this outcome, and considering O'Connor (1993) animal dismembering through time, we can confidently assume that multiple activities were carried out in this space.

In the R1, activities related to primary processing for meat acquisition from young individuals' ribs and secondary processing related to marrow extraction from elements such as metatarsals, radioulna, tibia, and femur would have been carried out. Additionally, it is important to mention that 64.91% of the sample consists of fragments measuring 0–20 mm (young camelid); where few elements with cut marks, impacts, and fresh fractures were recovered, such as ribs, ischium, femur, tibia, radioulna, and diaphysis of long bones; as well as whole and fragmented elements (cervical and lumbar vertebrae, mandibles, phalanges, carpals, tarsals). This is consistent with models proposed for the extraction of components such as bone fat (De Nigris 1999; Outram 2001; Church and Lyman 2003). Furthermore, the presence of one hearth, which were cleaned before leaving the dwelling, was recorded. Therefore, based on the above, these remains are the result of waste from domestic food processing and consumption activities, i.e., activities related to culinary techniques, as suggested by O'Connor (1993).

In R2, elements with cut marks related to leather extraction from the skull of juvenile *Lama guanicoe* and meat from long bones and ribs were observed; while impact marks and fresh fractures present on long bones and metatarsals of juveniles and adults are associated with marrow extraction. Therefore, this space shows a mixture of secondary processing waste and primary processing waste for crafting bone tools. This indicates that R2 functioned similarly to RC, allowing for an expansion of areas dedicated to various activities.

Macroscopic and microscopic analyses of bone artifacts have revealed technical, morphological, and selection similarities with those analyzed by Izeta et al. (2013) from Valle del Cajón and Santa María. These authors associate these artifacts with the manufacture of wool and basketry. This suggests shared knowledge in the crafting techniques of such tools for animal fiber processing at a regional level. The presence of bone tools for spinning animal fibers, dating back to *ca.* 3800 years BP (in TPV1), indicates indirect practices related to camelid management. Wool production has persisted over time, initially practiced in communal spaces and later within the family sphere.

The isotopic values of llamas indicate a diet based on C_3 photosynthetic pathway plants, similar to that of guanacos. This can be attributed to the predominance of C_3 plants with good forage capacity, such as *Stipa* sp. (Martín et al. 2002), and the availability of permanent water sources—such as Los Corrales river and nearby wetlands—in the study area. This suggests a feeding strategy that utilized resources both from the study area and from nearby regions like Tafí del Valle or Amaicha. On the other hand, it is important to highlight that the productive structures likely used as corrals are located in areas associated with productive activities and are separate from the dwellings. This raises questions about whether each structure was shared by multiple families, or if the "productive" space (animals and crops) represented an area of shared communal responsibilities.

Based on the evidence presented above, it is possible to propose that the QDLC village assemblage was likely composed of several key components: the herders, the dwelling inhabitants, domestic animals (particularly camelids), wild animals, pastures, and productive spaces. Through the analysis of dwelling U3, we can suggest the presence of three potential mechanisms: affective/sensory, mnemonic/temporal, and political. Animals (both wild and domestic) function as an affective/sensory mechanism within the community, fostering affective interactions among its members through the sharing of knowledge during resource processing, as well as sensory experiences linked to the aromas of food preparation. Another affective/sensory mechanism is evident in the presence of neonatal and adult domesticated camelids, reflecting "networks of ownership" (Dransart 2011) within the family unit, where the responsibility for the care and protection of the herd in productive structures plays a central role. This also suggests the mobility of herders to feed and manage their herds.

Domestic spaces act as mnemonic/temporal mechanisms, as they encapsulate practices from the past (such as hunting) and the present (such as domestication). At the same time, these spaces function as political mechanisms, as the shared architecture with neighboring societies (reflecting a cultural identity) and daily practices transmit social, economic, and religious norms, demonstrating the capacity and power of the involved agents.

On the other hand, productive spaces could function as mnemonic/temporal mechanisms, as they evoke the communal use of spaces in the past, now transformed into areas of exclusive use for the villagers. Furthermore, these spaces might also act as political mechanisms, serving as areas for negotiation over the use and preservation of pastures, the maintenance of cultivation terraces, and other social matters.

Regarding stable isotope analyses, it is possible to propose that herding spaces would represent another political mechanism within this assemblage. The pastures might reflect social agreements among families regarding the communal use of resources, both within the village and in interaction with other social groups in the surrounding areas.

Based on this, we can begin to think of a landscape shaped collectively by herders, camelids, grazing spaces, and productive spaces. The herders would leave their dwellings to gather their herds from the productive structures, allowing the herds to graze within the village while moving among the cultivation terraces. At the same time, they would move to nearby pastures, where they exchanged information (such as tool-making techniques) and raw materials (possibly including specimens of *Vicugna pacos*) with other herders. The camelids, while grazing and fertilizing the land, played a role in preserving the communal pastures. Later, they became a source of food and raw materials for tools within each dwelling. The productive spaces would have served as spaces of social cohesion, as they required care and participation from each member of the village, fostering the creation of social bonds.

5.6 Conclusions

Pastoralism has been and continues to be a widely analyzed topic though most studies have focused on highland environments such as the *Puna*. Therefore, the study of areas such as Quebrada de Los Corrales in the sector of mesothermal valleys, contributes valuable information to this field. The landscape of QDLC, understood as a socially constructed space filled with meanings and interactions, is shaped by the interrelation between herders, domestic animals, and the environment. These relationships, as seen through the lens of assemblage theory, create a unique pastoral identity specific to this region.

Among the notable findings, the presence of a possible *Vicugna pacos* individual is a significant archaeological discovery for pastoralism within the first village settlements of northern Argentina, as this species typically inhabits higher altitude and wetter areas. This discovery not only highlights the unique environmental conditions of QDLC, but also reflects how the assemblage of herders and camelids adapted to these local landscapes, forming a distinct way of life.

Regarding pastoral models, it has been established that these can be diverse. Based on zooarchaeological, isotopic, and productive structure information (*corrales*), the hypothesis for QDLC is that camelid management strategies involve using local pastures and meadows, with access to nearby lower areas like Tafí del Valle or Amaicha. This local adaptation of pastoral practices, viewed through assemblage theory, shows how the various components (herders, animals, landscapes) interrelated in a stable yet flexible system, influenced by both local and broader interaction networks.

Therefore, the Puesto Viejo village reflects a different way of inhabiting this landscape. This local adaptation is more evident in the use of local raw materials, ceramic-making techniques (Gramajo Bühler and García Rosselló 2020), and camelid management strategies with local pastures (during the spring-summer seasons). Despite this localized adaptation, the people of QDLC were still part of broader interaction networks that allowed access to foreign goods and information. Furthermore, the houses should be understood as private spaces where multiple activities were shared solely by family members, as indicated by their spatial distribution. However, these spaces were also sites of reproduction for shared social structures, which, like the assemblages described, were both stable and dynamic.

These spaces reflect the strong social, economic, and religious ties, as proposed by Tarragó (1999) for the central-western region of Tucumán province, in Tafí del Valle and its subsidiary quebradas of El Mollar, Anfama, and El Pedregal. These ties persisted and were transmitted over 1000 years across 350 km^2, establishing regional identity (Oliszewski 2017; Oliszewski and Di Lullo 2020). The circulation spaces between structures might have functioned as exchange areas, with the river as a meeting space and agricultural lands as negotiation fields (Oliszewski and Di Lullo 2020, p. 331).

The integration of landscape and assemblage theories in the study of pastoral practices at QDLC opens new avenues for understanding the dynamic relationships between humans, animals, and their environments in ancient Andean societies. Future research could further explore how these assemblages evolved over time, particularly in response to environmental changes, social transformations, and broader regional interactions. Additionally, applying these theoretical frameworks to other archaeological sites within the Andes could provide comparative insights, revealing both the uniqueness and commonalities of pastoral strategies across different landscapes. By continuing to investigate these interconnections, we can gain a deeper appreciation of how Andean communities adapted to their environments and maintained resilient social and economic networks that have persisted over millennia.

Acknowledgments This work was carried out within the framework of the following research projects: PICT 2018-03150, PIP-CONICET 0434, and PIUNT G624. Mercedes Izeta contributed to the final English version of the manuscript.

Competing Interest Declaration The authors have no conflict of interest to declare that are relevant to the content of this chapter.

References

Aschero C, Hocsman S (2011) Arqueología de las ocupaciones cazadoras-recolectoras de fines del Holoceno Medio de Antofagasta de la Sierra (Puna Meridional Argentina). Chungara, Rev Antropol Chilena 43(1):393–411

Aschero C, Izeta A, Hocsman S (2012) New data on south American camelid bone size changes during middle-late Holocene transition. Osteometry at Peñas Chicas 1.5 (Antofagasta de la sierra, Argentinian Puna). Int J Osteoarchaeol 24:492–504. https://doi.org/10.1002/oa.2235

Berberián E, Nielsen A (1988) Sistemas de asentamiento prehispánico en la etapa Formativa del valle de Tafí (Pcia. de Tucumán- Rep. Arg.). In: Berberián E (ed) Sistemas de Asentamiento Prehispánicos en el Valle de Tafí. Córdoba, Editorial Comechingonia, pp 21–51

Binford LR (1981) Bones. Ancient men and modern myths. Academic, New York

Brain CK (1981) The hunters or the hunted? The University of Chicago Press, Chicago

Cabrera A (1976) Regiones fitogeográficas argentinas. Enciclopedia Argentina de Agricultura y Jardinería (Segunda Edición, Tomo II, Fascículo 1). Editorial ACME, Buenos Aires

Carbonelli JP, Winocur D, López B, de Medina C, Carminatti N, Peisker V (2021) Cueva Abra del Toro, registro de la ocupación más antigua para el Valle de Yocavil, Catamarca y de la erupción de mayor intensidad conocida para el Holoceno en la Zona Volcánica Central de los Andes. Rev Museo Antropol 14(1):123–138. https://doi.org/10.31048/1852.4826.v14.n1.29889

Cardich A, Izeta AD (1999-2000) Revisitando Huargo (Perú). Análisis cuantitativos aplicados a restos de Camelidae del Pleistoceno Tardío. Anal Arqueol Etnol 54(5):29–40

Caria M, Oliszewski N, Gómez Augier J, Pantorrilla M, Gramajo Bühler M (2010) Formas y espacios de las estructuras agrícolas prehispánicas en la Quebrada del río de Los Corrales (El Infiernillo-Tucumán). In: Korstanje MA, Quesada M (eds) Arqueología de la agricultura: casos de estudio en la región andina Argentina. Magna, Aurora, pp 144–165

Cartajena I (2009) Explorando la variabilidad morfométrica del conjunto de camélidos pequeños durante el Arcaico Tardío y el Formativo Temprano en Quebrada Tulán, norte de Chile. Rev Museo Antropol 2:199–214. https://doi.org/10.31048/1852.4826.v2.n1.5418

Cartajena I, Núñez L, Grosjean M (2007) Camelid domestication on the western slope of the Puna de Atacama, northern Chile. Anthropozoologica 42(2):155–173

Church RR, Lyman RL (2003) Small fragments make small differences in efficiency when rendering grease from fractured artiodactyl bones by boiling. J Archaeol Sci 30:1077–1084. https://doi.org/10.1016/S0305-4403(03)00010-4

Correa I, Carrasco C, Ballester B, Gallardo F (2018) Efectos colaterales de la transición de la transición hacia el Formativo: una nueva culinaria entre los Cazadores-recolectores marinos del desierto de Atacama. Chungara Rev Antropol Chilena 50(1):87–106. https://doi.org/10.4067/S0717-73562018005000101

De Nigris M (1999) Lo crudo y lo cocido: Sobre los efectos de la cocción en la modificación ósea. Arqueología 9:239–264. http://repositorio.filo.uba.ar/handle/filodigital/6415

DeLanda M (2006) A new philosophy of society. Assemblage theory and social complexity. Continuum, London

Deleuze G, Guattari F (1987) A thousand plateaus. Capitalism and Schizophrenia. University of Minnesota Press, Minneapolis

Di Lullo E (2010) El Espacio Residencial Durante el 1er Milenio D.C. en la Quebrada de Los Corrales (El Infiernillo, Tucumán). Bachelor Thesis. Facultad de Ciencias Naturales e IML, UNT, Tucumán. Unpublished

Dransart P (1999) La domesticación de los camélidos en los andes centro-sur. Una reconsideración. Revista. Relac Soc Argentina Antropol XXIV:125–138

Dransart P (2011) Social principles of Andean camelid pastoralism and archaeological interpretations. In: Albarella U, Trentacoste A (eds) Ethnozooarchaeology: the present and past of human-animal relationships. Oxbow Books, Oxford, pp 123–130. https://doi.org/10.2307/j.ctvh1dwvg.17

Elkin DC (1996) Arqueozoología de Quebrada Seca 3: indicadores de subsistencia humana temprana en la Puna Meridional Argentina. Ph.D. Dissertation. Facultad de Filosofía y Letras, Buenos Aires, Argentina. Unpublished

Gasco A, Marsh E, Kent J (2014) Clarificando variables osteométricas para la primera falange de camélidos sudamericanos. Intersec Antropol 15(1):131–138

Gómez Augier J, Oliszewski N, Caria M (2008) Altitude cultivation: phytolith analysis in archaeological farming structure of Quebrada del Río de Los Corrales site El Infiernillo, Tucumán, Argentina). In: Osterrieth M, Fernández Onaire M, Borelli N (eds) abstracts, 7th international meeting on Phytolith research and 4th south American meeting Phytolith research, p 64

González AR (1998) Cultura La Aguada: arqueología y diseños. Filmediaciones Valero, London

Grant Lett-Brown J (2008) El recurso Camelidae en sitios de la Puna Meridional Argentina: Una aproximación osteométrica. Bachelor Thesis. Facultad de Filosofía y Letras, Universidad de Buenos Aires. Unpublished

Gramajo Bühler, M and García Rosselló, J (2020). Aproximaciones Traceológicas a la Cerámica del Primer Milenio d. C. de Quebrada de Los Corrales (El Infiernillo, Tucumán, Argentina). Rev Museo Antropol 13(2):349–358. https://doi.org/10.31048/1852.4826.v13.n2.27183

Haber A (1999) Una arqueología de los oasis puneños. Domesticidad, interacción e identidad en Antofalla, primer y segundo milenios d. C. Ph.D. Dissertation, Facultad de Filosofía y Letras, Universidad de Buenos Aires. Unpublished

Hocsman S (2010) Cambios en las puntas de proyectil durante la transición de cazadores-recolectores a sociedades agro-pastoriles en Antofagasta de la Sierra (Puna Argentina). Arqueología 16(1):59–86. https://doi.org/10.34096/arqueologia.t16.n1.1728

Hocsman, S and Babot, MP (2018) La transición de cazadores-recolectores a sociedades agro-pastoriles en Antofagasta de la Sierra (Puna de Catamarca, Argentina): perspectivas desde la agencia y las prácticas. Chungará Revista de Antropología Chilena, 50(1), 51–70. https://doi.org/10.4067/S0717-73562018005000202

Ingold T (1993) The temporality of the landscape. World Archaeol 25(2):152–174. https://doi.org/10.1080/00438243.1993.9980235

Izeta AD (2004) Zooarqueología del sur de los valles Calchaquíes. Estudio de conjuntos faunísticos del Período Formativo. Ph.D. Dissertation. Facultad de Ciencias Naturales y Museo, Universidad Nacional de La Plata, Argentina. https://doi.org/10.35537/10915/4560

Izeta AD (2007) Zooarqueología del sur de los valles Calchaquíes (Provincias de Catamarca y Tucumán, República Argentina): Análisis de conjuntos faunísticos del primer milenio A.D. B.A.R. International Series S1612. John and Erica Hedges https://suquia.ffyh.unc.edu.ar/handle/suquia/14922

Izeta AD (2008) Late Holocene camelid use tendencies in two different ecological zones of northwestern Argentina. Quat Int 180:35–144. https://doi.org/10.1016/j.quaint.2007.08.026

Izeta AD, Otaola C, Gasco A (2009) Osteometría de falanges proximales de camélidos sudamericanos modernos. Variabilidad, estándares métricos y su importancia como conjunto comparativo para la interpretación de restos hallados en contextos arqueológicos. Rev Museo Antropol 2(1):169–180. https://doi.org/10.31048/1852.4826.v2.n1.5416

Izeta AD, Srur MG, Labarca Encina R (2012) Guía osteométrica de camélidos sudamericanos. Museo de Antropología IDACOR, Córdoba. https://suquia.ffyh.unc.edu.ar/handle/suquia/4746

Izeta AD, Cattáneo R, Scattolin MC, Cortés LI (2013) Changed into tools. Camelid bones from the southern Calchaquíes valleys (formative period, North-Western Argentina). In: Choyke A, O'Connor S (eds) From these bare bones: raw materials and the study of worked osseous objects. Oxbow Books, Oxford, pp 50–58

Kaufmann C (2004) Metodologías para construir perfiles de mortalidad de guanaco (Lama guanicoe) en sitios arqueológicos: el Sitio Paso Otero 1 como caso de estudio. In: Mengoni Goñalons GL, Olivera DE, Yacobaccio HD (eds) El manejo de los camélidos sudamericanos: perspectivas multidisciplinarias. San Salvador de Jujuy, Buenos Aires, pp 151–176

Kaufmann C, L'Heureux L (2009) El dimorfismo sexual en guanacos (Lama guanicoe). Una evaluación osteométrica de elementos poscraneales. Rev Museo Antropol 2:181–198. https://doi.org/10.31048/1852.4826.v2.n1.5417

Kent JD (1982) The domestication and exploitation of the south American camelids: methods of analysis and their application to circum-lacustrine archaeological sites in Bolivia and Peru. Ph.D. Dissertation. Washington University. St. Louis, Missouri, USA. Unpublished

L'Heureux GL (2005) Variación morfométrica en restos óseos de guanaco de sitios arqueológicos de Patagonia austral continental y de la Isla Grande de Tierra del Fuego. Magallania 33(1):81–94. https://doi.org/10.4067/S0718-22442005000100006

Lavallée D, Julien M, Karlin C, García LC, Pozzi-Escot D, Fontugne M (1997) Entre desierto y quebrada. Tomayoc, un alero en la Puna. Av Arqueol 3:9–40. http://repositorio.filo.uba.ar:8080/xmlui/handle/filodigital/5808

Le Neün M, Dufour E, Goepfert N, Neaux D, Wheeler JC, Yacobaccio H, Mengoni Goñalons G, Elkin D, Gasco A, Cucchi T (2023) Can first phalanx multivariate morphometrics help document past taxonomic diversity in south American camelids? J Archaeol Sci Rep 47:103708. https://doi.org/10.1016/j.jasrep.2022.103708

López Campeny SML (1999) Actividades domésticas y organización del espacio intrasitio. El sitio Punta de la Peña 9 (Antofagasta de la Sierra, Prov. de Catamarca). Bachelor Thesis Facultad de Ciencias Naturales e Instituto Miguel Lillo, Universidad Nacional de Tucumán. Unpublished

López B, de Medina CR (2011) Zooarqueología del sitio formativo Soria 2, valle de Yocavil (Catamarca), siglo I d.C. Rev Museo Antropol 4:3–16

Madero C (1992) Análisis faunístico de Huachichocana III (Jujuy): identificación arqueológica de la caza y el pastoreo. Palimpsesto Rev Arqueol 2:107–122

Marsh E (2015) The emergence of agropastoralism: accelerated ecocultural change on the Andean altiplano, 3540–3120 cal BP. Environ Archaeol 20(1):13–29. https://doi.org/10.1179/1749631414Y.0000000036

Martín GO, Raya F, Lucas J, Fernández D, Colombo MB, De Marco de Roncaglia N (2002) Gradiente de distribución de la diversidad florística, en la transecta Tafí del Valle—Amaicha (Prov. de Tucumán, Argentina). Actas de las XIX Jornadas de la Asociación de Biología de Tucumán. Asociación de Biología de Tucumán, pp 212–218

Martínez JG, Funes Coronel J (2020) Trayectorias cruzadas de gente y de materia prima. Movilidad entre Puna y Valles en el Noroeste de Argentina durante el Holoceno Medio. Rev Museo Antropol 13(1):37–44. https://doi.org/10.31048/1852.4826.v13.n1.23838

Martínez JG, Mauri E, Mercuri C, Caria M, Oliszewski N (2013) Mid-Holocene human occupations in Tucuman (NW Argentina). Quat Int 307:86–95. https://doi.org/10.1016/j.quaint.2013.04.027

Martínez JG, Oliszewski N, Arreguez GA, Backwell LR, Luna LH, Molar R, Naharro ME (2020) Prácticas funerarias y ritualidad en la Quebrada de Los Corrales, Tucumán-Argentina (3.800-3.500 a.p.). Rev Chilena Antropol 42:290–318. https://doi.org/10.5354/0719-1472.2020.60494

Menegaz A, Goin FJ, Ortiz Jaureguizar E (1989) Análisis morfológico y morfométrico multivariado de los representantes fósiles y vivientes del género Lama (Artiodactyla, Camelidae). Sus implicancias sistemáticas, biogeográficas, ecológicas y biocronólogicas. Ameghiniana 26(3–4):153–172

Mengoni Goñalons GL (1988) Análisis de materiales faunísticos de sitios arqueológicos. Xama 1:78–120

Mengoni Goñalons GL (1999) Cazadores de guanacos de la estepa patagónica. Sociedad Argentina de Antropología, Colección Tesis Doctorales, Buenos Aires

Mengoni Goñalons GL, Yacobaccio HD (2006) The domestication of south American camelids: a view from the south-Central Andes. In: Zeder MA, Bradley DG, Emshwiller E, Smith BD (eds) Documenting domestication: new genetic and archaeological paradigms. University of California Press, Los Angeles, pp 228–244. https://doi.org/10.1525/9780520932425-019

Mercuri C (2017) Introducción a los artefactos compuestos del 1° milenio DC en El Infiernillo, Tucumán. Intersec Antropol 18:261–270

Mercuri C, Mauri E (2013) El 1° milenio de Quebrada de Los Corrales desde los materiales líticos: análisis del conjunto de PV2 Estructura 1. In: Bárcena J, Martín S (eds) Arqueología Argentina en el Bicentenario de la Asamblea General Constituyente del año 1813. UNLaR-CONICET, Buenos Aires, pp 347–348

Miyano JP (2018) El uso de animales por las sociedades agropastoriles tempranas: análisis zooarqueológico de un basural de la aldea de Palo Blanco (valle de Fiambalá, Catamarca). Arqueología 24(1):77–101

Molar R, Oliszewski N, Arreguez G (2024) Tras los rastros de los alimentos: prácticas de procesamiento de vegetales en contextos aldeanos tempranos (Quebrada de los Corrales, Tucumán, Argentina). Rev Museo Antropol 17(2):287–302. https://doi.org/10.31048/1852.4826.v17.n2.44290

Morello J, Matteucci S, Rodríguez A, Silva M (2012) Ecorregiones y complejos sistémicos argentinos. Orientación Gráfica Editora, Buenos Aires

Muntaner A (2012) Vida y muerte en Puesto Viejo 1: Estudio de un entierro humano del 1° milenio en La Quebrada de los Corrales, El Infiernillo, Tucumán. Bachelor Thesis, Universidad Nacional de Tucumán

Naharro ME (2019) Osteobiografía y prácticas mortuorias prehispánicas en El Infiernillo (Tucumán) Bachelor Thesis. Universidad Nacional de Tucumán. Unpublished

Naharro ME (2020) Osteobiografía en la quebrada de Los Corrales (El Infiernillo, Tucumán) durante el Holoceno tardío. La Zaranda de Ideas 18(2):59–79

Núñez L, Perlès C (2018) Ceremonialismo y congregación durante la transición arcaico tardío y formativo temprano en la circumpuna de Atacama (norte de Chile). Rev Geografía Norte Grande 70:183–209. https://doi.org/10.4067/S0718-34022018000200183

Núñez L, Cartajena I, Carrasco C, de Souza P, Grosjean M (2006) Emergencia de comunidades pastoralistas formativas en el sureste de la Puna de Atacama. Estud Atacameños 32:93–117

O'Connor T (1993) Process and terminology in mammal carcass reduction. Int J Osteoarchaeol 3:63–67. https://doi.org/10.1002/oa.1390030202

Oliszewski N (2017) Las aldeas "Patrón Tafí" del sur de Cumbres Calchaquíes y norte del Sistema del Aconquija. Comechingonia Rev Arqueol 21(1):205–232. https://doi.org/10.37603/2250.7728.v21.n1.19391

Oliszewski N, Di Lullo E (2020) Puesto Viejo, una aldea Tafí en las alturas (Quebrada de Los Corrales, El Infiernillo Tucumán). Rev Museo Antropol 13(2):325–338. https://doi.org/10.31048/1852.4826.v13.n2.27183

Oliszewski N, Martínez JG, Caria M (2008) Ocupaciones prehispánicas en una quebrada de altura: el caso de Cueva de los Corrales 1 (El Infiernillo, Tafí del Valle, Tucumán). Rev Relac 33:209–221

Oliszewski N, Martínez JG, Di Lullo E, Gramajo Bühler M, Arreguez G, Cruz H, Mauri E, Mercuri C, Muntaner A, Srur G (2015) Contribuciones al estudio de sociedades aldeanas en el noroeste argentino: el caso de la quebrada de Los Corrales (El Infiernillo, Tucumán). In: Korstanje A, Lazzari M, Basile M, Bugliani F, Lema V, Pereyra Domingorena L, Quesada M (eds) Crónicas materiales precolombinas: arqueología de los primeros poblados del Noroeste Argentino. Sociedad Argentina de Antropologíam, Buenos Aires, pp 51–79

Oliszewski N, Martínez JG, Arreguez G, Gramajo Bühler M, Naharro ME (2018) "La transición" vista desde los valles intermontanos del noroeste argentino: nuevos datos de la Quebrada de Los Corrales (El Infiernillo, Tucumán, Argentina). Chungara Rev Antropol Chilena 50(1):71–86. https://doi.org/10.4067/S0717-73562017005000110

Oliszewski N, Galván VK, Srur G, Olivera DE, Martínez JG (2020) Human paleodiet studies between ca. 3300–1500 years BP in Quebrada de Los Corrales (Tucumán, Argentina). J Archaeol Sci Rep 32:1–8. https://doi.org/10.1016/j.jasrep.2020.102429

Oliszewski N, Martínez JG, Arreguez G, Coronel A, Di Lullo E, Gramajo Bühler M, Molar L, Naharro ME, Nasif N (2022) Cueva de Los Corrales 1 (Quebrada de Los Corrales, El Infiernillo, Tucumán): un sitio multifuncional de altura en el Norte de las Sierras del Aconquija (ca. 3000-600 años AP). Relac Soc Argentina Antropol 47(2):188–208. https://doi.org/10.24215/18521479e036

Olivera D (1997) La importancia del recurso Camelidae en la puna de Atacama entre los 10.000 y 500 años AP. Estud Atacameños 14:29–41. https://www.jstor.org/stable/25674670

Olivera D (1998) Cazadores y Pastores tempranos de la puna Argentina, past and present in Aandean prehistory and early history. Etnol Stud 42:153–180

Outram AK (2001) A new approach to identifying bone marrow and grease exploitation: why the "indeterminate" fragments should not be ignored. J Archaeol Sci 28:401–410. https://doi.org/10.1006/jasc.2000.0619

Puig S (1988) Craneología y craneometría de camélidos: diferenciación interespecífica y determinación de la edad. Xama 1:43–56

Raffino R (1977) Las aldeas del Formativo inferior en la Quebrada del Toro, (Pcia. de Salta. Argentina). Obra Centenario Museo La Plata II:253–299

Raffino RA, Tonni EP, Cione AL (1977) Recursos alimentarios y economía en la Región de la Quebrada del Toro, Provincia de Salta, Argentina. Relac Soc Argentina Antropol XI:9–30. http://sedici.unlp.edu.ar/handle/10915/25246

Scattolin MC (2010) La organización del hábitat precalchaquí (500 a.C. - 1000 d.C.). In: Korstanje MA, Albeck ME, Scattolin MC (eds) El hábitat prehispánico. Arqueología de la arquitectura y de la construcción del espacio organizado. Editorial de la Universidad Nacional de Jujuy, San Salvador de Jujuy, pp 13–51
Scattolin MC (2019) Yutopián, donde los objetos se sublevan. Rev Museo La Plata 4(1):69–102. https://doi.org/10.24215/25456377e070
Scattolin MC, Lazzari M (1997) Tramando redes: Obsidianas al oeste del Aconquija. Estud Atacameños 14:189–209. https://www.jstor.org/stable/25674681
Scattolin MC, Bugliani MF, Izeta AD, Pereyra Domingorena L, Lazzari M, Martínez LA (2001) Conjuntos materiales en dimensión temporal. El sitio Formativo "Bañado Viejo" (Valle de Santa María, Tucumán). Relac Soc Argentina Antropol 26:167–192
Scattolin MC, Cortés LI, Bugliani MF, Calo CM, Domingorena LP, Izeta AD, Lazzari M (2009) Built landscapes of everyday life: a house in an early agricultural village of North-Western Argentina. World Archaeol 41(3):396–414. https://doi.org/10.1080/00438240903112310
Sistema de Información de Biodiversidad de la Administración de Parques Nacionales de Argentina (n.d.) Información general sobre *Dicotyles tajacu*. https://sib.gob.ar/especies/dicotyles-tajacu?tab=info-general
Somonte C, Baied CA (2013) Edad mínima de exposición de superficies en canteras-taller: reflexiones en torno a las primeras dataciones mediante microlaminaciones de barniz de las rocas (VML) para el noroeste argentino. Chungara Rev Antropol Chilena 45(3):427–445
Tarragó MN (1994) El Formativo en el Noroeste argentino y el alto Valle Calchaquí. Actas del XI Congreso Nacional de Arqueología Argentina. Rev Museo Hist Nat 27(1–4):103–119
Tarragó MN (1999) El Formativo y el surgimiento de la complejidad en el noroeste argentino. In Ledergerber-Crespo P (ed) Formativo sudamericano: una revaluación. Abya-Yala, pp 302–313
Tilley C (1994) A phenomenology of landscape. Berg, Oxford
Todd LC (1987) Taphonomy of the Horner II Bonebed. In: Frison GC, Todd LC (eds) The Horner site: the type site of the Cody cultural complex. Academic Press, New York, pp 107–198
Wheeler JC (1982) Aging llamas and alpacas by their teeth. Llama World 1:12–17
Yacobaccio HD (2001) La Domesticación de Camélidos en el Noroeste Argentino. In: Berberián E, Nielsen A (eds) Historia Argentina Prehispánica, vol. 1, pp 7–40
Yacobaccio HD, Madero CM, Malmierca MP, Reigadas MC (1997) Caza, domesticación y pastoreo de camélidos en la Puna Argentina. Relac Soc Argentina Antropol 22:389–418. http://sedici.unlp.edu.ar/handle/10915/25596
Yacobaccio HD, Escola PS, Pereyra FX, Lazzari M, Glascock MD (2004) Quest for ancient routes: obsidian sourcing research in northwestern Argentina. J Archaeol Sci 31(2):193–204. https://doi.org/10.1016/j.jas.2003.08.001

Chapter 6
Humans and Prey. Holocene Hunter Occupation of High-Altitude Spaces in the Puna of Salta, Argentina

Hernán Muscio, Juan Pablo Orsi, and Guido Carballo

6.1 Introduction

Human dispersal in the South Andean Highlands, the high-altitude areas of Northern Chile and the Puna of Argentina, was a process that began during the Pleistocene to Holocene transition about 12,000 years ago (Yacobaccio and Morales 2011). For about 9000 years after their initial dispersal, humans maintained a forager way of life adapted to the extreme conditions of the high-altitude Puna environments, where South American camelids hunting was the predominant human subsistence strategy (Yacobaccio and Vilá 2013). As many processes of behavioral and cultural evolution in the South Andean highlands arose from the increased human dependence on camelid hunting, the ecological relation of predator-prey can be useful as a theoretical frame of reference for illuminating these processes. This work examines the implications of studying human-camelid interactions in terms of a predator-prey ecological relationship that began early in the history of humans in the Andes and not ended with the transition to food production economies. To this end, we discuss the archaeological record of two high-altitude gorges, Quebrada Nacimiento and Quebrada Cavi, which document early specialization in hunting and long-term habitat construction patterns resulting from the human selectivity of spaces focused on the hunting of wild camelids.

H. Muscio (✉)
Consejo Nacional de Investigaciones Científicas y Técnicas (CONICET), Instituto de Arqueología, Facultad de Filosofía y Letras, Universidad de Buenos Aires, Buenos Aires, Argentina

J. P. Orsi (✉)
Instituto de Arqueología, Facultad de Filosofía y Letras, Universidad de Buenos Aires (UBA), Buenos Aires, Argentina

G. Carballo
Departamento de Antropología, Universidad de Buenos Aires (UBA), Buenos Aires, Argentina

R. Frontini et al. (eds.), *Non-human Animals and People: Interactions, Agents and Materialities*, The Latin American Studies Book Series,
https://doi.org/10.1007/978-3-031-97495-3_6

6.1.1 Predator-Prey Relationship and High Andean Hunters

In this work, our goal is to apply the predator-prey model as a theoretical framework to examine the emergence of a hunting landscape in the Puna region of Salta. We aim to document how the distribution of archaeological sites across broad spatial and temporal scales, alongside faunal patterns from localized and buried sites, can offer insights into past predator-prey interactions in which humans were the predators. Our primary dataset includes the spatial distribution of archaeological sites across two high-altitude gorges, Quebrada Nacimiento and Quebrada Cavi (Fig. 6.1). Also, we discuss the zooarchaeological record of the site Cueva Nacimiento 2 (CN2), located in Quebrada Nacimiento with an occupational sequence ranging from the Late Early Holocene to the onset of the Late Holocene.

From an evolutionary-ecological perspective (Pianka 1994), one possible interaction between humans and animals is the predator-prey relationship (Abrams 2001; Burak et al. 2018). This is a biological interaction between two species in which one of them, the predator, feeds on the other species, the prey. In this type of relationship, predators obtain food by hunting and consuming the prey (Abrams 2001). Classically, the population effects of these dynamics were described by the oscillatory trajectories of prey and predator populations over time, captured by the Lotka-Volterra model (Hou et al. 2013).

There are different strategies and adaptations in predators and prey that evolve over time increasing their chances of survival. These might imply the evolution of more or less specialized phenotypic adaptations for hunting or avoidance (Lima 2002). Predators often have specialized physical and behavioral abilities to capture their prey, such as speed, agility, camouflage, or ambush techniques. On the other hand, prey can develop defensive mechanisms such as speed, camouflage, protective structures (such as shells or spines), group formation to reduce the risk of capture, fleeing, and habitat abandonment. In predator-prey interactions, other important changes can arise through niche construction, the process by which an organism alters its own (or another species) local environment producing feedback on selection (Odling-Smee et al. 2003). Niche construction can lead to a predator benefiting its prey. This may occur when predators reduce prey mortality by protecting them against other predators through interference or predation of alternative predators, or controlling prey's competitors, or when a predator helps its prey's dispersal and reproduction by actively managing both (Piccot et al. 2019).

The predator-prey interaction has many interesting effects to consider on archaeological-evolutionary timescale (Hoare 2019). In this work, we discuss some implications of applying the logic of the predator-prey relationship to account for the archaeological record from the early Holocene to the beginning of the Late Holocene in the highlands of the southern Andes, particularly in the Puna region of Salta Province, Argentina.

In demographic terms, the predator-prey dynamics predict correlated trajectories, where human population growth decreases prey populations due to hunting pressure, leading to a reduced hunting selectivity and dietary expansion (Kelly

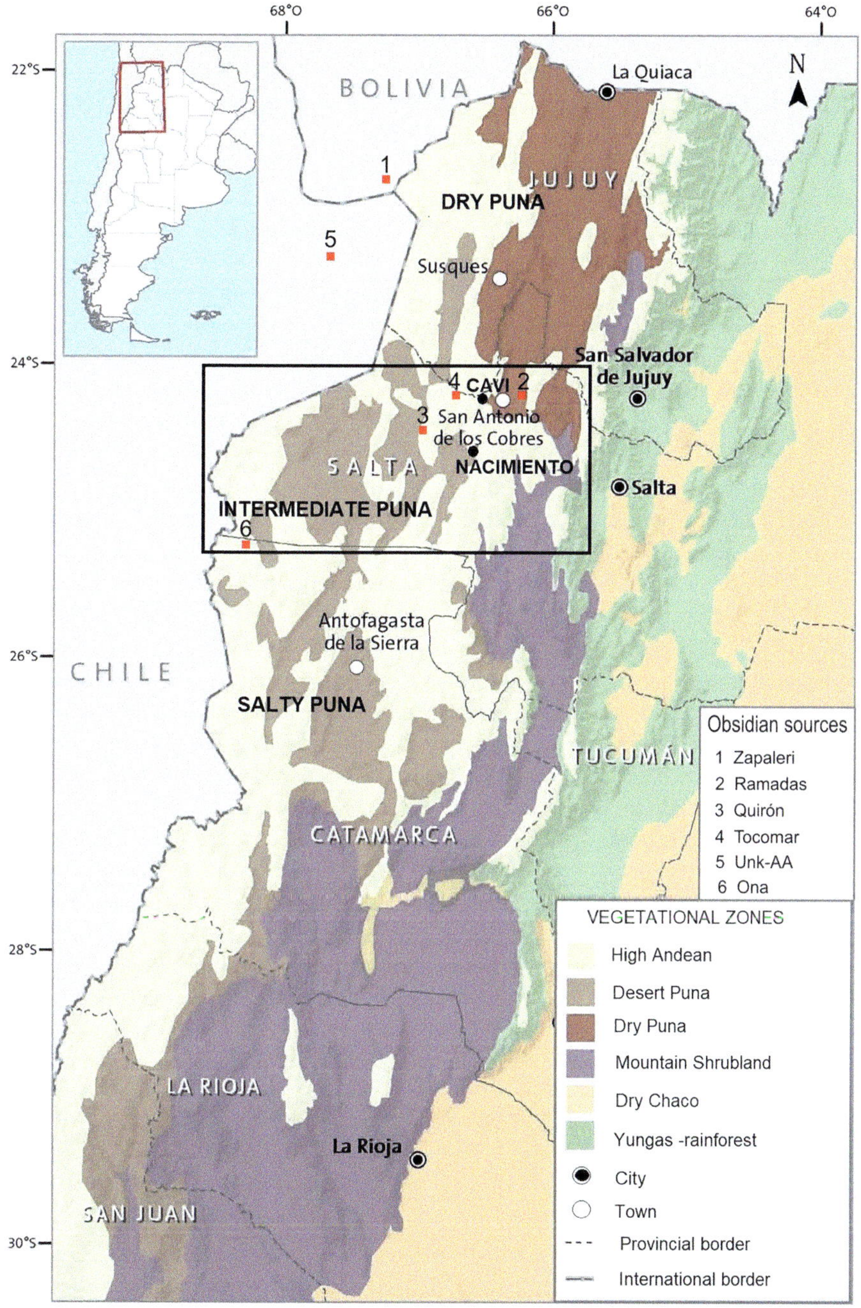

Fig. 6.1 Map of the Puna of Argentina, main divisions, plant zoning, and location of the Nacimiento and Cavi gorges, indicating the obsidian sources mentioned in the text. Modified from Carilla et al. (2018)

1995; Stiner et al. 2008). This process of adaptation to human modified environments qualifies as niche construction (Laland and O'Brien 2010). Other expected processes include the evolution of phenotypic adaptations, as specialized hunting strategies for a specific type of prey, particular hunting technologies including weapon systems, and hunting architecture (Lemke 2021).

Additionally, hunter-gatherer patterns of mobility are closely related to the structure of resources in their environment. As a group heavily dependent on hunting exploits a less dense food source, the area of land exploited by hunter-gatherers increases as dependence on hunting increases (Kelly 1983, 1995). A hunting strategy targeting high-quality prey, typically artiodactyls (Broughton et al. 2011), may lead to the expansion of the human distributional range to new spaces by occupying prey-specific habitats. Hence, hunting is a powerful incentive for human expansion into unoccupied spaces (see Gamble 2009). Furthermore, among specialized hunters, other dynamics are expected that may involve the human abandonment of habitats. For example, in ecologically highly heterogeneous spaces like deserts, with a localized distribution of prey available in specific patches, human mobility will be structured among prey-specific habitats. In these cases, decreasing prey populations can lead to the restructuring of mobility with a less use or abandonment of impoverished habitats. The latter is applicable to the dynamics of predatory human use of sectors of the Puna of Argentina. This is an environment where ecological productivity is low and correlates with altitude, herbivore diversity is scarce and the species with the highest energy return for human hunting are wild camelids, guanacos *(Lama guanicoe)* and vicuñas *(Vicugna vicugna)*. Both species exhibit adaptations for browsing xerophilic vegetation and are distributed heterogeneously in the high Andean landscape (Vilá 2000). While the vicuña is an "obligate drinker" (Koford 1957), and therefore its distribution is more sensitive to the presence of water, the guanaco is an occasional drinker and possesses special physiological mechanisms to cope with water scarcity (Fowler 2010). Particularly for the vicuña, changes in habitat use resulting from changes in aridity have been documented in the Puna of Salta Province, Argentina (Mosca Torres 2010).

In ecologically poor and fragile environments, such as the higher altitude gorges with lower availability of water and pasture in the Puna, the human use of these spaces will be conditioned by the presence of prey. Therefore, any cause with the potential for eradicating prey populations from those habitats, such as climate change, ecological impoverishment, hunting pressure, or the concurrent action of both, would also extirpate (sensu Smith-Patten et al. 2015) human populations that depend on them (Ripple et al. 2013). On this basis over a long timescale, it is expected that in the Puna region the archaeological record will document predator-prey dynamics of habitat use and abandonment of certain habitats, particularly the ecologically most fragile ones such as some very arid and high-altitude gorges.

Furthermore, in the Puna of Argentina, the existence of artificial hiding places made with rock walls for hunting vicuñas on mountain slopes has been documented (Moreno 2009, 2013). Particularly, the construction of hides and traps are important sources of habitat modification with the potential for long-term transgenerational inheritance of modified landscapes. In this work, our hypothesis is that the long-term human occupation of Quebrada Nacimiento and Quebrada Cavi resulted in the construction of archaeological landscapes characterized by heterogeneous distribution patterns. The repeated use of localized spaces for hunting and animal processing likely led to the formation of low-resolution temporal palimpsests, shaped by the accumulation of averaged archaeological discards and the establishment of architectural facilities for hunting. Furthermore, we hypothesize that the zooarchaeological record of CN2 will document the predominant use of camelids, driven by a highly efficient hunting strategy focused on high-ranking prey.

6.1.2 Paleoenvironmental Framework and Archaeological Background in the Puna of Argentina

Human dispersal to the Puna of Argentina began with the environmental conditions of the Coipasa event (12,615–10,556 cal yr. BP), characterized by improved climatic conditions and a lowering of the snow line in a warmer environment compared to the preceding period (Yacobaccio and Morales 2005).

In the Puna of Argentina the early Holocene, around 11,700–8200 cal yr. BP (sensu Walker et al. 2012), was a period characterized by a cooler and more humid climate compared to the current environment (Yacobaccio and Morales 2005). The higher relative humidity would have been associated with a landscape of greater ecological homogeneity. Archaeological evidence documents that the first human who dispersed in the area made an extensive use of the space in a context of low demography, absence of density-dependent pressures, and with a highly mobility strategy mainly oriented towards the exploitation of wild camelids (Aschero 1994; López and Restifo 2012; Muscio 2009; Yacobaccio 2001, 2012). During this period, there is consensus that long-range hunting predominated with mostly triangular projectile points, and a handheld dart-throwing weapon system (Núñez 1992; Aschero and Martínez 2001; De Souza 2004; Martínez 2003;Ratto 2003).

During the middle Holocene, around 8200–4200 cal yr. BP (sensu Walker et al. 2012), a change towards a less humid and warmer climate with local variations is documented (Yacobaccio and Morales 2005). It is suggested that from this point onward, a warmer, drier, and more stable period started, particularly between 6800 and 4200 cal yr. BP. This timeframe witnessed persistent extreme regional aridity with the most severe drought period occurring around 5700 cal yr. BP (Núñez and Grosjean 1994; Morales et al. 2022). Throughout the area, the drier climate was accompanied by a more heterogeneous resource distribution, in a landscape of greater ecological contrast, segmented into patches. This increased the spatial predictability of camelids for hunting and decreased prey encounter frequency (Muscio 2009). In this context, the residential mobility of human populations decreased and was oriented towards the use of more localized spaces with availability of key resources for subsistence, primarily water and access to wild camelid populations (Yacobaccio and Morales 2005). This led to specialization in camelid hunting, in a process that some authors characterize as intensification (López and Restifo 2012; Yacobaccio and Vilá 2013), and around 3400 cal yr. BP, resulted in the full development of food-producing niches, based on pastoralism and high-altitude agriculture (Escola 1996; Yacobaccio 2001; Yacobaccio and Vilá 2013).

6.1.3 Case Studies: Quebradas Nacimiento and Cavi

The Puna region of Argentina comprises the arid highlands located between 19° and 27° south latitude, ranging from 3000 to 4800 m a.s.l. (Fig. 6.1). This area is defined as a high-altitude desert biome crossed by several mountain ranges oriented from southeast to northwest. It is characterized by high solar radiation due to its altitude, broad daily thermal amplitude, marked seasonality in precipitation, and low atmospheric pressure. Primary productivity is very low, as is animal biomass. Ecological productivity is mainly concentrated in stable hydrological systems and high-altitude gorges and valleys with meltwater streams that develop highly localized wetlands (named as *vegas*) and springs, within endorheic basins, forming salt flats and brackish lagoons (Reboratti 2006; Matteucci 2012).

Although precipitation is variable, there is a decreasing gradient of rainfall that distributes from north to south and east to west. Two zones have been defined: the Dry Puna with annual precipitation of 300 mm forms a band to the east, covering parts of southern Peru, western Bolivia, northern Chile, and northwestern Argentina and the Salty Puna, with annual precipitation of less than 100 mm, covers areas of northeastern Chile and northwestern Argentina (Paoli 2002; Bianchi et al. 2005). Between both zones lies the intermediate Puna, around 24° south latitude comprising the Puna of the Province of Salta. This is an extremely arid environment, with very low precipitation ranging between 115 and 85 mm annually, and intermontane valleys at high altitudes, above 3700 m above present sea level (Bianchi and Yánez 1992; Muscio 1999). From an ecological point of view, this sector presents vegetation of Dry puna and Salty Puna (Carilla et al. 2018). Quebrada Nacimiento and Quebrada Cavi are located in the intermediate Puna. Both are high-altitude gorges situated over 4200 m.a.s.l. In these gorges hypoxia and seasonality are extreme, leading to the concentration of ecological productivity in patches where water is present as *vegas*, which are formed by the precipitations in the mountain headwaters.

Quebrada de Cavi is a high-altitude gorge located above 4300 m.a.s.l. (Fig. 6.2a). The gorge features a watercourse that meanders through the mountainous massifs on the western side of the Eastern Cordillera. The *vega* describes an inverted L-shaped trajectory, with a 2 km south-north course before turning eastward for a stretch of 0.9 km (Fig. 6.1). The landscape in this sector of the Puna corresponds to high Andean grassland (Cabrera 1976; Carilla et al. 2018). In the *vega*, there are cushion plants and grasses, characteristic vegetation of wetlands with high humidity and productivity (Izquierdo et al. 2022).

On the other hand, Quebrada Nacimiento (Fig. 6.2b) integrates a series of high-altitude gorges, located about 18 km southwest of San Antonio de los Cobres. The

Fig. 6.2 The landscapes of Quebrada Cavi (**a**) and Quebrada Nacimiento (**b**)

altitude of the gorge averages above 4200–4300 m.a.s.l. Currently, it lacks permanent watercourses, but it does have some narrow streams and vegas with seasonal flow, limited to the wet season, with sources located above 5000 m.a.s.l. The geomorphological structure of Quebrada Nacimiento features a narrow alluvial bottom of Quaternary origin, in a U-shape, sculpted by the seasonal watercourse extending approximately 4 km in length and nestled between mountain massifs running north to south. Phytogeographically, the landscape corresponds to high Andean steppe vegetation (sensu Cabrera 1976) and high-altitude shrubs (Carilla et al. 2018). Within the gorge, there is a primary vega with greater development and a secondary *vega* located at a higher elevation, both with perennial grass cover and soil and paleosol formation. In the intermediate sector of the gorge, there is a series of three rock shelters: Nacimiento 1, 2, and 3.

6.2 Methodology

In this work, we discuss both surface and subsurface archaeological evidence. The latter includes the archaeofaunal record and radiocarbon evidence from the site Cueva Nacimiento 2 (CN2). The anatomical and taxonomic identification of the specimens was performed using reference material and comparative samples from the Instituto de Arqueología (Universidad de Buenos Aires). Additionally, bone quantification was carried out using the following taxonomic abundance measures: total number of specimens (NSP), number of specimens identified per taxon (NISP), and number of unidentified specimens (NUSP) (Grayson 1984; Binford 1984; Lyman 2008). The zooarchaeological description of each assemblage was done using taxonomic categories at the order, family, and species level when possible.

In order to obtain information about the archaeological landscape of Quebrada Nacimiento and Quebrada Cavi, we conducted surface surveys in both gorges. The methodology involved surveying the margins of the vegas and the mountain slopes through pedestrian transects where each operator covered an area of 2 m on each of their sides. In Cavi, four transects were surveyed, oriented along a south-to-north axis to align with the natural layout of the meadow. The plain in this area spans approximately 1.1 km^2, of which 0.038 km^2 was systematically surveyed. In Nacimiento, four transects were established, oriented along a northwest-to-southeast axis to cover the main plain and its two forks. The plain's total area is approximately 1.93 km^2, with the surveyed area encompassing 0.075 km^2. During the surveys, isolated finds were documented, GPS points were recorded for structures and artifact concentrations, and only temporally diagnostic artifacts were collected. To estimate the density of artifact concentrations, counts were conducted within 1 x 1 m grids at specific locations, following the methodology proposed in previous studies (Muscio 2001, 2009; Muscio and López 2007).

To document patterns on a large spatial scale, we geolocate surface sites, including into this category both concentrations and architectural sites. For this analysis, we used Google Earth Engine 2024 (Luo et al. 2018) and complementary global positioning systems (GPS) points. Based on locational and altitudinal data, we calculated the distribution area of the surface archaeological record by determining the area of the polygon that encloses all the sites within each gorge. We then calculated the density of sites within each gorge based on the area of the polygons. To analyze the isolation-concentration patterns of sites within each gorge, we measured the site-to-site distance, or Inter-Site Distance (ISD), the geodesic distance of each site to every other site within the polygon (Ragheb and Ragab 2015). On this basis, we calculated the Site Net Distancing (SND) as the averaged distance of a single site with all the other sites in the polygon. Besides, we analyzed the effects of altitude on the distribution of sites within each gorge.

Time-sensitive artifacts, whether isolated or associated in concentrations, were collected during the surveys. Time-sensitive artifacts serve as chronometers to segment time and to assign hypothetical temporality to surface palimpsests. In the Southern Central Andes, and particularly in the Puna of Argentina, there are projectile point classes that function as chronometers, with known appearance, disappearance, and duration dates, indicating cultural transmission (Cardillo and Muscio 2021).

6.3 Results

6.3.1 Occupational History and Faunal Patterns of Cueva Nacimiento 2

The information presented here comes from five excavation units located near the walls of CN2. These units are 1 m^2 grids, excavated down to the bedrock. The excavation of CN2 was conducted through a stratigraphic survey of layers from which samples were obtained for radiocarbon dating. The stratigraphic units were determined based on their sedimentary characteristics (Muscio et al. 2020).

The earliest signal of human presence in CN2 is documented in the early Holocene to the middle Holocene transition, within the range 9451–8651 cal y BP. The calibration of the C^{14} dates was made using Oxcal (Bronk Ramsey 2009) with the Shcal20 southern hemisphere curve (Hogg et al. 2020). The distribution of calibrated ages for CN2, two-time blocks, located separately within the stratigraphy (Fig. 6.3), where two temporal blocks with the absence of calibrated dates are evident. These temporal blocks span approximately 1500 years each.

Additionally, the absence of radiocarbon signal correlates with specific stratigraphic layers, namely S1 and S. Both stratigraphic units are sedimentary accumulations of aeolian sediments that are associated with arid climates that favored the deposition into the rock shelter of light sediments transported by wind and which are characteristics in the study region (Ledesma et al. 2019). Consequently, the lack of radiocarbon signal and the genesis of these layers documents periods of non-human occupation of the rock shelter. The initial period of radiocarbon loss, spanning approximately from 7600 to 6000 cal BP, coincides with the onset of the aridification process during the mid-Holocene in the Puna region. Subsequently, the second period without human presence, spanning approximately from 5500 to 4000 cal yrs. BP, coincides with the peak of regional hyperaridity around 5700 cal yrs. BP (Tchilinguirian and Morales 2013) and persisted until the establishment of wetter conditions in the Late Holocene in the Puna.

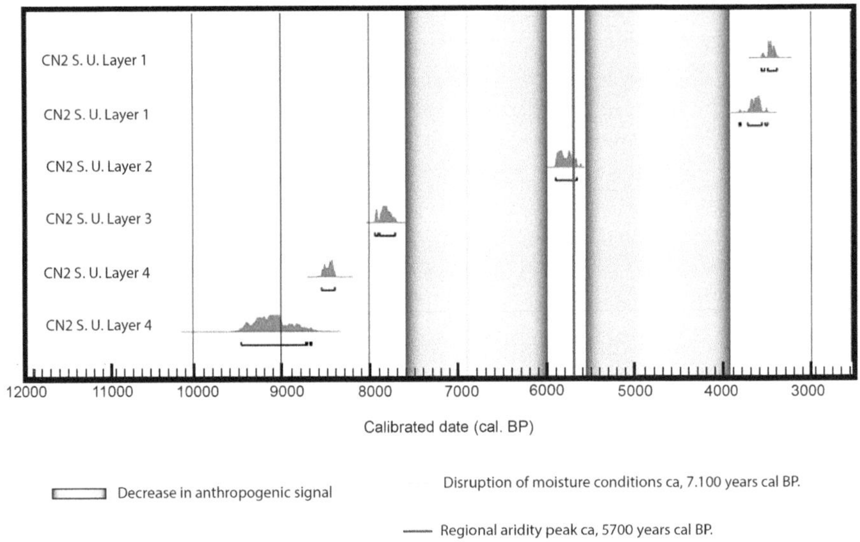

Fig. 6.3 Stratigraphic units and calibrated age distribution of site CN2. Note the spans where radiocarbon signals are absent

Table 6.1 Weathering stages of the zooarchaeological record from Cueva Nacimiento 2

Layer	Stage 0 (%)	Stage 1 (%)	Stage 2 (%)	Stage 3 (%)
1	31.82	62.03	4.64	1.52
2	65.94	13.76	15.43	4.89
3	81.56	14.08	3.17	1.21
4	47.16	45.88	4.91	2.32

6.3.2 *Zooarchaeological Patterns of the Stratigraphic Record*

As we mentioned above, we are interested in discussing the taxonomic composition of the zooarchaeological record from the CN2 site. To do this, we first evaluate the preservation of the assemblage. In Table 6.1, the varying degrees of weathering of the zooarchaeological specimens can be observed. According to these results, the faunal remains of CN2 exhibits low percentages of weathering across all layers. Specifically, the early stages of weathering prevail in all layers, indicating that the zooarchaeological record has not been significantly altered by weathering-like processes. Figure 6.4 shows a comparison of taxa at the family level, where camelids predominate over chinchillidae throughout all layers of CN2. There is a slight peak of chinchillas in Layer 3. However, the representation of camelids progressively increases from Layer 2, corresponding to the interval 5659–5899 cal yrs BP. In other words, the faunal record shows the preferential consumption of high-return fauna, mostly camelids, throughout the entire archaeological sequence. As we observed, the camelid family, based on body size, constitutes the highest-ranking

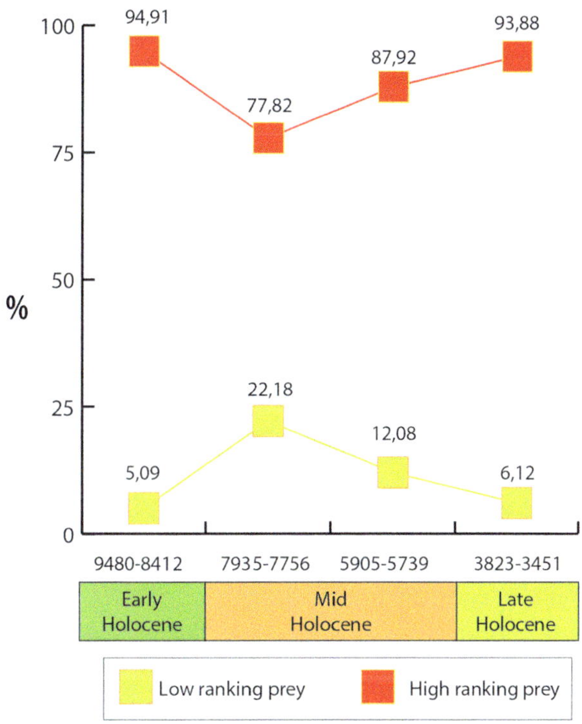

Fig. 6.4 Number of specimens identified by taxon in percentages from the site Cueva Nacimiento 2, Puna de Salta

faunal resource. In contrast, the Chinchillidae family, specifically *Lagidium viscacia*, represents animals with a low energy return. According to the diet breadth model (Broughton et al. 2011), this pattern of low use of small prey is expected by predators that optimize hunting. Also, this pattern aligns with findings documented in the archaeofauna record of the Puna de Salta from comparable ages, especially with the sequences of Alero Cuevas (López 2021) and Abrigo Pozo Cavado (Orsi 2018).

6.3.3 *Scales of the Archaeological Landscapes*

The first major difference is the scale of the spatial distributions (Fig. 6.5). While in Nacimiento the site distribution extends over 12 km, in Cavi it spans around 2 km. The scales of the archaeological distributions in both gorges do not differ significantly. In Nacimiento, it is approximately 3.5 km, while in Cavi, it is 2 km. In both cases, this coincides with the extent of the gorge bottoms. The notable difference lies in the site density of both gorges. In Nacimiento, there are 10 sites, with a

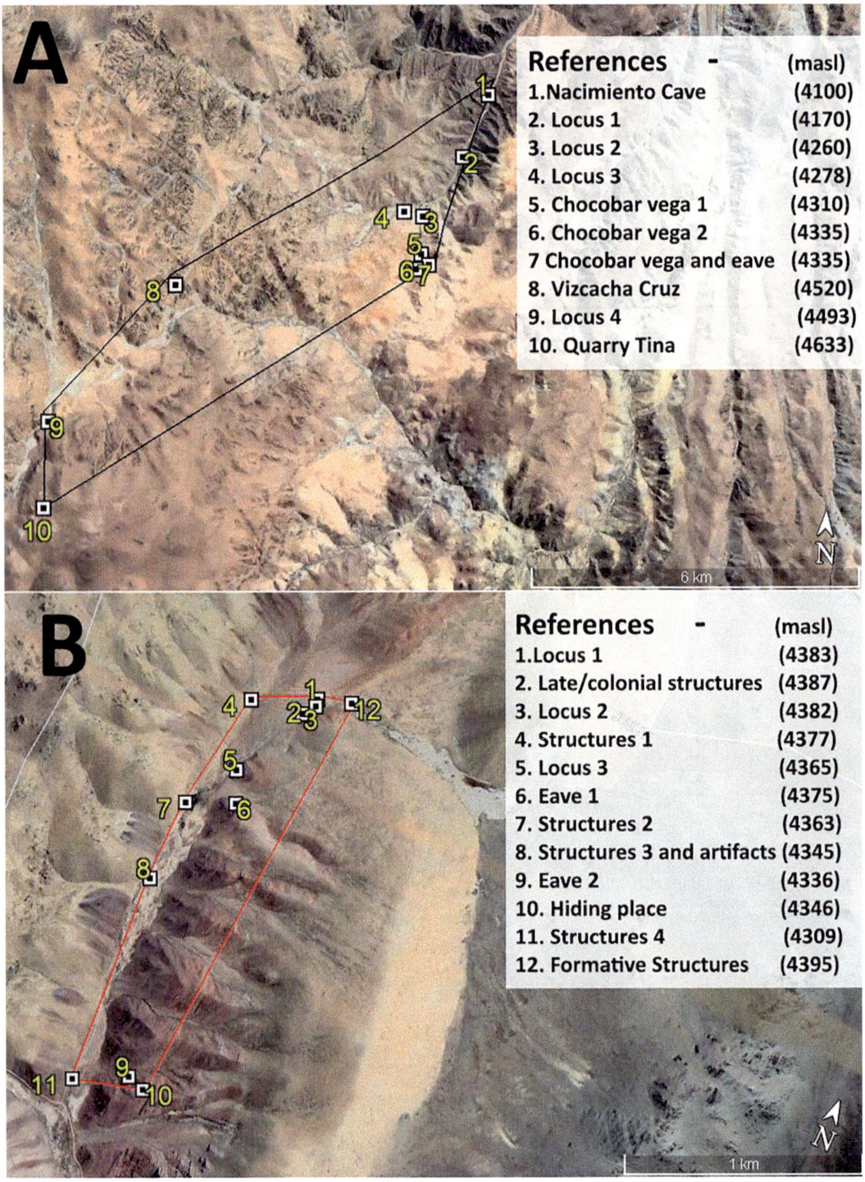

Fig. 6.5 Distribution of sites and polygons in Quebrada Nacimiento (**a**) and Quebrada Cavi (**b**). The column displays the altitudes of the sites

density of 0.51 sites/km^2 for a polygon area of 19.5 km^2. In Cavi, there are 12 sites, with a density of 17.14 sites/km^2 for a polygon area of 0.70 km^2.

In Nacimiento, the distribution of surface sites is markedly irregular, with the highest concentration occurring at the bottom of the gorge and in the secondary *vega*. The estimated archaeological densities do not exceed 2 artifacts/m^2 (Muscio et al. 2021). The sites with architectural structures are living spaces delimited by small walls in caves or rocky shelters and reclaimed archaeological corrals. Three of these sites are located outside the vega, in high-altitude spaces of the mountains that border the gorge (Fig. 6.5a, sites 8, 9, and 10).

At Cavi, the bottom of the gorge, often immediately on the margins of the *vega*, is the space with the highest density and diversity of artifacts. Here the archaeological densities vary between 2 and 5 artifacts/m^2 (Muscio and Carballo 2023). Within these spaces, the architectural structures appear more or less concentrated, with variously shaped floors, predominantly rectangular and some horseshoe-shaped hunting blinds or parapets. Additionally, on the mountain slopes, there are architectural structures and a hunting blind located at the top of the hill (Fig. 6.6a). In the

Fig. 6.6 Examples of sites in Quebrada Cavi. (**a**) Architectural structure with walls made from recycled flake cores, (**b**) Hunting blind located at the top of the hill, (**c**) Concentration of artifacts at *Locus* 1 and cores with laminar extraction

Puna of Argentina, it has been proposed that during the middle Holocene, the placement of hunting blinds on mountain slopes and hilltops was part of an ambush hunting strategy, as vicuñas often use these spaces as resting places (Aschero and Martínez 2001; López et al. 2023; Moreno 2009, 2013; Moreno et al. 2021).

Comparatively, Quebrada Cavi is wetter than Nacimiento. At Cavi, the slope of the ravine bottom is 2%, in Nacimiento it is 5%. This lower slope implies less water runoff and a greater extension and stability of the water body and the *vega*. These conditions must have made the Cavi gorge a more attractive space for wild camelid populations and a more predictable high-altitude hunting habitat for human hunters. Thus, leaving constant taphonomic bias, we propose that the higher frequency and density of sites at Cavi compared to Nacimiento document a greater frequency of human activity and archaeological discard in a more predictable high-altitude hunting space. It is interesting to note the use of flaked cores for the construction of walls in Quebrada Cavi. This provides evidence of reclamation (sensu Schiffer 1987, 1991) and indicates the utilization by pastoralist populations of previously deposited lithic resources.

On the other hand, the artifact concentrations in both gorges provide evidence of lithic knapping, indicated by the presence of flaked cores and debitage (Muscio et al. 2021; Muscio and Carballo 2023). At Quebrada Cavi, there is an abundance of flakes representing different stages of reduction with the characteristic pattern of the so-called quarry sites (Muscio and Carballo 2023). In Quebrada Nacimiento, stratigraphic data also document lithic reduction activities and edge retouching (Muscio et al. 2020). Although the sample sizes are small, some observations can be made regarding the lithic raw materials in the surface record. In Quebrada Nacimiento ($n = 33$), the raw materials and their percentage frequencies are Andesite (36.36%), Metaquartzite (48.49%), Quartz (3.03%), Obsidian (3.03%), and Silica (9.09%) (Muscio et al. 2021, Table 4). In contrast, at Cavi ($n = 61$), the percentages are Basalt (40.74%), Metaquartzite (29.6%), Obsidian (7.41%), Quartzite (13%), Silicified Tuff (7.4%), and Quartz (1.83%) (Muscio and Carballo 2023, Fig. 10). The rocks sourced from within the gorges are considered local (Muscio et al. 2021). Thus, in both cases, with the exception of obsidian, the remaining raw materials are local. In Quebrada Nacimiento, the identified obsidian sources are Tocomar (33.53%), Ona (17.92%), Zapaleri (16.19%), and Quirón (4.05%). Glascock (2022) recently identified obsidian from the Ramadas source, as well as from the called Unk-AA source, in materials recovered from Cueva Nacimiento 2. The Unk-AA source is thought to be located near Tulán in northern Chile. Additionally, the Quirón and Zapaleri obsidians are present at Cavi although their frequencies have not yet been determined. The diversity of these obsidian sources reflects the long-term participation of the societies inhabiting Quebrada Nacimiento and Quebrada Cavi in extensive obsidian circulation networks. These networks spanned a broad geographic area, covering the north-south Puna regions, the intermediate zone, and northern Chile (Fig. 6.1). There is consensus that obsidian was a resource whose procurement led to the emergence of extensive, circulation networks across the South Andean Highlands, including the Puna of Salta, starting from the early Holocene (Yacobaccio et al. 2002, 2004; Pintar et al. 2016; López 2020).

6.3.4 *The Temporality of the Archaeological Landscapes*

Time-sensitive artifacts serve as chronometers to segment time and assign hypothetical temporality to surface palimpsests. Figure 6.7 displays the temporal seriation of Cavi and Nacimiento time-sensitive projectile points. In Cavi, the specimens come from the surface record while those from Nacimiento come from the CN2 excavation (Muscio et al. 2020) and from the surface record (Muscio and Carballo 2023).

The chronology of these artifacts is known for contexts in the Puna of Salta and Antofagasta de la Sierra, Catamarca Province, Argentina. The chronological blocks cover calibrated years and are constructed based on bibliographic information. We calibrated the dates reported in the literature using Oxcal (Bronk Ramsey 2009) with the Shcal20 southern hemisphere curve (Hogg et al. 2020).

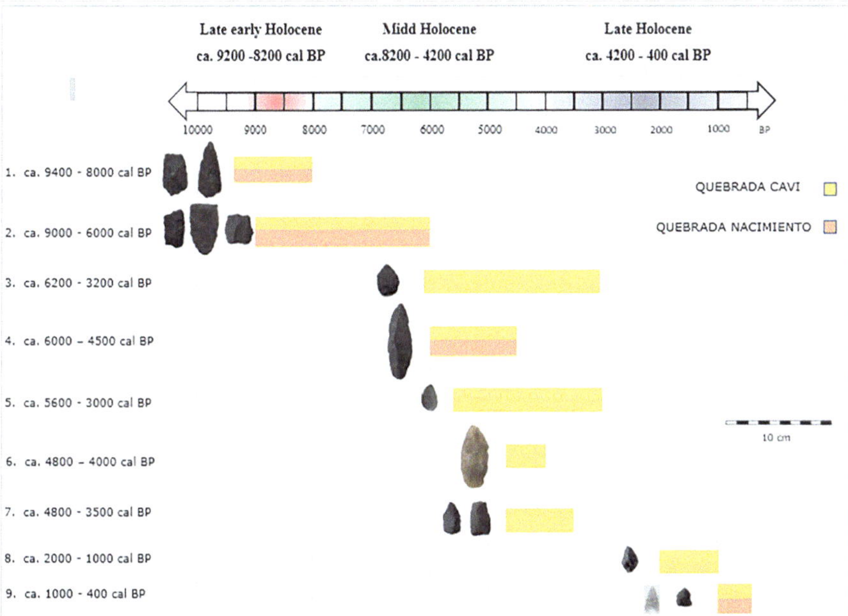

Fig. 6.7 Occurrence seriation of projectile points and temporally diagnostic artifacts from Quebrada Nacimiento and Quebrada Cavi. Temporal segments are expressed in calibrated years BP. From top to bottom: (1) San Martin PP, (2) Lanceolate PP, (3) Triangular limb PP, with shoulders, prominent peduncle, and straight base, (4) Saladillo artifact, (5) Small Limb Lanceolate PP, (6) Bipoint or Mandorla artifact, (7) Triangular limb PP with, straight edges, and convex or straight basecon (PCh F type), (8) Triangular limb PP with peduncle, obsidian, (9) Triangular limb projectile points with notched base and straight base (obsidian). The number of analyzed projectile points is 56. Of these, 12 come from the surface record in Quebrada Cavi, 36 specimens were recovered from excavations at sites CN1 and CN2 in Quebrada Nacimiento, and 8 specimens come from the surface record of Quebrada Nacimiento

For the end of the Early Holocene and the beginning of the middle Holocene, characteristic artifacts include San Martin projectile points (López and Restifo 2017; Muscio and Cardillo 2019) and lanceolate projectile points with lengths greater than 5 cm (López and Restifo 2017; Muscio and Cardillo 2019: Cardillo and Muscio 2021). For the middle Holocene, characteristic artifacts in the Puna de Salta and Jujuy include Saladillo artifacts, with the earliest date in the puna of Salta (Muscio et al. 2020), and other laminar artifacts (Restifo 2015; Restifo and Patané Araoz 2017). For the middle and late middle Holocene, smaller lanceolate projectile points and the Peñas Chica C.a y B.a projectile points are characteristic (Aschero and Hocsman 2011; Hocsman 2006) and the biconvex points called mandorla (Aguerre et al. 1975; Aschero and Yacobaccio 1999). Moving towards the Late Holocene, we find triangular leaf-shaped projectile points around 2500 years BP and later triangular projectile points with notched bases. On the other hand, in Cavi there are middle Holocene bifaces with a perimeter edge (Fig. 6.6c) that frequently occur in other surface contexts of Northwestern Argentina valleys (Cigliano et al. 1962; Carbonelli 2011).

Figure 6.7 illustrates the contrasting temporal representation of the archaeological record in Quebrada Nacimiento and Quebrada Cavi, suggesting diverse occupation histories. Specifically, in Quebrada Nacimiento, there is a notable absence of evidence between the end of the middle Holocene and the beginning of the late Holocene, consistent with radiocarbon data indicating a discontinuous occupation of this gorge. In contrast, Quebrada Cavi exhibits greater diversity of artifact classes and a higher representation of Holocene temporal segments. Furthermore, both gorges show a later human use, as evidenced by rectangular architectural structures distributed along the margins of the *vegas*, alongside corrals. Additionally, the presence of pedunculated triangular projectile points dated to the late Holocene and ceramics assignable to historical periods further supports this observation.

From this we conclude that in Quebrada Cavi and Quebrada Nacimiento the formation of the surface archaeological distributions began at least during the early to middle Holocene transition. This involved human populations using high-altitude *vegas* as a habitat for obtaining and processing hunting resources and lithic raw materials. The projectile points found in the surface record of both gorges document the use of these habitats as high-altitude hunting patches throughout the middle Holocene and the beginning of the late Holocene. Besides, in both gorges there is also evidence of resource processing, such as the retouched edges on flakes, some of which are laminar. Additionally, it has been suggested that Saladillo artifacts might have been used in resource processing (Fernández Distel 1978; Restifo and Patané Araoz 2017).

At Cavi, the location of hunting blinds in tophills of the gorge suggests ambush hunting strategies. In this gorge, the architectural distributions across the *vega* bottom document the development of a habitat construction process that modified the landscape through the accumulation of persistent architecture and lithic artifacts. This process of constructing a hunting landscape began minimally at the end of the early Holocene.

6.3.5 Site Spacing Patterns and Altitudinal Distribution

In this section, we examine the spatial distribution of sites within the gorges to identify patterns of aggregation. Our focus is on comparing site spacing and clustering across broad spatial scales, independent of their specific functions or chronology. As mentioned, the metrics we used are the site-to-site distance, or Inter-Site Distance (ISD), and the Site Net Distancing (SND), which is the average ISD of a single site. A site with a low SND is closer to all other sites in the polygon, whereas as SND increases, site isolation increases.

When comparing inter-site distances in both gorges, the results show that Nacimiento has the greatest ISD, and the larger spatial scale of the distribution of sites in the gorge (see Fig. 6.5). Meanwhile, in Nacimiento, the mean ISD is 4.39 km, whereas in Cavi, it is 0.87 km, with similar variation (Table 6.2).

It is important to note that the spacing between sites and the scale of the archaeological distribution of Nacimiento is strongly affected by the three sites located at extreme altitudes out of the *vega* (Table 6.3). These are sites 8, 9, and 10, located at 4535, 4496, and 4616 m a.s.l., respectively. When these sites are removed from the sample, the greatest distance is 3.46 km, between sites 1 and 7 and the mean ISD is 1.8 km. At this scale, the sites are located along the edges of the *vegas* and at an average altitude of 4255 m a.s.l. Nevertheless, the spacing of sites in Nacimiento remains much lower than at Cavi. Hence the more spaced distribution of the archaeological record of Nacimiento reflects the low archaeological density and the localization of sites near the water bodies. Conversely, the shorter distances between sites at Cavi reflect a greater spatial aggregation of the surface archaeological record, particularly in the final sector of the gorge.

Considering altitude in the distribution of sites, a common pattern emerges for both gorges. Altitude and SND consistently covary, with an R^2 coefficient of 0.5 in Nacimiento and 0.8 at Cavi. In Nacimiento, the R^2 coefficient excludes the high-altitude sites 8, 9, and 10 as outliers in the regression analysis. In short, both gorges exhibit a non-homogeneous distribution of sites, where, as altitude increases, the aggregation of sites also increases. When regressing altitude against SND in both gorges, that is, treating the two gorges as a single sample and excluding the outliers, the correlation is strong and negative ($r = -0.8766$; $F = 56.41, p < 0.05$). Recall that SND measures the net spatial proximity of each individual site to the rest of the sites

Table 6.2 Summary statistics of ISD for Quebrada Nacimiento and Quebrada Cavi

	Q. Nacimiento	Q. Cavi
N	45.00	66.00
Min	0.24	0.03
Max	11.10	1.94
Mean	4.39	0.88
Std. error	0.49	0.07
Variance	10.77	0.36
Coeff. var	74.68	68.68

Table 6.3 Altitude and SND for each site of Quebrada Nacimiento and Quebrada Cavi

Q. Nacimiento			Q. Cavi		
Site	Altitude m a.s.l.	SND (km)	Site	Altitude m.a.s.l.	SND (km)
1	4110	4.93	1	4386	0.78
2	4166	3.92	2	4387	0.76
3	4248	3.13	3	4382	0.73
4	4292	3.12	4	4376	0.76
5	4316	3.06	5	4365	0.65
6	4326	3.16	6	4375	0.66
7	4325	3.17	7	4363	0.70
8	4535	4.85	8	4343	0.81
9	4496	6.91	9	4337	1.22
10	4616	7.68	10	4346	1.24
			11	4309	1.33
			12	4395	0.88

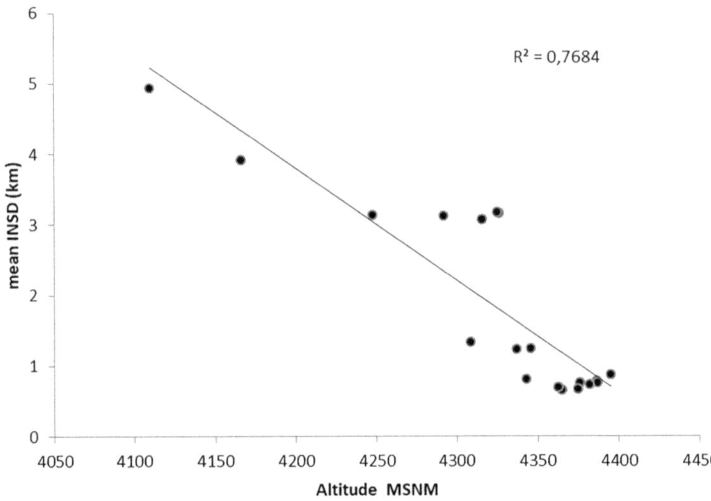

Fig. 6.8 Altitude and SND of Quebrada Nacimiento and Quebrada Cavi (adjusted R^2 0.7548)

in each stream. So, this correlation shows that as altitude increases, the proximity (concentration) between sites also increases in both gorges (Fig. 6.8). Taking into account the R^2 value, altitude explains 76% of the variation in the distance between sites.

In both gorges due to the slopes, the greatest width of the water bodies and the *vegas* are in the highest spaces. In both, the main water bodies are located at the inner ends of the gorges, at the highest altitudes of the geoforms. This makes an altitudinal gradient in the intensities of space use within the gorges expected and an increasing frequency of sites from the entrances to the interior of the gorges. So, the greater concentration of sites inside the gorges may document a more intense use of these highest sectors into the gorges as hunting habitats and in pastoral economies, as logistical high-altitude locations in the livestock mobility circuits. But this behavioral anthropogenic hypothesis may mask taphonomic bias, especially when dealing with sites on slopes (Borrazzo 2009; Borrazzo and Borrero 2015).

Gravity-driven processes might produce individual or mass movement of surface archaeological materials through downward displacement (Ozán 2017). These processes depend greatly on local contexts, including the microtopography of the landforms. Therefore, it is difficult to identify a taphonomic agent that operated in the same way in both gorges to produce an inverse covariation pattern between altitude and the aggregation of sites, including sites with architectural structures. We propose the possibility that mass transport gravity-driven processes acted with greater intensity in the middle sections and at the entrances of the gorges, thereby more effectively destroying and burying the archaeological record in these sectors. Especially flash floods events, which are the sudden increase in the flow of a mountain torrent (Carmichael et al. 2023). On a broad spatial scale, the intensity of these processes is a consequence of the slopes of the geoforms. So, a greater slope should be associated with a distribution of the archaeological record more correlated with altitude.

However, contrary to expectations, Quebrada Cavi, despite having less than half the slope of Nacimiento, exhibits a stronger correlation between altitude and the spacing of sites by regression analysis. Consequently, based on the available data, this pattern of correlation supports the behavioral hypothesis of more intensive use of the higher and inner sectors of the gorges, rather than the taphonomic hypothesis.

6.3.6 Information Summary

1. The occupation history of CN2 reveals a predominant pattern of camelid use from the beginning of the human occupation, documenting that the rock shelter was used by specialized hunters that optimized their niche breadth through focusing on camelids as the primary prey. The specialized strategy is observed throughout the entire occupational sequence of the cave.
2. A predator-prey relationship may explain the occupational dynamics of Quebrada Nacimiento, with a phase of human use beginning at least in the Early-Middle Holocene transition and two periods of radiocarbon signal loss coinciding with the onset of the middle Holocene aridification and with the regional aridization peak. These more arid environmental conditions must have resulted in a deterioration or retraction of the pastures and the water body in Quebrada Nacimiento, leading to a decrease in the frequency of camelid encounters.
3. The surface archaeological record of Quebrada Cavi and Quebrada Nacimiento documents a long-term human occupation within the framework of a hunting strategy, where high-altitude gorges were primarily used as hunting patches localized above 4200 m a.s.l. In both gorges, the archaeological record documents lithic reduction activities and resource processing during the middle

Holocene. The obsidians from the northern Puna, southern Puna, intermediate Puna, and northern Chile in both gorges document the participation of the human groups that inhabited these spaces in large-scale social interaction networks over time.

4. At Quebrada Cavi, the frequency of sites and the density of artifacts are greater than in Nacimiento, as well as the diversity of temporally diagnostic artifacts, suggesting human presence across more temporal blocks. This supports the hypothesis of an unequal occupation of high-altitude gorges, highly conditioned by the greater or lesser local presence of pastures and water bodies.
5. In Quebrada Nacimiento and Quebrada Cavi, the archaeological record documents a selective habitat use with a preference for high-altitude spaces located at the inner ends of the gorges where the largest expanses of vega grasslands are found. These spaces should have concentrated populations of wild camelids, particularly vicuñas, due to their dependence on water, making these areas predictable for camelid hunting. Additionally, in both gorges these areas show the highest aggregation of sites and the least spacing between them. At Cavi, the construction of hunting blinds in the high hillside areas suggests ambush hunting strategies.
6. In both gorges, the human use of these spaces led to the formation of a high-altitude hunting landscape upon which overlapped the archaeological record of pastoral economies in the late Holocene. At Quebrada Cavi, this resulted in the formation of persistent places associated with lithic reduction activities and resource processing during the middle Holocene.
7. The evidence from Quebrada Nacimiento and Quebrada Cavi agree with macro-regional information documenting processes of specialization in camelid use, suggesting that the high-altitude gorges of the Puna of Salta were suitable habitats for human use but might have been abandoned during some periods at the onset of the arid conditions in the middle Holocene (Orsi 2018). Nevertheless, an approach based on regional taphonomy studies (Borrero 2001) is necessary for a more refined explanation of the archaeological variability of the Puna of Argentina.

6.4 Discussion

Predator-prey dynamics have been invoked in archaeology to study the behavioral evolution and demographics of hunting human populations and their prey in different areas of the world (Harpending and Bertram 1975; Winterhalder and Goland 1993; Stiner et al. 2008; Marom and Wolkowski 2024). Furthermore, it has been proposed that predator-prey dynamics played a decisive role in the evolution of Plio-Pleistocene hominids (Hoare 2019).

In this work, we have shown that predator-prey dynamics offer a useful theoretical framework for explaining the spatial distribution of archaeological sites and the faunal record of the Puna de Salta from an evolutionary perspective. An early evolved predator-prey relationship between humans and camelids appears to be at the basis of many behavioral and cultural evolutionary processes that occurred from the end of the Pleistocene and throughout the Holocene in the South Andean Highlands. Human dispersal in the area occurred during the Pleistocene-Holocene transition and coincided with the onset of the Coipasa event (12,615–10,556 cal year BP), which involved an improvement in climatic conditions and the lowering of the snow line in a warmer environment than the preceding period (Yacobaccio and Morales 2011). These new conditions should have brought about greater accessibility to high-altitude populations of wild camelids. Human dispersal to the Puna can then be modeled by the expansion of hunting ranges to include prey-specific habitats in the highlands. As we have shown in this work, Quebrada Cavi and Quebrada Nacimiento are examples of high-altitude spaces used as hunting patches. Another case in the Puna de Salta is Alero Cuevas, a site that also exhibits greater temporal depth (López 2021).

Furthermore, a pattern of cladistic diversification of lanceolate projectile points from the southern Andean highlands has been proposed as a consequence of a growing specialization of hunting (Muscio and Cardillo 2019; Cardillo and Muscio 2021). Thus, this process of technological evolution can also be framed in the predator-prey dynamics. In this regard, it is interesting to highlight the high diversity of lithic projectile points in spaces such as Quebrada Cavi. Hunting in these high-altitude spaces with natural and artificial enclosures and more open spaces, must have involved different hunting contexts, promoting the evolution of different technological solutions. Another case of construction of a hunting landscape with hunting architecture in the Puna of Salta is found in the Salar de Pocitos, at an altitude of 3800 m a.s.l. at the Cazadero site (López et al. 2023). This site shows the location of a hunting patch in flatter sectors of lower altitude. Therefore, the variability of hunting contexts that occur in the high-altitude gorges and the lower plains but with water courses in the Puna may be the key to explaining the evolution of lithic point diversity (Aschero and Martínez 2001).

On the other hand, predator-prey relationship can lead to predator niche construction processes. One way is by excluding competitors and monopolizing prey populations. The puma (*Felis concolor*) is the only carnivore that systematically hunts large ungulates in the Puna, making it a potential competitor with human hunters (Mondini 2004). However, a strong avoidance of habitats impacted by humans was observed in pumas (Piccot et al. 2019), along with changes in predatory behavior (Guerisoli et al. 2019). This reduces predation pressure on prey, as was observed in deer (Piccot et al. 2019). Hence, in the Argentine Puna, the anthropization of habitats could have meant some benefit for the camelids by reducing hunting pressure from the pumas, leaving humans as their main predators. In fact, the earliest zooarchaeological record from the Puna of Argentina contains no evidence of competition with pumas (Mondini 2004, 2008). This could indicate an early exclusion of competitors by human hunters focused on camelids and manipulating prey habitats.

Additionally, the cases studied here illustrate past processes involved in the construction of hereditary habitats. These processes form the foundation of ecological inheritance and niche construction (Odling-Smee et al. 2003). The archaeological record of both gorges reflects the construction of inherited environments over millennia, emerging from the dynamic relationship between humans and camelids in a predator-prey context. Initially, hunters selected these spaces as hunting habitats, and later, shepherds reused them as grazing areas. The architectural distributions in Quebrada Nacimiento and Quebrada Cavi—some of which can be attributed to hunting blinds—document the development of habitat construction practices that transformed the local landscapes through the accumulation of persistent architecture and lithic artifacts, likely beginning at the end of the Early Holocene.

A landscape modified in this way facilitates reoccupation across generations, as the presence of pre-constructed facilities, whether for hunting or herding, makes it more attractive for human use. Additionally, the discarded materials and lithic accumulations produced by human activity in the habitat left artifacts available for reuse as raw material. At Quebrada Cavi, reclaimed basalt cores were used for constructing walls. At Quebrada Nacimiento, entire rock shelters were reused as corrals during the late Holocene, with walls built to enclose animals. Therefore, the human construction of habitats in Quebrada Cavi and Quebrada Nacimiento involved endowing these gorges with cultural elements, in addition to natural ones, making them suitable spaces for human use and reuse over time. The intentional creation of landscapes by these human groups contributed to an adaptive management of the environment. In this way, rather than being mere passive users of their environment, the hunter-gatherers who occupied these high-altitude spaces in the Puna were active landscape builders.

More generally, the framework of predator-prey dynamics can also shed light on the development of a specialized human niche based on camelid exploitation, leading to a domestication process in what Zeder (2012) referred to as the path from prey. As a strong human-camelid predator-prey relationship is defensible since the beginning of human dispersion in the South Andean Highlands, we suggest that a predator-prey interaction was the critical selective framework explaining many technological adaptations, behavioral processes, demographic changes, and the construction of human habitats.

Acknowledgments This work was funded by the National Scientific and Technical Research Council of Argentina (CONICET). We would like to thank Sebastián Frete for his collaboration in the fieldwork.

Competing Interest Declaration The authors have no conflict of interest to declare that are relevant to the content of this chapter.

References

Abrams PA (2001) Predator-prey interactions. In: Fox CW, Roff DA, Fairbairn DJ (eds) Evolutionary ecology: concepts and case studies. Academic, Oxford, p 2020. https://doi.org/10.1093/oso/9780195131543.003.0028

Aguerre AM, Fernández Distel AA, Aschero CA (1975) Comentarios sobre nuevas fechas en la cronología arqueológica precerámica de la Provincia de Jujuy. Relac Soc Argentina Antropol IX:211–214

Aschero CA (1994) Reflexiones desde el Arcaico Tardío 6.000 - 3.000 A.P. Rumitacana 1:13–17

Aschero CA, Hocsman S (2011) Arqueología de las ocupaciones cazadoras-recolectoras de fines del Holoceno Medio de Antofagasta de la Sierra (Puna meridional Argentina). Rev Antropol Chilena 43:393–411

Aschero CA, Martínez JG (2001) Técnicas de caza en Antofagasta de la Sierra, Puna Meridional Argentina. Relac Soc Argentina Antropol 26:215–241

Aschero CA, Yacobaccio H (1999) 20 años después: Inca Cueva 7 reinterpretado. Cuad Ins Nac Antropol Pensamiento Latinoamericano 18:7–18

Bianchi A, Yánez C (1992) Las Precipitaciones en el Noroeste Argentino. INTA (Instituto Nacional de Tecnología Agropecuaria), Secretaría de Agricultura, Ganadería y Pesca de la Nación. Estación Experimental Agropecuaria Salta. Segunda Edición

Bianchi AR, Yáñez CE, Acuña LR (2005) Bases de datos mensuales de las precipitaciones del Noroeste Argentino. Informe del Proyecto Riesgo Agropecuario. INTA-SAGPYA

Binford LR (1984) Faunal remains from Klasies River mouth. Academic Press, Orlando

Borrazzo K (2009) El uso prehistórico de los afloramientos terciarios en la bahía San Sebastián (Tierra del Fuego, Argentina). In: Salemme M, Santiago F, Álvarez M, Piana E, Vázquez M, Mansur ME (eds) Arqueología de Patagonia: Una mirada desde el último confín. Editorial Utopías, Ushuaia, pp 291–305

Borrazzo K, Borrero LA (2015) Taphonomic and archaeological perspectives from northern Tierra del Fuego, Argentina. Quat Int 373:96–103

Borrero LA (2001) Regional taphonomy. The scales of application to the archaeological record. In: Buitenhuis H, Prummel W (eds) Animals and man in the past, vol 41. ARC Publication, Groningen, pp 17–20

Bronk Ramsey C (2009) Bayesian analysis of radiocarbon dates. Radiocarbon 51(1):337–360

Broughton JM, Cannon MD, Bayham FE, Byers DA (2011) Prey body size and ranking in zooarchaeology: theory, empirical evidence, and applications from the northern Great Basin. Am Antiq 76(3):403–428. https://doi.org/10.7183/0002-7316.76.3.403

Burak MK, Monk JD, Schmitz OJ (2018) Eco-evolutionary dynamics: the predator-prey adaptive play and the ecological theater. Yale J Biol Med 91(4):481–489. https://doi.org/10.26031/PMC6302632

Cabrera A (1976) Regiones fitogeográficas argentinas. In: Kugler W (ed) Enciclopedia Argentina de agricultura y jardinería. Acme, Buenos Aires, pp 1–85

Carbonelli JP (2011) El sitio Ampajango, cincuenta años después. Arqueología 19(1):41–64. https://doi.org/10.34096/arqueologia.t19.n1.1693

Cardillo M, Muscio H (2021) Filogenia, innovación y recambio tecnológico en las puntas de proyectil del Holoceno medio de las Tierras Altas Surandinas. Mundo Antes 5(2):13–38. https://doi.org/10.59516/mda.v15.249

Carilla J, Grau A, Cuello AS (2018) Vegetación de la Puna Argentina. In: Grau R, Babot J, Izquierdo A, Grau A (eds) La Puna Argentina. Naturaleza y cultura. Fundación Miguel Lillo, pp 146–156. http://lillo.org.ar/publicaciones/serie-conservacion-de-la-naturaleza

Carmichael B, Daly C, Fatorić S, Macklin M, McIntyre-Tamwoy S, Pittungnapoo W (2023) Global riverine archaeology and cultural heritage: flood-risk management and adaptation for the anthropogenic climate change crisis. Climate 11(10):197. https://doi.org/10.3390/cli11100197

Cigliano E, Bereterbide S, Carnevali B, Lorandi AM, Tarragó M (1962) El Ampajanguense. Publ Inst Antropol 5:7–104

De Souza P (2004) Cazadores-recolectores del arcaico temprano y medio en la cuenca superior del río Loa: sitios, conjuntos líticos y sistemas de asentamiento. Estudios Atacameños 27:7–42

Escola P (1996) Riesgo e incertidumbre en economías agropastoriles: consideraciones teórico-metodológicas. Arqueología 6:9–23

Fernández Distel A (1978) Nuevos hallazgos precerámicos en la región de Salinas Grandes, Puna de Jujuy, Argentina. Rev Inst Antropol VI:15–62

Fowler M (2010) Medicine and surgery of south American camelids: llama, alpaca, vicuña, guanaco, 3rd edn. Wiley, Hoboken

Gamble C (2009) Human display and dispersal: a case study from biotidal Britain in the middle and upper Pleistocene. Evol Anthropol 18(4):144–156. https://doi.org/10.1002/evan.20209

Glascock MD (2022) Provenance for obsidian artifacts from the northwestern Argentina sites of Abrigo Pozo Cavado and Cueva Nacimiento 2 by X-ray fluorescence. ANIDs: APC001-28, CN2029-36

Grayson D (1984) Quantitative zooarchaeology: topics analysis of archaeological faunas. Academic Press, Orlando. https://doi.org/10.1016/b978-0-12-297280-5.50009-4

Guerisoli MM, Caruso N, Luengos Vidal E, Lucherini M (2019) Habitat use and activity patterns of Puma concolor in a human-dominated landscape of Central Argentina. J Mammal 100(1):202–211

Harpending H, Bertram J (1975) Human population dynamics in archaeological time: some simple models. Memoirs Soc Am Archaeol 30:82–91. https://doi.org/10.1017/S0081130000003828

Hoare S (2019) The possible role of predator-prey dynamics as an influence on early hominin use of burned landscapes. Evol Anthropol 28(6):295–302. https://doi.org/10.1002/evan.21807

Hocsman S (2006) Tecnología lítica en la transición de cazadores-recolectores a sociedades agro-pastoriles en la porción meridional de los Andes Centro-Sur. Estud Atacameños Arqueol Antropol Surandinas 32:59–73

Hogg AG, Heaton TJ, Hua Q et al (2020) SHCal20 southern hemisphere calibration, 0–55,000 years cal BP. Radiocarbon 62(4):759–778. https://doi.org/10.1017/RDC.2020.59

Hou Z, Lisena B, Pireddu M, Zanolin F (2013) Lotka-Volterra and related systems: recent developments in population dynamics. De Gruyter, Alemania

Izquierdo AE, Blundo C, Carilla J, Foguet J, Navarro CJ, Casagranda E, Chiappero MF, Vaieretti MV (2022) Floristic types of high-Andean wetlands from Northwest Argentina and their remote-sensed characterization at a regional scale. Appl Veg Sci 25(2):12658. https://doi.org/10.1111/avsc.12658

Kelly RL (1983) Hunter-gatherer mobility strategies. J Anthropol Res 39(3):277–306. https://doi.org/10.1086/jar.39.3.3629672

Kelly RL (1995) The foraging spectrum: diversity in hunter-gatherer lifeways. Smithsonian Institution Press, Washington

Koford CB (1957) The vicuña and the puna. Ecol Monogr 27:153–219

Laland KN, O'Brien MJ (2010) Niche construction theory and archaeology. J Archaeol Method Theory 17:303–322. https://doi.org/10.1007/s10816-010-9096-6

Ledesma J, del Papa C, Payrola P (2019) Revisión de la estratigrafía Eocena-Pliocena de la cuenca intermontana de San Antonio De Los Cobres, Puna Salteña, Argentina. Latin Am J Sedimentol Basin Anal 26(1):57–73

Lemke A (2021) Literal niche construction: built environments of hunter-gatherers and hunting architecture. J Anthropol Archaeol 62:1–18. https://doi.org/10.1016/j.jaa.2021.101276

Lima SL (2002) Putting predators back into behavioral predator–prey interactions. Trends Ecol Evol 17(2):70–75

López G (2020) Macro-regional interaction processes in the south-Central Andes along the Holocene (ca.10000-500 BP): obsidian circulation, cultural transmission and evidence of allochthonous materials at archaeological sites in the Puna of Salta, North-Western Argentina. J Archaeol Sci Rep 29:1–11

López G (2021) El sitio Alero Cuevas, Pastos Grandes, Puna de Salta: Actualización y síntesis arqueológica de los cambios registrados a lo largo del Holoceno en perspectiva regional. Rev Museo Antropol 14:151–166. https://doi.org/10.31048/1852.4826.v14.n3.33842

López G, Restifo F (2012) The middle Holocene domestication and intensification of camelids in North Argentina, tracked by zooarchaeology and lithics. Antiquity 86(334):1041–1054. https://doi.org/10.1017/s0003598x00048237

López G, Restifo F (2017) El sitio Alero Cuevas, Puna de Salta, Argentina: secuencia de cambio en artefactos líticos y resolución cronológica macrorregional durante el Holoceno temprano y medio. Chungara Rev Antropol Chilena 49(1):49–63

López G, Orsi J, Seguí S, Araya S, Solá P, Coloca F (2023) Cazadores de la Puna de Salta en escala cronológica amplia (ca. 11000-3000 años AP): Paisaje arqueológico cinegético en el sector de Pozo Cavado, cuenca de Pocitos. Rev Museo Antropol 16(2):51–66. https://doi.org/10.31048/1852.4826.v16.n2.40183

Luo L, Wang X, Guo H, Lasaponara R, Shi P, Bachagha N, Li L, Yao Y, Masini N, Chen F (2018) Google earth as a powerful tool for archaeological and cultural heritage applications: a review. Remote Sens 10(10):1558. https://doi.org/10.3390/rs10101558

Lyman RL (2008) Quantitative paleozoology. Cambridge University Press, Cambridge

Marom N, Wolkowski U (2024) A note on predator-prey dynamics in radiocarbon datasets. bioRxiv. https://doi.org/10.1101/2023.11.12.566733

Martínez J (2003) Ocupaciones humanas tempranas y tecnología de caza en Antofagasta de la Sierra (10000-7000 AP). Ph.D. Dissertation, Facultad de Ciencias Naturales e Instituto Miguel Lillo, Universidad Nacional de Tucumán

Matteucci SD (2012) Ecorregión puna. In: Morello J, Matteucci SD, Rodriguez A, Silvia M (eds) Ecorregiones y Complejos Ecosistémicos Argentinos. Orientación Gráfica Editora S.R.L., pp 87–127

Mondini MN (2004) La comunidad de predadores en la puna durante el Holoceno. Interacciones bióticas entre humanos y carnívoros. Relac Soc Argentina Antropol 29:183–209

Mondini M (2008) La acción tafonómica de mamíferos carnívoros en el sitio arqueológico Inca Cueva-cueva 4, borde oriental de la Puna Argentina. Geobios 41(1):101–112

Morales M, Hoguin R, Oxman B, Pirola MM, Ruan Sirolli J, Merler Carbajo J, Bustos S, Tchilinguirian P, Álvarez L, Samec C, Kohan P, Yacobaccio H (2022) Evolución ambiental y registro arqueológico de la cuenca del río Barrancas, provincia de Jujuy. Argentina Rev Museo Antropol 15(1):97–116. https://doi.org/10.31048/1852.4826.v15.n1.34596

Moreno E (2009) El paisaje cazador en la Quebrada de Antofalla (Antofagasta de la Sierra, Catamarca). La Zaranda de Ideas. Rev Jóvenes Invest Arqueol 5:101–120

Moreno E (2013) Estrategias de caza y paisajes culturales en Antofagasta de la Sierra. Cat Com 17(2):165–190

Moreno E, Martínez JG, Aschero CA (2021) Ancient hunting strategies of wild camelids through the study of multiple lines of archaeological evidence at southern Argentine Puna. In: Belardi JB, Bozzuto DL, Fernández PM, Moreno EA, Neme GA (eds) Ancient hunting strategies in southern South America. The Latin American studies book series. Springer, Cham, pp 1–25. https://doi.org/10.1007/978-3-030-61187-3_1

Mosca Torres ME (2010) Selección de alimento y hábitat por vicuñas silvestres de la reserva provincial Los Andes (Salta, Argentina). Ph.D. Dissertation, Universidad de Buenos Aires. Facultad de Ciencias Exactas y Naturales

Muscio H (1999) Tendencias en la variabilidad ambiental de la Puna Argentina: Implicancias para la ecología humana prehistórica y para los paisajes arqueológicos. Cuad Inst Nac Antropol Pensamiento Latinoamericano 18:271–296

Muscio H (2001) Arqueología de San Antonio de los Cobres (SAC): Primeras interpretaciones del registro de superficie. Estud Soc NOA 2:121–165

Muscio H (2009) Nicho y estrategia predominante: Dos conceptos necesarios en Arqueología Evolutiva. In: Barboza M, Avila J, Piccoli C, Cornaglia Fernández J (eds) 150 años después... La vigencia de la teoría evolucionista de Charles Darwin. Universidad Nacional de Rosario, Rosario, pp 83–105

Muscio HJ, Carballo GE (2023) El registro arqueológico superficial de la quebrada de Cavi: Una aproximación inicial. Relac Soc Argentina Antropol 48(2):168–188. https://doi.org/10.24215/18521479e085

Muscio HJ, Cardillo M (2019) Filogenias andinas: análisis cladístico de las puntas líticas lanceoladas del Holoceno Medio de la Puna de Argentina y Norte de Chile. Relac Soc Argentina Antropol 44(2):213–232

Muscio H, López G (2007) Unidades de análisis arqueológicas en el estudio evolutivo de adaptaciones con economías de producción de alimentos: Un análisis de las arqueofaunas de la Quebrada de Matancillas (Puna de Salta). Rev Shincal 7:111–134

Muscio H, Varde M, Restifo F (2020) El conjunto lítico de las ocupaciones del Holoceno medio en Cueva Nacimiento 2, Puna de Salta, Argentina. Cuad Inst Nac Antropol Pensamiento Latinoamericano 29(2):39–58

Muscio HJ, Seguí S, Varde M, Coloca FI, Rucci L, Araya S (2021) Escalas, espacio y patrones del registro arqueológico de las Quebradas Matancillas y nacimiento, San Antonio de los Cobres (Puna de Salta, Argentina). Rev Relac Soc Argentina Antropol 46:449–481

Núñez L (1992) Ocupación arcaica en la Puna de Atacama: secuencia, movilidad y cambio. In: Meggers B (ed) Prehistoria sudamericana: nuevas perspectivas. Taraxacum, Washington, pp 283–307

Núñez L, Grosjean M (1994) Cambios ambientales Pleistoceno-Holocénicos: Ocupación humana y uso de recursos en la Puna de Atacama (Norte de Chile). Estud Atacameños 11:11–24

Odling-Smee J, Laland KN, Feldman MW (2003) Niche construction: the neglected process in evolution. Princeton University Press, Princeton

Orsi J (2018) Explotación de recursos faunísticos y procesos de cambio en el uso de los camélidos en la Puna de Salta ca. 10.000-2.500 años AP: Análisis del registro arqueofaunístico de los sitios Alero Cuevas y Abrigo Pozo Cavado. Ph. Dissertation, Facultad de Filosofía y Letras, Universidad de Buenos Aires

Ozán IL (2017) Gravity and the formation of the archaeological record: main concepts and methodological tools. Geoarchaeology 32:646–661. https://doi.org/10.1002/gea.21620

Paoli H (2002) Recursos Hídricos de la Puna, Valles y Bolsones Áridos del Noroeste Argentino. Instituto Nacional de Tecnología Agropecuaria

Pianka ER (1994) Evolutionary ecology, 5th edn. Harper Collins, New York

Piccot A, Monnin T, Loeuille N (2019) From apparent competition to facilitation: impacts of consumer niche construction on the coexistence and stability of consumer-resource communities. Funct Ecol 33:1746–1757

Pintar E, Martínez J, Aschero CA, Glascock M (2016) Obsidian use and mobility during the early and middle Holocene in the salt Puna, NW Argentina. Quat Int 422:93–108

Ragheb A, Ragab A (2015) Enhancement of Google earth positional accuracy. Int J Eng Res Technol 4(1):627–630

Ratto N (2003) Estrategias de caza y propiedades del registro arqueológico en la Puna de Chaschuil (Departamento de Tinogasta, Catamarca, Argentina). Ph. Dissertation, Facultad de Filosofía y Letras, Universidad de Buenos Aires

Reboratti C (2006) Situación ambiental en las ecorregiones Puna y Altos Andes. In: Brown A, Martínez Ortíz U, Acerbi M, Corcuera J (eds) La situación ambiental Argentina 2005. Fundación Vida Silvestre Argentina, Buenos Aires, pp 33–39

Restifo F (2015) Tecnología de hojas líticas en tierras altas andinas: perspectivas desde la Puna de la provincia de Salta (Argentina). Estud Atacameños 51:33–51

Restifo F, Patané Araoz CJ (2017) Artefactos Saladillo de la Puna de la provincia de Salta (Argentina): interpretación del proceso de cambio tecnológico hacia fines del Holoceno medio a partir de su clasificación funcional macroscópica. Rev Relac Soc Argentina Antropol 42:13–34

Ripple WJ, Wirsing AJ, Wilmers CC, Letnic M (2013) Widespread mesopredator effects after wolf extirpation. Biol Conserv 160:70–79. https://doi.org/10.1016/j.biocon.2012.12.033

Schiffer MB (1987) Formation processes of the archaeological record. University of New Mexico Press, Albuquerque

Schiffer MB (1991) Los procesos de formación del registro arqueológico. Bol Antropol Am 23:39–45

Smith-Patten BD, Bridge ES, Crawford PHC, Hough DJ, Kelly JF, Patten MA (2015) Is it extinction forever? Public Underst Sci 24(4):481–495

Stiner MC, Beaver JE, Munro ND, Surovell TA (2008) Modeling paleolithic predator-prey dynamics and the effects of hunting pressure on prey 'choice'. In: Bocquet-Appel JP (ed) Recent advances in palaeodemography. Springer, Dordrecht, pp 143–178

Tchilinguirian P, Morales M (2013) Holocene paleoenvironments in northwestern Argentina: main patterns and discrepancies. Quat Int 307:14–23

Vilá BL (2000) Los camélidos y sus pastores. Rev Cien 2:30–41

Walker MJ, Berkelhammer M, Björck S, Cwynar LC, Fisher DA, Long AJ, Lowe JJ, Newnham RM, Rasmussen SO, Weiss H (2012) Formal subdivision of the Holocene series/epoch: a discussion paper by a working group of INTIMATE (integration of ice-core, marine and terrestrial records) and the subcommission on quaternary stratigraphy (international commission on stratigraphy). J Quat Sci 27(7):649–659

Winterhalder B, Goland C (1993) On population, foraging efficiency, and plant domestication. Curr Anthropol 34(5):710–715

Yacobaccio H (2001) La domesticación de camélidos en el Noroeste Argentino. In: Berberián E, Nielsen A (eds) Historia Argentina prehispánica. Editorial Brujas, Córdoba, pp 7–40

Yacobaccio H (2012) Intercambio y caravanas de llamas en el sur andino (3000-1000 AP). Comechingonia 16:9–29

Yacobaccio H, Morales M (2005) Mid-Holocene environment and human occupation of the Puna. Quat Int 132:5–14

Yacobaccio H, Morales MR (2011) Ambientes pleistocénicos y ocupación humana temprana en la Puna Argentina. Bol Arqueol PUCP 15:337–356. https://doi.org/10.18800/boletindearqueologiapucp.201101.013

Yacobaccio H, Vilá B (2013) La domesticación de los camélidos andinos como proceso de interacción humana y animal. Int Antropol 14:227–238

Yacobaccio H, Escola P, Lazzari M, Pereyra F (2002) Long-distance obsidian traffic in Northwestern Argentina. In: Glascock M (ed) Geochemical evidence for long-distance exchange, scientific archaeology for the third millenium. Bergin & Garvey, Westport, pp 167–203

Yacobaccio H, Escola P, Pereyra F, Lazzari M, Glascock M (2004) Quest for ancient routes: obsidian sourcing research in northwestern Argentina. J Archaeol Sci 31:193–204

Zeder MA (2012) Pathways to animal domestication. In: Gepts P, Famula TR, Bettinger RL et al (eds) Biodiversity in agriculture: domestication, evolution, and sustainability. Cambridge University Press, Cambridge, pp 227–259

Chapter 7
Communal, Collective, and Massive Guanaco Hunts: The Regional Archaeological Contribution to Its Discussion in Southern Patagonia

Juan Dellepiane, Josefina Flores Coni, Agustín Agnolin, Gisela Cassiodoro, Rafael Goñi, Francisco Guichón, and Anahí Re

7.1 Introduction

Collective, communal and/or massive hunts (CCMH) in the southern Patagonia region have raised a series of questions and approaches from different authors. Some argue that they existed in all their forms, and others question their existence or, to put it more clearly, whether there are aspects of the archaeological record that support or prove the idea of their realization. Borrero (1990, 2013) was one of the main authors to discuss the existence of this type of practice in Patagonia prior to contact with Europeans, who provided local populations with horses as facilitators of a collective hunting practice. This author argues that collective hunts (large numbers of hunters—see Sect. 7.3) may have occurred, but, conversely, regarding

J. Dellepiane (✉)
USI, Instituto Nacional de Antropología y Pensamiento Latinoamericano, Buenos Aires, Argentina

J. Flores Coni · G. Cassiodoro · A. Re
Consejo Nacional de Investigaciones Científicas y Técnicas (CONICET) - Universidad de Buenos Aires (UBA) - Instituto Nacional del Antropología y Pensamiento Latinoamericano (INAPL), Buenos Aires, Argentina

A. Agnolin
Consejo Nacional de Investigaciones Científicas y Técnicas (CONICET) - Instituto Nacional de Antropología y Pensamiento Latinoamericano (INAPL), Buenos Aires, Argentina

R. Goñi
Instituto Nacional de Antropología y Pensamiento Latinoamericano (INAPL) - Universidad de Buenos Aires (UBA), Buenos Aires, Argentina

F. Guichón
Consejo Nacional de Investigaciones Científicas y Técnicas (CONICET) - Comisión Nacional de Actividades Espaciales (CONAE), Buenos Aires, Argentina

R. Frontini et al. (eds.), *Non-human Animals and People: Interactions, Agents and Materialities*, The Latin American Studies Book Series,
https://doi.org/10.1007/978-3-031-97495-3_7

massive hunts, which imply a high number of prey obtained, there would be no solid evidence. His arguments can be summarized as ecological, based on his perspective of the ethology of the main prey (guanaco—*Lama guanicoe*), and archaeological since he maintains that no archaeological record is known in southern Patagonia that supports this presumption. In particular, he emphasizes the zooarchaeological record as a main characteristic and the absence of the so-called bone beds known for other parts of the world, especially for the Great Plains of North America (Frison 1978; Speth 2017). Thus, this author initially raises the discussion on the specificity of a type of archaeological record as proof of this practice. A perspective that does not coincide with the previous one is that proposed by Santiago and Salemme for Tierra del Fuego (Santiago and Salemme 2010, 2016; Colasurdo et al. 2012). The analyses carried out at the Las Vueltas 1 site led to defining it as a guanaco killing and processing site, where at least three occupation moments have been identified. For the third and most recent occupation (ca. 500 years BP), the zooarchaeological analysis has allowed them to consider the possibility of being faced with a killing event of dozens of animals (Santiago and Salemme 2010; Colasurdo et al. 2012).

Whatever the type of archaeological evidence available to date, in these works the discussion is focused on a site scale, mainly prioritizing the zooarchaeological record. In this work, we propose to demonstrate that, under an exhaustive regional analysis of the archaeological record, we can identify a series of unequivocal patterns of the presence of CCMH in Patagonia in pre-contact times. We will focus spatially on the central-western Province of Santa Cruz (COSC) (south-central Patagonia, Argentina) and, chronologically, on the Late Holocene, especially the middle Late Holocene (ca. last 2000 years BP). Thus, we will direct the discussion toward two aspects: on the one hand, the relevant climatic, environmental, geographical, and ecological factors for the purpose of analyzing CCMH, and, on the other hand, the social and cultural factors to take into account. The central argument is that from an eminently regional perspective, with different temporal/spatial micro, meso, and macro scales (sensu Dincauze 2000), identifiable patterns in the archaeological record are recognizable as a correlate of the CCMH hunting practices. In summary, we seek to establish that, during the Late Holocene, vast geographical spaces of seasonal use, such as the COSC plateaus, were the propitious scenario for the prosecution of these CCMH strategies and that they left a rich archaeological record referable to such practices. To do this, we discuss the variability of the archaeological record behind these CCMH present in the region under study.

First, we present the environmental and archaeological characteristics of the region under study (Sect. 7.2); the proposed settlement model for the mentioned sector of Santa Cruz (Patagonia) where the discussion of this work is framed is described; then, in Sect. 7.3, the concepts discussed here are defined; thus, specifications are provided about what we understand by the different plural hunting strategies, differentiating between collective, communal and massive. Section 7.4 presents the archaeological evidence of the COSC that allows us to discuss the presence of CCMH strategies; for this, different lines of evidence from two areas within the study region are described. Finally, Sect. 7.5 integrates all the information related to the regional analysis presented through the discussion of the fundamental factors we have considered to address CCMH.

7.2 The Late Holocene and Human Settlement of the Study Region in South-Central Patagonia

7.2.1 *The Study Region*

The region under study is located in the central-western portion of Santa Cruz province (Patagonia, Argentina) (Fig. 7.1). It comprises lowlands (below 400 masl)—essentially lake basins—and highlands—primarily basaltic plateaus (over 700 masl), in an approximate range of 250 km in a North/South axis, and 120 km in an East/West axis. It is characterized by a semi-desert climate (Paruelo et al. 1992) and a prevalence of shrub and grass steppe. The region has a heterogeneous distribution of water resources, with all year permanent freshwater sources constrained to river valleys, lake basins and wetlands (*mallines or vegas*). The region exhibits a strong seasonality with harsh winters characterized by heavy snowfall, mainly in the highlands. Although lowlands and highlands are environmentally similar, they differ in their resource availability, exhibiting significantly different effective temperatures (9.8 °C in the highlands vs 12.6 °C in the lowlands). The basaltic plateaus are a prominent feature of the landscape and share geological, geomorphological, ecological, and environmental characteristics (Mazzoni and Rabassa 2010). These have been described as part of extensive basaltic flows, whose origin dates back to the

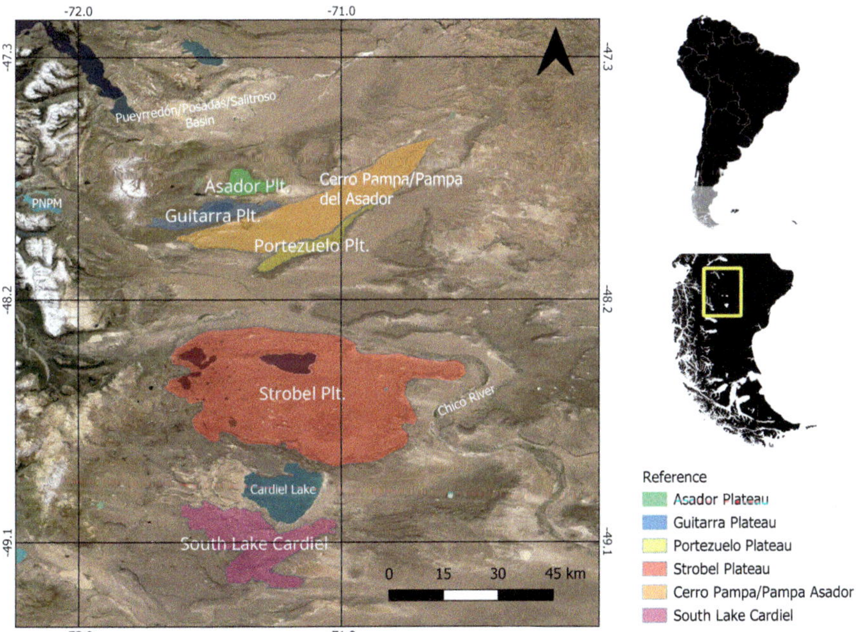

Fig. 7.1 Map of the region under study in central-western Santa Cruz province (Patagonia, Argentina). Several plateaus and highlands over 700 m.a.s.l. are shown. During the Late Holocene and related to climatic and environmental changes, these areas were fully incorporated into hunters' mobility ranges

Upper Miocene (Ramos 2002). During the winter, they accumulate a large amount of snow, which inhibits their occupation. Snow tends to accumulate in the few shelters and rock walls that exist, lasting well into spring. Given that these places are the ones with the main evidence of human occupation (Goñi 2010), we can assume that until summer or late spring the possibilities of occupying these spaces must have been very limited. In late spring and summer, when the snow melts, the plateaus become an important grazing place for the guanaco. Winds are constant in a west-easterly direction, with average speeds of 40/60 km/h and gusts often exceeding 100 km/h. This phenomenon is especially intense during spring and summer.

Regarding animal species, the diversity is quite low. The guanaco, an ungulate of up to 120 kg, has been the most important resource for hunting populations in the past (Borrero 1990; Mengoni Goñalons 1999; Rindel 2009) (Fig. 7.2). The guanaco has migratory habits, which in the region are characterized by the use of high plateaus to feed and give birth in spring/summer and the descent to the lower basins for the winter (Candino et al. 2022).

The plasticity observed in the migratory characteristics of this species (Candino et al. 2022) stands out. The different behavioral patterns of the guanaco profoundly influence human mobility, creating an isomorphic hunter-prey dynamic in southern Patagonia during the Late Holocene.

Other species that can be found are carnivores like foxes (*Pseudalopex griseus* and *P. culpaeus*) and pumas (*Puma concolor)*, edentata mammals like hairy armadillos (*Chaetophractus villosus*), dwarf armadillos (*Zaedyus pichiy*), and numerous rodents. There is also a variety of birds like cauquén (*Chloephaga picta*) and flightless birds like the choique (*Rhea pennata*), a prominent species. All of these minor resources, however, made up a minimal part of the diet of local hunting populations throughout the Holocene (Rindel 2009; Dellepiane 2019), until the introduction of horses in the region, during the twenty-seventh century.

According to various paleoclimate studies, the present landscape was established during the Late Holocene. Changes in the direction and intensity of winds, the

Fig. 7.2 Herd of guanacos in the Patagonian steppe. These ungulates have seasonal migratory habits: during spring and summer they occupy plateaus and highlands while in winter they inhabit lower basins. This behavioral characteristic influenced human mobility during the Late Holocene

so-called Southern Westerlies, were recorded from 6000 years BP and particularly after 1800 years BP (Gilli et al. 2001). Consequently, fluctuations in the distribution of hydric resources were observed as well a deep decreasing of environmental humidity at a regional scale since 2500 years BP (Stine and Stine 1990; González 1992; Gilli et al. 2001; Ariztegui et al. 2010; Horta et al. 2017; Quade and Kaplan 2017). The presence of "epic droughts" circa 929–797 and 787–722 cal BP was established during the global scale episode known as Medieval Climatic Anomaly (MCA) (Stine 1994). Drier conditions generated hydric retractions and therefore new lands became available. These fluctuations resulted in a progressive fragmentation of the landscape into spaces with different availability of water, vegetation, and fauna.

7.2.2 Human Peopling

The region was occupied by hunter-gatherer groups from the Early Holocene in some areas until the beginning of the twentieth century. Based on the environmental and paleoclimate conditions detailed above, combined with the results of an extensive program of archaeological work, a peopling model for the central-western region of Santa Cruz province was presented (Goñi 2000, 2010; Goñi et al. 2011-2012, 2019). The central argument is that the significant climatic changes recorded during the Holocene, particularly Late Holocene, directly affected the environment of the study region. As mentioned above, drier conditions generated hydric retractions and therefore new lands became available in the low lake basins. These new spaces were gradually colonized by species from the shrub steppe, mainly molle (*Schinus polygamus*), creating suitable living conditions for hunter-gatherer populations.

As water is a critical resource for humans, the increasing aridity during the Late Holocene generated dramatic changes in hunter-gatherer mobility. Thus, a strong reduction in residential mobility took place reaching its critical point during and after the MCA (Goñi 2010). In this scenario, lowland basins such as lakes Salitroso and Cardiel were ideal to establish more stable residential camps due to their year-round resource availability (i.e., water, firewood, plants, and animals), and better sheltering conditions during winter, due to less snowfall. We have also suggested that this gradual reduction in residential mobility, together with the potential spatial constraint originated by the new arid conditions, resulted in larger concentrations of human populations in specific areas (Goñi 2010). Considering that resources were unevenly distributed in time and space, it became necessary for hunter-gatherers to undertake logistical movements (sensu Binford 1980). We have proposed an extension of action ranges that included the surrounding highlands (i.e., Strobel and Guitarra Plateaus, high plains from Pampa del Asador, and the basins from the Perito Moreno National Park) (Fig. 7.1). Thus, an extensification strategy (Flores Coni et al. 2021) was established to allow for the complementary use of highlands and lowlands with different effective temperatures (Goñi 2010; Goñi et al. 2019). In this process, high plateaus were fully incorporated into hunters' mobility ranges, during the Late Holocene.

Given that the diet of these populations was based on the guanaco, it is considered that during this period was adopted a mobility strategy strongly linked to the migrations of this ungulate. In this way, populations were organized around the use and monitoring of herds, establishing isomorphic mobility with this ungulate (Goñi 2000, 2010; Goñi et al. 2011-2012; Rindel et al. 2017). Predictability must have been achieved by observing herds of guanacos, when and where they are, a logistical task in order to inform the rest to plan the hunt. In this context, animal behavior offers multiple opportunities to plan massive procurement of prey. In summary, human behavior is guided by the behavior of the guanaco, but, given that it is a versatile animal, with seasonal variations and flexible behavior, it allows for a variety of alternatives for its acquisition by hunters. Within this framework, the different CCMH alternatives are discussed.

The populations of this period can be characterized as groups with a very low population density (Goñi 2010), given the limited carrying capacity of the environment (Oliva et al. 2001) (Goñi et al. 2019). The presence of a sparse and dispersed population would not have prevented the circulation of information or periodic groupings (Group 2 and 3 on Binford 2001), but could have stimulated them as a necessity for survival. Thus, based on the characteristics of the rock art record (see below), it has been suggested that the high plateaus, especially the Strobel, would have functioned as spaces of population convergence, attracting groups from distant places and acting as spaces for sharing information (Belardi and Goñi 2006; Re et al. 2017, 2021).

Previous works show the complementary use of various types of weapons toward the late Holocene, especially the last 1000 years cal. BP (Cassiodoro et al. 2014; Flores Coni 2018; Flores Coni et al. 2021; Agnolin et al. 2019). Thus, the use of spear/dart, bow and arrow, and bola is recorded in diverse contexts that include hunting in open spaces, topographic traps and equipping the space with hunting blinds (Belardi et al. 2005; Cassiodoro et al. 2014; Agnolin et al. 2019; Flores Coni et al. 2021). These same groups, after contact with Europeans and the acquisition of horses, became mounted hunters between the twenty-eight and twenty-ninth centuries. During this period, the development of CCMH using bolas is well attested by numerous sources (Onelli 1904; Musters 2005).

In summary, the dynamics of regional settlement toward the late Holocene involves populations of terrestrial hunters dependent on the guanaco, with an important repertoire of hunting strategies and high logistical mobility.

7.3 Plural Hunts: Definitions and Their Archaeological Evidence

7.3.1 Communal, Collective, and Massive Hunts

For the purposes of this work, it is necessary to operationally define the strategies encompassed under the concept of plural hunting, referred to as communal, collective, and mass hunting. A first aspect to highlight is that all three alternatives mentioned here involve a high degree of planning. Moreover, the first two, communal and collective hunting, refer to the number and social organization of the hunters, whereas mass hunting refers to the number of prey obtained (hunting productivity) (Driver 1990).

Collective Hunting Collective hunting requires the presence of a plural number of hunters, enabling the implementation of various strategies generally aimed at obtaining a greater number of prey than would be possible individually. The focus is on the strategies implemented during the hunt, such as enclosure or herd management, rather than other characteristics. Following Frison (1978, pp. 147–149), these groups can range from three hunters to a band (12 or more hunters), allowing for the containment of multiple animals. Meanwhile, Driver (1990), adopting a broader perspective, differentiates collective from individual hunting, indicating that with two or more individuals, a hunt qualifies as collective.

Collective hunting thus includes an adequate number of hunters, allowing for planned or programmed hunts, minimizing opportunistic or contingent conditions, though not excluding them entirely. The emphasis is placed on the strategy rather than the prey. What matters most is achieving the programmed objective, regardless of the energy yield.

Communal Hunting Communal hunting, on the other hand, refers to plural hunts involving the participation and/or presence of different social segments, meaning not only hunters but also entire family groups and/or aggregates, with a variety of age and gender groups present. It can be cyclical and not necessarily annual. These groups were characterized as Group 2 and Group 3 by Binford (2001), including contexts where members of the group who do not usually participate in hunting are incorporated (e.g., the involvement of women and children in driving animals toward a group of hunters in position) (see Kelly 2013, p. 214).

Mass Hunting Finally, mass hunting refers to the number of prey obtained, emphasizing the energy return. In this sense, the quantity of prey is key, and there can even be mass hunts involving a small number of hunters. Traditional archaeological examples include bison hunting in North America (Frison 1978; Speth 2019), and, ethnohistorical, Inkaic Chakus (Aguilar Meza 2018; Ratto & Orgaz 2002).

The proposal developed here aims to differentiate concepts that are often treated as synonyms and which, in this context, require distinction to avoid confusion. Therefore, we have defined three types of plural hunting (collective, communal, and massive), which, while different, may complement or overlap without contradiction.

7.3.2 CCMH and Archaeological Record Expectations

The definitions provided here cannot be understood as rigid categories with unequivocally associated archaeological records. The identification of CCMH practices in archaeology has traditionally been complex due to the absence of unambiguous indicators (Lubinski 2013). Below, we present aspects of ethnographic, ethnoarchaeological, and archaeological records that have been associated with CCMH practices in specialized literature. These include spatial provisioning technologies (hunting blinds and rock alignments), prey acquisition and processing technologies (artifacts), processing products (zooarchaeology), and rock art as associated information transmission.

It is in this context that a series of archaeological features gain significance, describing highly informative patterns when considered together. These mainly include technologies such as stone structures (hunting blinds), tools (projectile points, scrapers, waste, etc.), and even rock art representations. Their spatial, temporal, frequency, and distributional combinations allow us to ascribe meanings to derived patterns. This type of archaeological record, combined with associated environmental and geographical variables, allows us to analyze it analogically using evidence from other regions and sources presented below.

7.3.2.1 Stone Structures: Hunting Blinds

In international literature, one of the most widely accepted indicators of CCMH is the presence of stone structures (hunting blinds and alignments) associated with the management and stalking of prey (Benedict and Cassells 2000; Grønnow 2009; Holzer et al. 2010; Friesen and Stewart 2013; La Belle and Pelton 2013; Arkush 2014). Our archaeological research has also yielded a significant amount of information regarding stone structures and their relation to hunting strategies (Flores Coni et al. 2021).

From the perspective of ethnoarchaeology, as proposed by Binford, based on comparisons among the technologies of various groups of hunter-gatherers: "…heavy investment in technology of food procurement…" (Binford 2001, p. 395), and Proposition 10.21: "Tended facilities are increasingly successful in settings in which artiodactyl biomass is moderate to high and animals are at least periodically aggregated…" (Binford 2001, p. 487). As we shall see, the conditions described by Binford completely align with what is observed in the regional archaeological record.

Ethnographic information mentions the use of structures for hunting by diverse groups, such as the Australians of the Western Desert (Gould 1968), the Hadza (Marlowe 2010), the Nunamiut (Binford 1983, 2001), the San (Crowell and Hitchcock 1978), the Havasupai (Spier 1928), the Copper Eskimo (Jenness 1922), and the Shoshone (Thomas 1983). In these cases, stone structures were used as part of both communal and individual hunting strategies (Fig. 7.3). In cases of CCMH, the areas used tend to be very large, including lookouts, lines of rocks to guide prey, stone structures (or structures made of other materials) from which to finish off the animals, and spaces for processing carcasses, usually spread across areas of hundreds or thousands of meters (Fig. 7.4).

Fig. 7.3 Image of the current use of a hunting blind in Kalahari, Botswana to show the present use of these kinds of structures. Photo by courtesy of L. R. Binford

For the purposes of this study, ethnographic or ethnoarchaeological patterns serve as a bridge to build the argument between hunter dynamics and the resulting archaeological record.

7.3.2.2 Lithic Technology

In CCMH contexts, the artifact record would be associated with high frequency and density. The presence of hunting weapons was strongly linked to these activities—in our case, projectile points—as well as their discard and the byproducts of their manufacture and rejuvenation. Additionally, processing tools (knives, scrapers, grinding tools, etc.) are to be expected in these contexts. This processing toolkit may also include ceramic containers, necessary for fat processing.

Despite some artifact diversity, the tasks in hunting plateaus, butchering, and processing would require a discrete toolkit of lithic artifacts (Gould 1968; Binford 1983, 2001); in our case, projectile points, possibly bolas, knives, scrapers, and some retouched flakes. This situation contrasts with processing assemblages in semi-permanent camps in lower basins, where artifact richness is significantly greater (Goñi 2010).

Furthermore, several travelers' accounts and ethnographic information associate various resource processing tasks with women, supporting the idea of communal hunts (Aguerre 2000; Musters 2005).

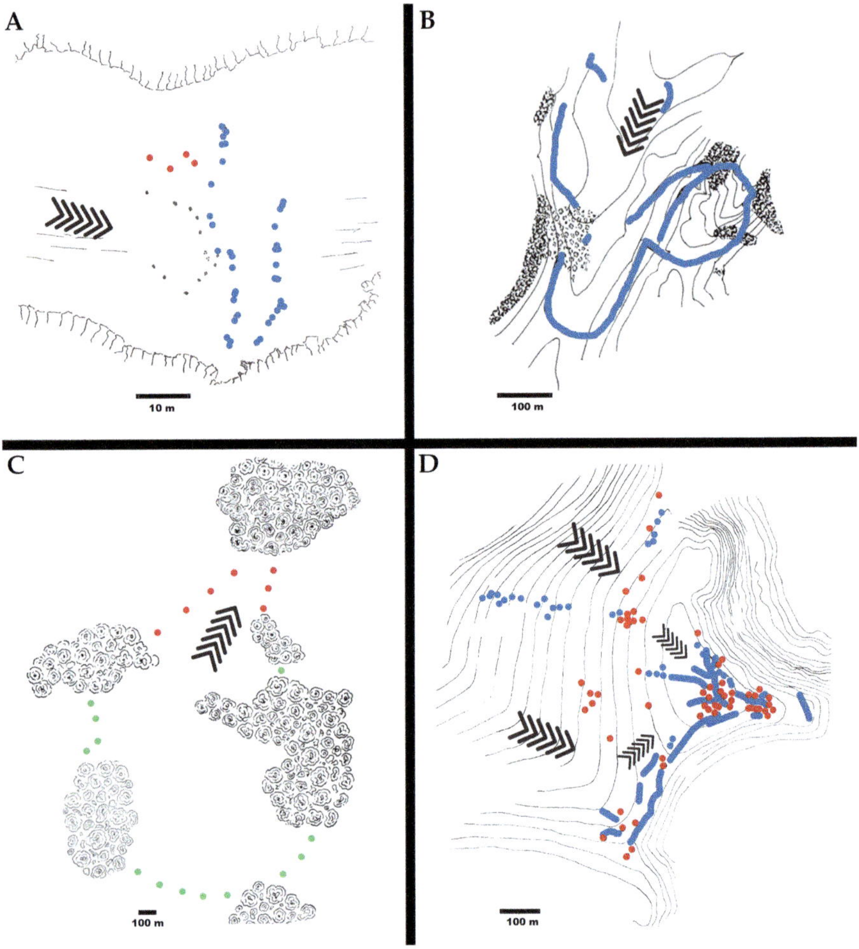

Fig. 7.4 Ethnographic and archaeological examples of CCMH site plans. (**a**) Caribou kill ethnographic site modified from Binford (2001); (**b**) antelope hunt archaeological site, modified from Frison (1978); (**c**) Selk'nam guanaco kill ethnographic site, modified from Furlong (1912); (**d**) wild sheep or deer hunt archaeological site, modified from Cassells (2000). References: green dots: drivers, red dots: structures or hiding places with hunters, blue lines and dots: structures to drive prey, arrows: observed or supposed direction of prey

7.3.2.3 Zooarchaeology

The zooarchaeological record is often at the center of discussions about CCMH. It is important to highlight that various ethnographic and ethnoarchaeological records of CCMH illustrate how faunal assemblages tend to be processed and deposited under such conditions. In both known cases of bison and caribou hunts, whole carcasses are not transported; instead, there is a first step of butchering on-site and a second step of processing, which may or may not coincide with the first. Both, however, result in bone assemblages being left in the open (Binford 1978; Frison 1978; see specifically the caribou mass-kill processing site in Binford 1983, fig. 59). The outcome is a record featuring a heterogeneous distribution of skeletal parts, combined with differential evidence of processing and distinct depositional contexts, including open-air discard.

Since in our study region the only prey capable of being mass-hunted is the guanaco (*Lama guanicoe*) given their gregarious behavior, it is possible that this type of kill and processing record would be dispersed across the field (see Claraz 2008), given the low or nonexistent preservation conditions present in our study area (Belardi et al. 2007). It should be noted that, circumstantially and especially recorded for equestrian moments, other species may have been hunted in large numbers.

7.3.2.4 Rock Art

The accumulation and transmission of information among hunter-gatherers is a key element in the development of these group hunting modalities, as suggested by Halstead and O'Shea (1989) and various other studies on the topic. Aunger (2009) proposes that communication acts as a form of niche construction, altering the behavior of other individuals. Within this context, this work operates on the premise that rock art can be understood as stored and/or preserved information, playing an important role in the circulation of information (Hartley 1992; Whallon 2006), particularly in our region (Goñi 2010; Re 2010).

These hunting strategies can be considered two-step processes: first, information is collected, and subsequently, it is planned and executed. Rock art may have assumed various roles and functions within this framework.

In this way, variables such as the frequencies of motif production and the identified variability contribute to discussions about the scenarios in which CCMH hunting strategies were developed. This correlation is observed in three dimensions: spatial (location and frequency of sites and motifs), temporal (temporal depth, reuse of spaces, and intensity of occupation), and formal (diversity of communicated information). Rock art at hunting sites might have been part of a provisioning process, analogous to the concept proposed for stone structures (Flores Coni et al. 2021).

7.4 General Characteristics of the Archaeological Record of Basaltic Plateaus from Central-Western Santa Cruz Province

According to what has been outlined so far, the discussion in our case study focuses on two main spheres: the geographic/environmental/ecological on the one hand, and the patterns of distribution, composition, and frequencies of the archaeological record on the other. In the first, the basaltic plateaus and high alluvial plains (Pampa del Asador—PDA) are the focus of analysis, and in the second, the stone structures and the technology associated with shelters in rock walls are the other (Fig. 7.5).

For over 25 years, as part of projects carried out by this research team, surveys have been conducted in plateau areas exceeding 700 m.a.s.l., such as PDA (Espinosa and Goñi 1999; Goñi et al. 2011-2012), the Strobel Plateau (Belardi and Goñi 2006; Re et al. 2017), the Guitarra Plateau (Goñi et al. 2010; Cassiodoro 2016), the Portezuelo Plateau, the Asador Plateau, and those located south of Lake Cardiel (Cassiodoro 2021; Cassiodoro et al. 2024). Significant information has been obtained regarding the chronology of occupation of the plateaus, indicating an initial exploratory process at least 5000 years BP (5319–5487 calibrated years BP) (Goñi et al. 2010) and a more established presence around 2000 years BP (Goñi 2000-2002; Cassiodoro et al. 2013; Re et al. 2017) (Fig. 7.6).

In these plateaus, various types of sites have been identified that were used complementarily, reflecting the seasonal and/or logistical use of these spaces (Goñi 2010).

Fig. 7.5 Types of sites mentioned in the text. (**a**) Photograph of basaltic walls—shelters. (**b**) Sketch of a generic hunting blind; (**c**) Photograph of a hunting blind in site K205 in the Strobel Plateau

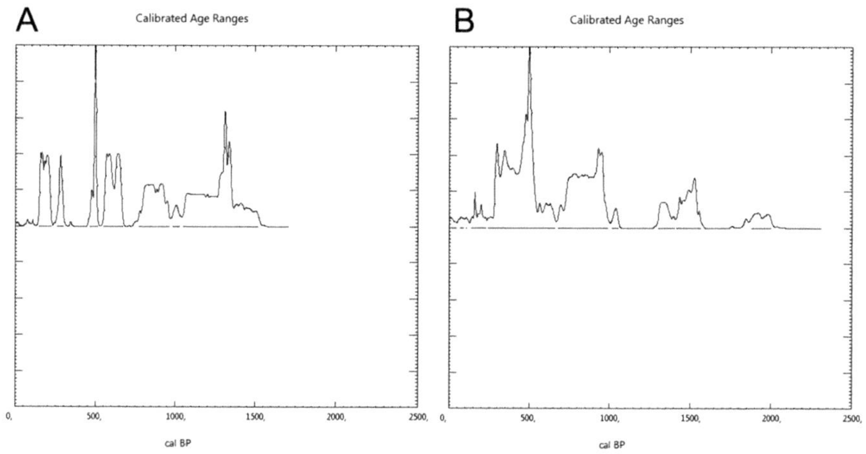

Fig. 7.6 (**a**) Graphic showing the sum of probabilities of dates from archaeological sites on rock walls and surface concentrations in central-western Santa Cruz province (Patagonia, Argentina). (**b**) Graph showing the sum of probabilities of dates from hunting blinds. Calibrations were performed using the Calib 8.2 software (Stuiver and Reimer 1993) with a Southern Hemisphere calibration curve (Hogg et al. 2013) and two-sigma ranges

7.4.1 Shelter in Rock Walls and Open-Air Concentrations

The distribution of shelters in basaltic plateaus in the region under study mainly consists of rocky walls (Fig. 7.5a) associated with both lagoons and river channels. These walls vary in size, ranging from 5 m to 1 km in length and 1–8 m in height. The western walls provide protection from strong winds and form the only possible living spaces in these extensive plains. Due to its geological and geomorphological characteristics, the Pampa del Asador lacks such shelters.

These walls show evidence of occupation (lithic materials and rock engravings) dating back to the Middle Holocene (Goñi et al. 2010; Re et al. 2017).

During the recent Late Holocene (last 2000 years, Fig. 7.6a), the number of sites increased, and the associated archaeological evidence diversified. The Strobel Plateau stands out among others and has been classified as a social aggregation point or population convergence area (Belardi and Goñi 2006). The Guitarra Plateau and its adjacent PDA are also significant.

The following chart presents the basic chronology of the colonization process of these plateaus, revealing a strong late signal of effective occupation of these spaces.

Regarding lithic technology, some sites stand out due to the high density of materials and a comparatively wide variety of artifacts (Cassiodoro et al. 2013; Flores Coni 2018). Evidence of ceramic technology is also present, albeit at low frequency, in some rock walls (Cassiodoro 2008; Cassiodoro et al. 2013). This points to the introduction of new techniques for resource cooking and storage in these spaces.

It is worth noting that sites associated with basaltic walls show very limited organic or zooarchaeological evidence due to the chemical characteristics of the sediments (Belardi et al. 2007; Dellepiane 2019).

On the other hand, western-facing rock walls feature predominantly engraved rock art representations (Fig. 7.7 and Table 7.1). The distinct characteristics of the rock art archaeological record suggest various strategies for information storage in these elevated spaces. While rock art motifs from the Middle Holocene have been identified, the proposed execution sequences and temporal trends indicate a strong connection to the recent Late Holocene (Re 2010; Re et al. 2017, 2021, 2024; Guichón 2018; Guichón et al. 2025).

Only in these COSC spaces are sites recorded with more than 500 motifs, which have been defined as key communication points in the region (Re et al. 2021). These coincide with extensive rock walls and associated flat surfaces, allowing the simultaneous presence of large groups of people.

It is noteworthy that the Strobel Plateau stands out for its negative handprints attributed to different age groups (adults, adolescents/young people, and children) (Fig. 7.8a) (Guichón 2018; Guichón and Re 2020). Moreover, human footprints engraved on the various plateaus come in different sizes: small (less than 100 cm^2), medium (between 100 and 500 cm^2), and very large (more than 500 cm^2) (Guichón 2012).

These plateaus also feature scenes created during the middle Late Holocene, including anecdotal scenes such as childbirth and hunting scenes (Fig. 7.8b, c) (Re 2010). The latter are particularly significant as they depict specific local situations (anecdotes) as well as hunting strategies. They are composed of hunters (represented by their footprints) who pursue and/or encircle guanacos (represented either by their figures or footprints) (Cassiodoro et al. 2014; Re 2016).

In summary, the archaeological record associated with basaltic rock walls illustrates a variety of activities and situations. These are also spatially linked to several of the hunting blind sites (see next section).

Lastly, it is worth mentioning the presence of open-air concentrations in these areas. As it will be seen in the next section, CP2B OA (in PDA) stands out due to its clear connection between the stone structures and the rest of the associated archaeological record.

7.4.2 Stone Structures: Hunting Blinds

In our study region, an exceptional amount of hunting blinds was registered: 1051 structures have been recorded so far (Table 7.2 and Fig. 7.5b, c). This includes both structures recorded in the field (summarized in Flores Coni et al. 2021) and those detected through remote tools (Dellepiane and Guichón 2023) (Fig. 7.9 and Table 7.2).

Fig. 7.7 Photos showing the context of rock art panels in the Strobel plateau. (**a**) Rock art panel with high density of motifs and superimpositions from K25 (called Gradin panel); (**b**) Panoramic of the southeast edge of the Strobel plateau; (**c**) Panel with high density of motifs from K145; (**d**) Panel with high density of motifs from K185; (**e**) Lagoon with rock art wall from the K145 Laguna del Puente

Table 7.1 Rock art sites in COSC plateaus. Bibliography from Strobel plateau: Guichón (2018), Re (2010), Re et al. (2019, 2021, 2024); from Guitarra plateau: Guichón et al. (2025) and Re (2010)

	Strobel plateau	Guitarra plateau
N rock art sites	118	27
N motifs	10,982 (in 94 sites)	1058 (in 19 sites)
N motif types (max 31)	31	24
Sites with more than 500 motifs	K25, K26, K27, K127, K145, K185-S2, K265	GUI42
N superimpositions	1863 (in 94 sites)	285 (in 19 sites)

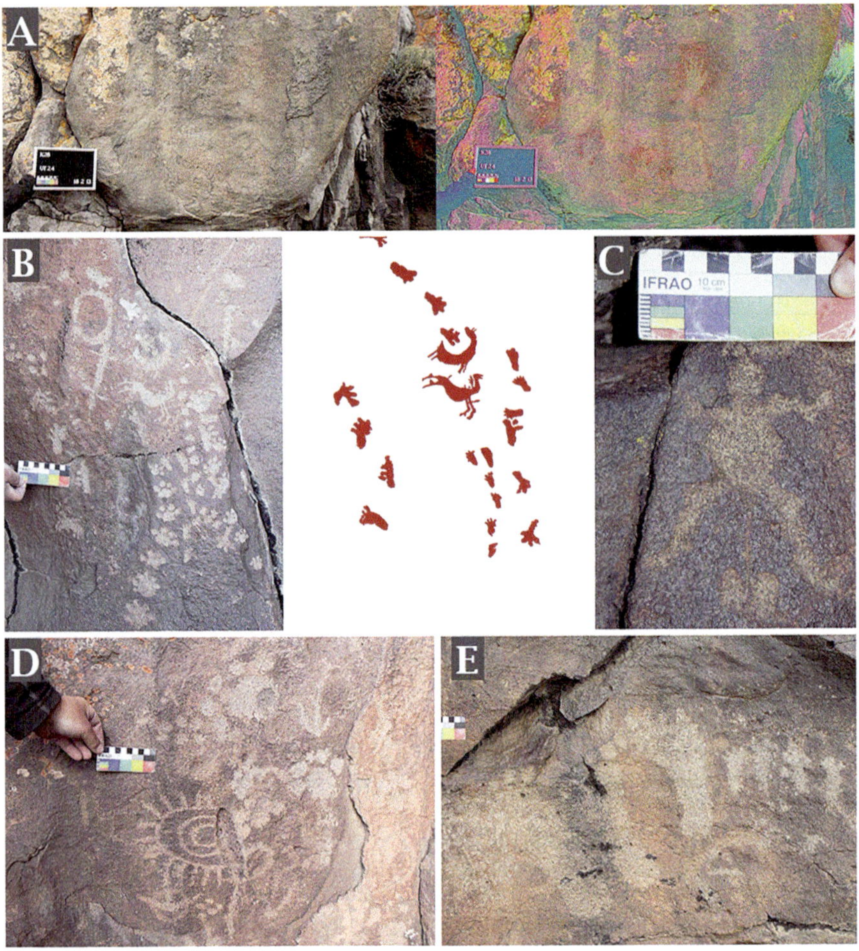

Fig. 7.8 Photos showing the variability of rock art motifs in Strobel Plateau. (**a**) Handprints with a variety of sizes (infants, adolescents, and adults) from site K28 (left original, right DStretch filter); (**b**) Panel with a hunting scene where two guanaco motifs and several human footprint trails are identified in site K145 (left original, right digital tracing of the hunting scene); (**c**) Human figure motif associated with a childbirth scene from site K25; (**d**) Variety of zoomorphic motifs (feline tracks, bird and abstract motifs (concentric circles with radial lines) from site K145; (**e**) Human footprints in site K25

Table 7.2 Total amount of stone structures recorded until now in the study region

Area	N structures (GPS)	N structures (remote sensing)	Total
PDA/Cerro Pampa	34	No data	34
Guitarra plateau	84	68	152
Strobel plateau	468	400	868
Total	586	468	1054

PDA Pampa del Asador

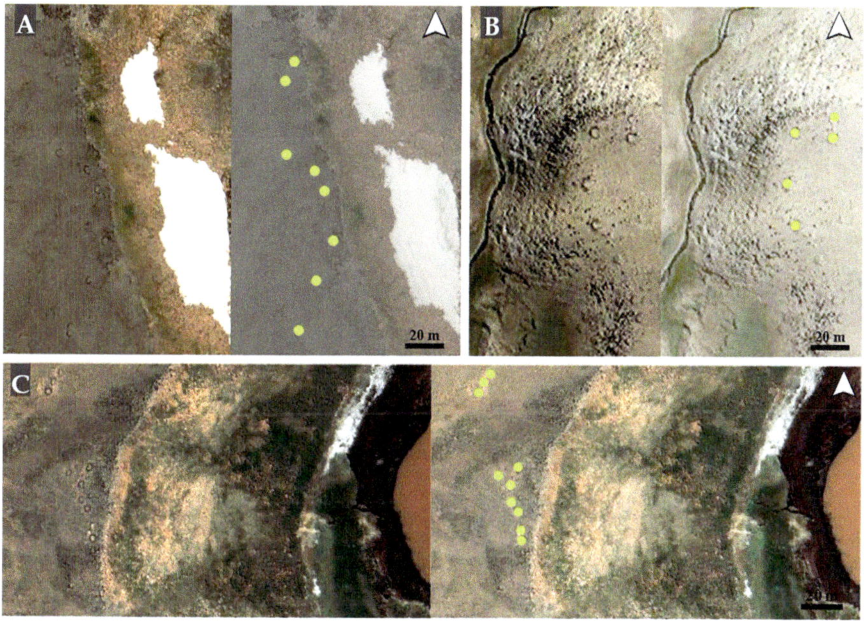

Fig. 7.9 Images showing the hunting blinds detected through remote sensing techniques in COSC, after recorded in field work. (**a**, **b**) Hunting blinds identified in Guitarra Plateau; (**c**) Hunting blinds identified in Strobel Plateau

They were built by the accumulation of rocks forming a low wall, without using mortar. Their walls are usually oriented against the prevailing westerly winds. Their shape is mainly semi-circular though circular and straight constructions were also registered (Cassiodoro 2011; Goñi et al. 2011-2012; Flores Coni 2018; Flores Coni et al. 2021). Different archaeological evidences are associated with these structures. Mainly lithic materials though also zooarchaeological evidence as well as ceramic sherds were recorded (Aragone and Cassiodoro 2005-2006; Dellepiane 2019). The evidence suggests that these structures were mainly used as hunting blinds, with less frequent examples of their use as residential structures (Flores Coni 2018; Flores Coni et al. 2021; Flores Coni and Cassiodoro 2023).

The earliest chronology for these sites begins 2000 years BP (1817–2021 cal BP, Fig. 7.6b). However, from 1000 years BP onwards, a persistent use of stone structures is observed. The cumulative process of constructing these structures indicates that by 500 years BP, the highlands featured a landscape equipped with hunting blinds (Flores Coni et al. 2021).

Hunting blinds can be found either isolated or in groups of two or more structures, with variable distances between them. For example, Table 7.3 presents sites composed of a large number of structures. It is worth noting that, solely for the purpose of this presentation, sites with more than 10 hunting blinds—out of the 586 studied—were arbitrarily selected to illustrate significant hunter gatherings, indicative of possible CCMHs.

Table 7.3 Examples of sites with clusters of 10 or more structures in the study region

Area	Sites	N structures
Guitarra plateau	GUI10	11
PDA/Cerro Pampa	CP6	13
	CP2	13
Strobel plateau	K36	15
	K47	11
	K107	17
	K116	55
	K123	13
	K130	11
	K168	10
	K205	25
	K268	10
	K280	31
Total	13	235

In the region, 13 sites have been detected in the field, each with 10 or more hunting blinds (Table 7.3). These have been identified in several of the elevated areas of the region under study.

The number of structures varies by site; while most have between 10 and 15 hunting blinds, there are notable clusters of more than 20 and 30 units, along with an exceptional case of 55 (site K116 on the Strobel plateau—see details in Sect. 7.4.2.2).

To discuss the presence of CCMH in the region, specific case studies are developed below, demonstrating the variability of the RA associated with these strategies, focusing on hunting blind sites. For this purpose, specific data is presented for two sites located in Pampa del Asador/Cerro Pampa (Fig. 7.10) and six sites on the Strobel plateau (Fig. 7.10).

7.4.2.1 Pampa del Asador/Cerro Pampa

La Pampa del Asador (Fig. 7.1) consists of fluvioglacial sedimentary deposits, situated between 900 and 1100 m above sea level, where various archaeological sites and lithic raw material deposits, especially black obsidian, are located (Espinosa and Goñi 1999; Cassiodoro 2016). Specifically, in the southern sector of Cerro Pampa lies the locality known as Cerro Pampa 2, and in the western sector is site Cerro Pampa 6, approximately 3.5 km away (Fig. 7.11).

The locality Cerro Pampa 2 (CP2) (S 47°56′41.90″ W 71°21′41.27″) comprises two groups of hunting blinds (CP2 sectors A and D on one side, and sector C on the other) and a significant surface concentration of archaeological material (CP2 sector B Ojo de Agua) (Espinosa and Goñi 1999; Goñi et al. 2011-2012) (Fig. 7.11). Altogether, this forms an associated area occupying approximately 410,400 m^2.

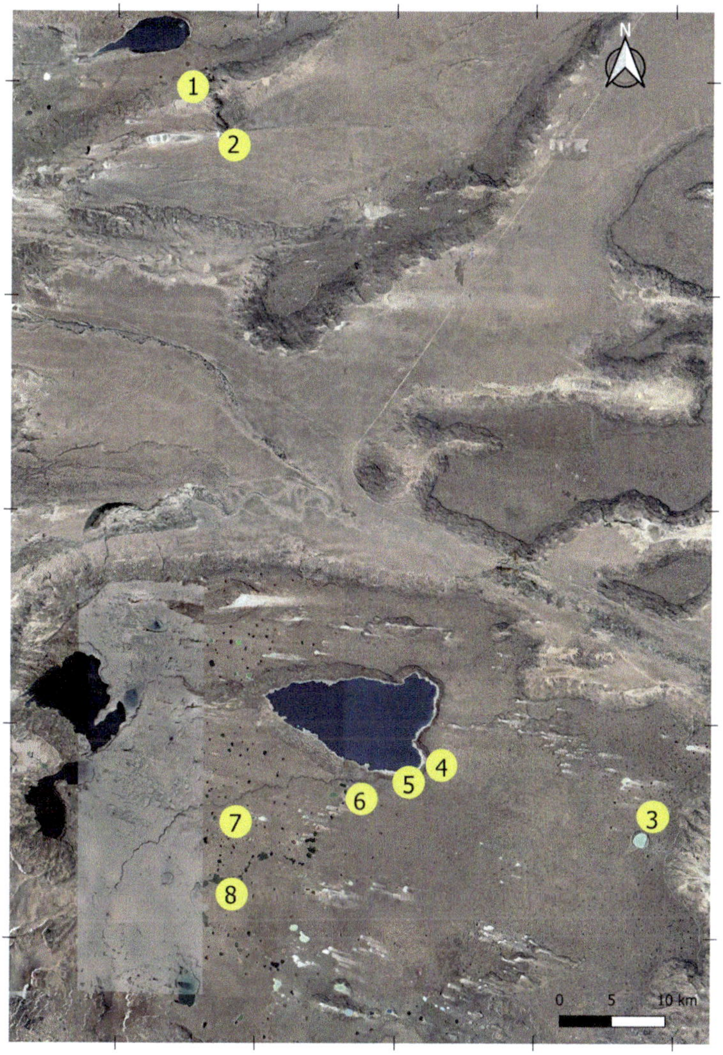

Fig. 7.10 Map with sites mentioned in the text: In Pampa del Asador: (1) Cerro Pampa 6; (2) Cerro Pampa 2; In the Strobel Plateau; (3) K205; (4) K123; (5) K116; (6) K107; (7) K30; (8) K168

A first group of hunting blinds (CP2A and D) (Fig. 7.11a) consists of 13 structures arranged in steps along a gentle slope of Cerro Pampa. The maximum distance between the structures of the two groups is 570 m. The second group (CP2C) is located further west and consists of five hunting blinds situated on a rocky outcrop. The distance from these structures to those of CP2A is 720 m and to those of CP2D is approximately 500 m.

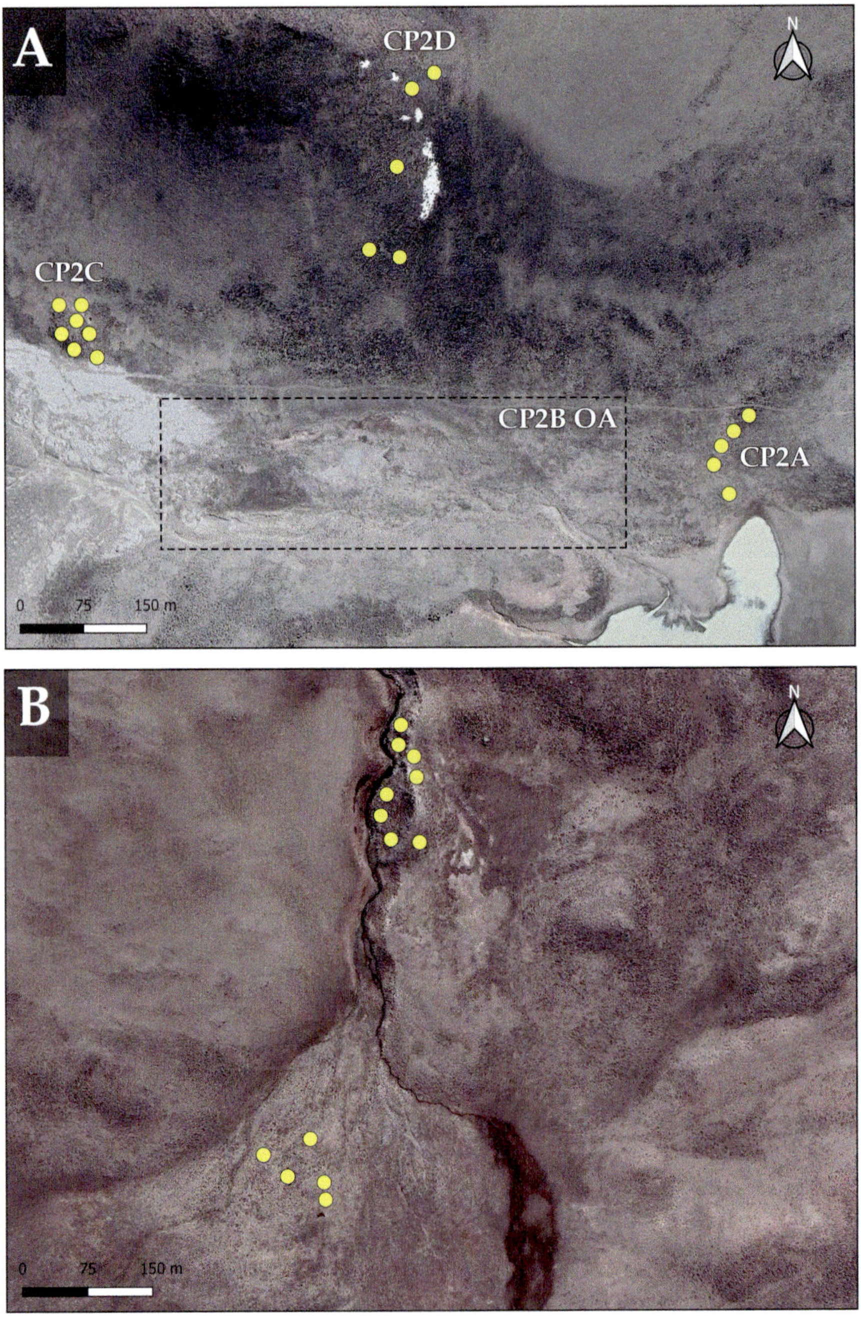

Fig. 7.11 Location of sites from Pampa del Asador/Cerro Pampa. (**a**) Locality CP2: sectors A, B, C, and D. (**b**) Site CP6

Meanwhile, the concentration called Cerro Pampa 2 sector B Ojo de Agua (CP2B OA) is an open-air site associated with a small freshwater body (Rindel et al. 2007). This concentration is characterized by a high richness of artifacts and archaeofaunal material (Tables 7.4 and 7.5). It is possible that the dynamics of this wetland over time favored the preservation of the recovered bone remains by burying the materials in a low-energy deposition context.

Notably, there is a considerable frequency of grinding artifacts and ceramic fragments (Table 7.4) that could be linked to the processing of animal resources (obtaining and storing fat). Specifically, isotopic and fatty acid analyses conducted on ceramic samples indicate an association with animal resources from the steppe (Chaile et al. 2018).

Dates indicate a chronology corresponding to the Late Holocene, between 2000 and 200 years BP (Goñi 2000-2002; Rindel 2009; Dellepiane 2019; Chaile et al. 2019). It is noteworthy that both for periods around 1400 years BP and 200 years BP, the occupation of CP2B OA aligns with that of the hunting blind structures (Fig. 7.12).

Indeed, no significant differences are established between the chronologies of CP2B surface and CP2A Hunting Blind 2 Level 3 (CP2A P2N3) (t:2.99 df: 1, α: 0.05) nor between CP2B surface and CP2A Hunting Blind 4 Level 2 (CP2A P4 N2) (t:1.78 df: 1, α: 0.05).

Regarding the zooarchaeological information available for both CP2B OA and the hunting blinds of CP2, guanaco dominates the bone assemblage (Table 7.5).

Table 7.4 Frequencies of technological AR recovered in surface and stratigraphy from sites in PDA/Cerro Pampa

	CP2		CP6
	Structures	CP2B OA	Structures
Debitage	49,090	2699	20,273
Tools	499	78	92
Cores	177	36	13
Projectile points	32	9	30
Bolas	1	4	2
Grinding artifacts	6	34	9
Ceramic sherds	40	390	40
Stone structures	13	0	13

Table 7.5 Zooarchaeological information for the Pampa del Asador/Cerro Pampa sites

	CP2		CP6
	Structures	CP2B OA	Structures
NSP	2966	1838	1364
NISP *Lama guanicoe*	420	1458	557
MNI	7	43	8
% juvenile individuals (MNE)	18.2	22.4	26.6

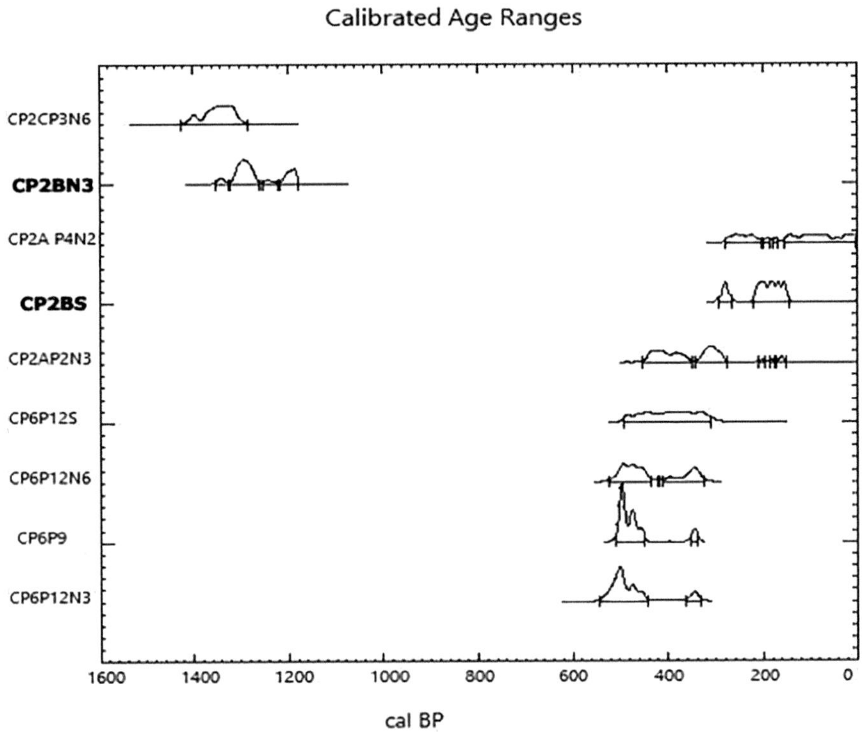

Fig. 7.12 Probability distributions for some dates in hunting blind and concentrations (in bold) from Pampa del Asador/Cerro Pampa using Calib 8.2 program (Stuiver and Reimer 1993), Southern Hemisphere calibration curve (Hogg et al. 2013)—two-sigma ranges were used

Cerro Pampa 6 site (CP6) (S 47°54′16.32″ W 71°24′09.67″) is a cluster of 13 hunting blinds (Fig. 7.11b), adjacent to a watercourse (Dellepiane and Cassiodoro 2019; Goñi et al. 2011-2012). This system of hunting blinds is arranged into two large groups. One group, consisting of eight structures, is located on the upper part of the terrain, while the second group, located 400 m to the south, comprises five structures distributed across a plain. The maximum distance between structures is approximately 480 m. Altogether, this sector occupies an area of 48,000 m^2.

In these structures, faunal, lithic, ceramic, pigment, and anthracological materials have been recovered (Cassiodoro and Tessone 2014; Cassiodoro 2016; Dellepiane 2019; Dellepiane and Cassiodoro 2019; Martínez and Re 2024) (Tables 7.4 and 7.5). Regarding the zooarchaeological information, guanaco is also the most represented species in CP6, and the overall taxonomic diversity is low (Table 7.5).

The recovered bone and ceramic materials provided six absolute chronologies, all corresponding to the Late Holocene (Cassiodoro and Tessone 2014; Dellepiane and Cassiodoro 2019). Notably, five of them date to the final Late Holocene (after 900 years BP), and similar chronologies exist for different structures (Dellepiane and Cassiodoro 2019).

In summary, the Cerro Pampa case shows that the topographic location, proximity to water bodies, frequency of projectile points, and characteristics of the zooarchaeological record of the hunting blinds establish a clear connection with activities focused on prey acquisition, mainly guanaco. Observations in CP2B OA recorded the recurring presence of large guanaco groups around the spring (groups of 90 and 36 animals were seen in January 2016 by the authors). This suggests that the structures were situated in locations suitable for interception hunting. Additionally, the hunting blinds are clearly linked to spaces where multiple activities were carried out. In other words, prey acquisition was likely combined with strategies for processing and planning deferred consumption. The development of all these activities would have necessarily required the involvement of different members of the social group.

7.4.2.2 Strobel Plateau

The Strobel plateau is bounded by the Cardiel Lake basin to the south, the Chico River valley to the north and east, and the Meseta de la Muerte to the west (Fig. 7.1) (Belardi and Goñi 2006). The plateau, with an approximate area of 3000 km^2, has a relatively flat landscape that slopes northwestward, reaching an elevation of approximately 1200 m a.s.l. at its western edge. There are low areas with shallow lakes exceeding 1000 (Lancelotti 2009), surrounded by basaltic walls.

Archaeological research suggests that the Strobel plateau has been systematically selected as a strategic location for hunting activities (Goñi 2010). In this regard, these structures serve as indicators of hunting territories that were used consistently over time (Flores Coni et al. 2021), with events of population convergence (Belardi and Goñi 2006). The largest number of hunting blind structures in southern Patagonia has been recorded here (Flores Coni et al. 2021) (Table 7.2). Rock art representations may also have been part of the process of equipping the space. In both cases, we observe behaviors that alter the natural and social environment, representing instances of niche construction recurrently occupied over time.

Below, several selected sites are described, notable for the number of structures they contain. These include sites K168, K130, K123, K107, K205, and K116 (Figs. 7.10 and 7.13).

Site K168

The set of structures known as K168 (S 48°34′15.3″W 71°23′06.3″) is located in the southwestern sector of the Strobel plateau, at 1003 m.a.s.l. This site comprises 10 hunting blinds arranged between two shallow lakes known as "Encadenadas" (Fig. 7.13a and Table 7.6).

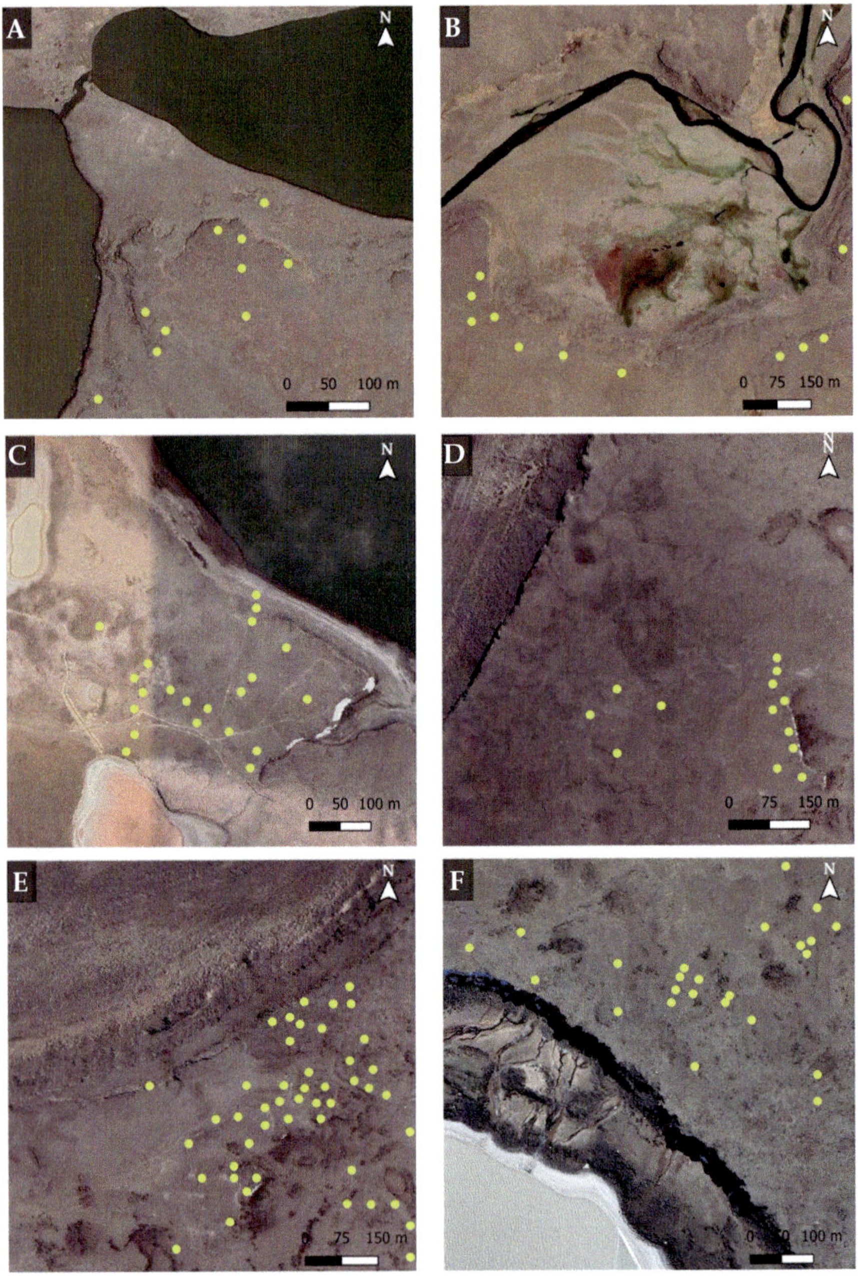

Fig. 7.13 Location of sites from the Strobel plateau mentioned in the text. (**a**) Site K168. (**b**) Site K130. (**c**) Site K107. (**d**) Site K123. (**e**) Site K116. (**f**) Site K205

Table 7.6 Characteristics of sites with a large number of structures in the Strobel plateau

Site	N. S.	Location	Masl	Covered area (m2)	Max. D B.S. (m)	Min. D. B.S. (m)	Zooarchaeological evidence
K168	10	Between two lagoons	1003	39,500	250	20	–
K130	11	In front of a wetland	963	54,780	900	1	–
K107	17	Next to lagoon	881	38,380	300	3	–
K123	13	Open pampa	833	26,505	400	30	–
K116	55	Open pampa	841	400,000	600	3	X
K205	25	Next to lagoon	768	178,000	500	10	X

NS number of structures, *Masl* meters above sea level, *Max.D.B.S.* maximum distance between structures, *Min.D.B.S.* minimum distance between structures

Site K130

It is located at S 48°30′28.5″ W 71° 21′45.9″, at 963 m above sea level, very close to the Barrancoso River (Fig. 7.13b). Some of the 11 hunting blinds are only a few meters away from a natural descent to the so-called Vega del Flaco, where site K127 is located, a prominent place on the plateau due to the large number of rock art representations and lithic material recorded (Flores Coni et al. 2019; Re et al. 2021).

Site K107

This site is located in the central eastern area of the plateau (S 48°29′51.9″W 71°12′33.5″). It is a concentration of 17 structures situated between two lagoons (Fig. 7.13c).

Site K123

It is located in the northeastern sector of the plateau (S 48°28′31.5″W 71°5′46.2″), at 833 m.a.s.l. It consists of a group of 13 structures situated on a high plain (Fig. 7.13d). From this vantage point, there is good visibility in all cardinal directions. Below the plain, there is a wall where 36 rock art motifs have been recorded (Re et al. 2019).

Site K116

Site K116 stands out on the Strobel plateau and in the region due to the large number of structures that make it up: a total of 55 (Fig. 7.13e). Five dates have been obtained from various hunting blinds that frame the occupation of this site between 944 and 1083 years BP (Re et al. 2017; Dellepiane 2019). Statistical analyses support the idea of a contemporary occupation of the structures and, therefore, the simultaneous use of the hunting blinds (Goñi et al. 2016; Flores Coni et al. 2021). Similarly, these chronologies are also similar to those recorded in some basalt walls (site K25 and site K26) (Fig. 7.14). This is particularly relevant because one of the highlighted aspects of this plateau is the number of both hunting blinds but also of basalt walls that served as residential shelters. The complementarity between these two types of sites is evident, not only due to their spatial proximity but also because of the same lithic technologies and homologous chronology obtained in both cases.

Site K205

This site consists of a group of 25 structures located on the northeastern shore of Laguna Hermosa (S 48°31′40.65″W 70° 49′10.94″) in the eastern sector of the plateau (Fig. 7.13f). It represents one of the groups with the largest number of hunting blinds on the Strobel plateau (Dellepiane 2019; Dellepiane and Flores Coni 2024). Field surveys identified the presence of five sectors within the site (A, B, C, D, and E), each comprising structures that exhibit a close spatial relationship. Three radiocarbon dates were obtained from guanaco bone samples recovered from hunting blind 2 (surface), 6 (stratigraphy), and 10 (stratigraphy) in Sector A, which frame the occupations in the Late Holocene (between 1600 and 200 years BP) (Re et al. 2017; Dellepiane 2019). It should be noted that the chronologies of hunting blind 2 do not differ from those of site K36, located on a nearby basalt wall (t:1.15, gl: 1, α: 0.05) (Fig. 7.14).

Table 7.6 summarizes spatial considerations for the hunting blinds, highlighting important aspects such as the total occupation area of the sets (in m^2), the maximum distance between structures (indicating an ideal length of the set), as well as the minimum distance (showing the proximity between some structures). These data are relevant for comparisons with ethnographic, ethnoarchaeological, and archaeological records cited in Sect. 7.3.

The following Table 7.7 summarizes some aspects of the lithic technology associated with the hunting blinds, highlighting the high frequency of projectile points.

Regarding the fauna recovered from sites on the Strobel plateau and similar areas, evidence is very scarce due to the general taphonomic conditions of these environments: low sedimentation rates, strong wind erosion, long periods of snow cover throughout the year, and so on. The recovery of faunal specimens in just a few cases (Table 7.8) has been highly significant, as it not only provided information about the prey hunted seasonally on the plateaus but also enabled the establishment of chronologies for the region, which was previously impossible due to the lack of

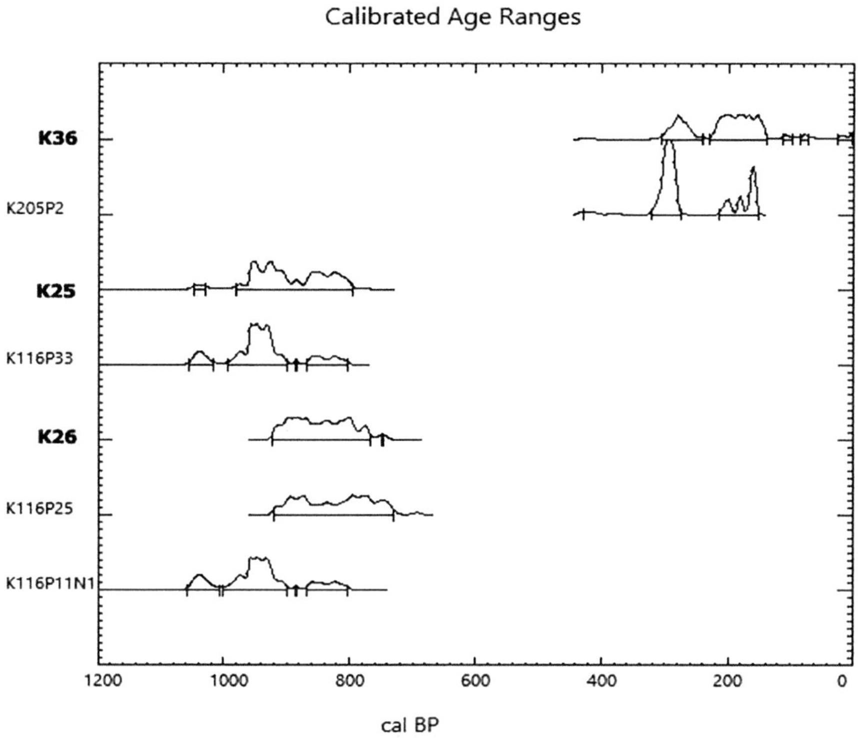

Fig. 7.14 Probability distributions for some dates in hunting blind and basaltic walls (in bold) from Strobel plateau using Calib 8.2 program (Stuiver and Reimer 1993), Southern Hemisphere calibration curve (Hogg et al. 2013) and two-sigma ranges were used

Table 7.7 Lithic technology from hunting blinds sites in the Strobel plateau mentioned in the text

Sites	K168	K130	K123	K107	K205	K116
Number of structures	10	11	13	17	25	55
Debitage	1629	407	48	961	+1000	4720
Tools	32	30	1	58	163[a]	408
Cores	1	1	0	2	11[a]	8
Projectile points	23	5	0	27	15[a]	203

[a] Sample

organic material (Belardi et al. 2007). In this instance, only sites K205 and K116 contained faunal remains.

In summary, the examples presented for the Strobel plateau highlight the presence of clusters of structures in the area, with clear evidence linking these to their use as hunting blinds. There is variability both in the number of hunting blinds and their placement. In most cases, there is a clear association with tasks related to the

Table 7.8 Zooarchaeological information of hunting blind sites from the Strobel plateau

Site	K205	K116
NISP	2598	1613
NISP *Lama guanicoe*	1703	1023
MNI	11	19
% juvenile individuals (MNE)	27.3	26

preparation and replacement of projectiles. On the other hand, the sites located on basalt walls are the most frequent and characteristic of the plateau (see Sect. 7.4.1). These locations have recorded the greatest technological diversity. Additionally, they are prominent local points due to the high frequency and diversity of registered rock art representations. Thus, they stand out as privileged sectors for the transmission of information within the information circulation system of the plateau (Goñi et al. 2014; Re 2010; Re et al. 2017, 2021).

7.5 Discussion

As a first step in this work, it was necessary to specify what we understand by communal, collective, and massive hunting. Sometimes, these concepts are not defined, and considering "massive" or "collective" as synonymous with "communal" can confuse the discussions (Borrero 2013; see Kelly 2013). Primarily, these should not be taken as rigid categories with unequivocal archaeological records associated with them.

Having established this, throughout this work, the aim has been to contribute to the discussion of whether CCMH (communal, collective, and massive hunting) could have taken place in southern Patagonia during the Late Holocene, before European contact. On the one hand, Borrero (2013) examines whether it is effectively possible to archaeologically identify large-scale (massive) guanaco hunting events based on solid zooarchaeological evidence. On the other hand, Santiago and Salemme (2010) (see also Colasurdo et al. 2012; Santiago and Salemme 2016) present intriguing evidence from a site in Tierra del Fuego with characteristics perfectly comparable to this type of hunting, specifically the site Las Vueltas 1. In both cases, the arguments primarily focus on zooarchaeological records, occasionally attributing to the so-called bone beds the characteristics they might have exhibited in the Great Plains of North America.

In this fascinating debate, we ask ourselves: Is it only through this type of evidence that CCMH sites can be identified? We propose that there are other ways to contribute to a much broader and more fruitful discussion, one that does not solely focus on locating specific sites, but rather addresses the organizational systems of hunter-gatherers in the past and their relationship with the omnipresent prey animal, the guanaco.

To discuss the issue at hand, centered on both CCMH and the characteristics and variability of the archaeological record behind it, we first posed a series of questions that can be understood as necessary geographical/environmental, ecological, and social factors. Secondly, our questions are based on the material correlates.

Geographical, environmental, ecological, and social factors required:

1. Does the behavior of the guanaco allow for the development of CCMH?

   There are aspects of the guanaco's behavior that establish it as a feasible prey for the development of CCMH. The fact that they are gregarious leads to regular situations where there are groups of guanacos with a high number of individuals. This is closely tied to the availability of grazing areas and water. Additionally, the seasonal migratory behavior of the guanaco makes it possible to anticipate its spatial and temporal location to some extent. In the same vein, recent information (Candino et al. 2022) regarding the guanaco's marked fidelity to specific routes emphasizes this possibility. The aggregation of animals and their patterned movement/transit make them predictable in terms of location-based behaviors. Predictability in behavior is a necessary step for cooperative planning in obtaining this resource.

   Thus, the social factor depends on the ethological factor of a singular prey species. Therefore, human social decisions cannot occur without the opportunity to act on a critical mass of animals. Unlike most international literature examples regarding CCMH hunts (see Morin et al. 2024), the case proposed here involves a single, absolutely predominant species in southern Patagonia. Consequently, it is the ethology of this animal that sets the conditions for the social behavior of human groups, which, in turn, translates into the hunting practices discussed here.
2. Are there areas in southern Patagonia that facilitate the development of CCMH?

   As discussed earlier, the plateaus of western southern Patagonia offer large expanses, shelter, and water, which are available to guanaco groups and human populations from November to March, outside the winter period. With the arrival of spring, new grazing areas sought by the guanaco begin to grow. Moreover, this late spring to early summer period coincides with the guanaco birthing season, from December to January, during which guanaco offspring of up to 1 month old are highly valued for their delicate fur used in making clothing.

   Thus, in a specific time of the year, the plateaus serve as the setting for large groups of guanacos, suitable for large-scale hunting. Simultaneously, these areas offer a broad visual range for managing/controlling groups, along with canyons and various water bodies that could enable the implementation of large-scale hunting techniques and strategies (Cassiodoro et al. 2014).

   A crucial aspect of these spaces is their seasonality. The idea of summer hunts underscores the crucial role that the temporal and seasonal components play in enabling CCMH in the region. In this regard, several authors (Frison 1978, 2004; Binford 1978) emphasize that, beyond the specific climatic issues inhibiting the transit of prey and hunters in winter, summer hunts involving large numbers of prey demand a different organization for processing and obtaining carcasses.

This is because, in this season, if multiple prey is obtained, butchering and processing must occur immediately to prevent decomposition.

Therefore, it becomes efficient for various social segments to associate in processing the hunted animals, thereby speeding up the acquisition of new specimens. Individual hunts mean that both tasks are carried out by the same actors simultaneously, reducing the likelihood of obtaining significant numbers of prey. Given that the use of plateau-like spaces is predominantly seasonal, ethnographic analogies (Frison 1978, 2004; Binford 1978) could be applied in our case.

An interesting point derived from these ethnographic or ethnoarchaeological sources is that large-scale hunting (massive) would need to be a communal activity, primarily in the spring/summer. In winter, while such hunts could still happen, immediate processing of prey would not be required since climatic conditions slow the decomposition process (Binford 1978; Frison 1978). However, if this were the case, it would have to occur in lower basins (such as lakes Cardiel or Salitroso), as the plateaus under study would not be available for animals or humans during winter.

3. Did the demographic conditions of hunter-gatherer populations require the development of strategies involving CCMH?

Various authors have identified three main reasons for the implementation of CCMH hunts: their greater economic utility (Binford 2001; Ugan 2005; Morin et al. 2024), their role as an effective means for obtaining storable resources in seasonal environments (Frison 1978; Speth 1997), and their capacity to sustain large groups of people for sharing information and establishing intergroup bonds (Driver 1990; Morin et al. 2024).

First, as initially presented, during the late Holocene, when environmental conditions tended toward increasing aridity, the plateaus provided vast spaces with good primary productivity marked by seasonality. Thus, the energetic return that these environments potentially produce and ensure has undoubtedly been a prime seasonal resource for Patagonian hunters. All the guanaco-derived products (meat, fat, and marrow, as well as raw materials like hides, bones, tendons, etc.), essential for the annual subsistence of human populations, justified this logistical and seasonal mobility of hunters (extensification sensu Goñi 2010; Flores Coni et al. 2021).

Second, from a social perspective, the argument relies on ethnoarchaeological and ethnographic information (see Binford 2001), which establishes that hunter-gatherer populations alternate over time between small, dispersed groups related to immediate subsistence activities (GRP1, Binford 2001), regular or medium-sized groups during phases of shared camps (GRP2), and larger social aggregations that occur periodically, assembled for reasons beyond mere subsistence (GRP3, Binford 2001). The latter two also respond to social needs such as exchanging information, forming partnerships, trading goods, and celebrations.

Thus, it is essential to consider that in some cases, CCMH may have been a consequence (and not a cause) of a form of aggregation (Group type 2 or 3, sensu Binford 2001). In other words, a social need requiring a suitable environment and timing. As previously discussed, on the Strobel plateau, we have evidence of population convergence (Belardi and Goñi 2006; Goñi et al. 2014; Re et al. 2017; Flores Coni 2019). This necessarily involved solving the food needs of aggregated groups in those instances.

Not only communal but also collective and very likely massive hunts could have been a consequence of logistics responding to varied motivations, regardless of the energetic return obtained. We consider that the low human population density is not an argument to rule out CCMH in Patagonia. On the contrary, aggregations are far more necessary on a regional scale (for genetic and cultural exchanges) for such low-density populations (Binford 1983, 2001), as in the case discussed here.

The archaeological record of the CCMH:

Following the points we have just presented, we establish how this perspective translates into the archaeological record (RA) we have presented in this work.

First, it has been essential to adopt a regional perspective where meso and macro scales (sensu Dincauze 2000) are utilized to explain the archaeological record. The micro scale is only used to explain or describe a particular archaeological site.

Starting from broader scales, the first element to consider is the extent of hunting territories, specific hunting areas, and their complementary alternatives (residential or transit areas). Historical examples, such as Yamnagoo (Miotti et al. 2004; Claraz 2008) or the descriptions by Musters (2005), give us a good idea of the spatial extent necessary to carry out these tasks. This need is not limited to the equestrian period but establishes a parameter for how much space is required for CCMH (see Furlong 1912, for hunts without the use of horses).

The cases and examples presented from the Strobel plateau and the Pampa del Asador/Cerro Pampa show that the distances and areas occupied by parapet groupings maintain values that exceed confined, small-scale surfaces. This is probably due to group hunting strategies involving direction, stalking, encircling, and killing, which would have required suitable locations with the qualities highlighted in the exemplified areas. Comparing our field records with those from ethnoarchaeological or archaeological investigations (Point 3.2.1, Fig. 7.3), we observe strong similarities in the spatial characteristics of the assemblages and the conditioning of space: distances between structures, number and location of structures per area, general utilized surfaces, etc.

Thus, like many human activities, CCMH requires the use of extensive spaces for its development. Therefore, in our research, we have addressed the archaeological record on the same scale. Considering that the high-plateau environments in the central-west region of Santa Cruz Province were likely the setting for CCMH and situations of human aggregation or population convergence (Belardi and Goñi 2006; Re et al. 2017; Flores Coni et al. 2021), these have been approached at scales exceeding the micro level (sensu Dincauze 2000).

Furthermore, as previously mentioned, ethnoarchaeological records illustrate the dispersion of faunal assemblages as they are processed (Binford 1978, 1983; Frison 1978). This process not only results in a broad spatial extent but also leaves bone assemblages exposed in open fields (see Binford 1983, fig. 59). Given the low preservation potential on basaltic plateaus (Belardi et al. 2007), in our case, this would have resulted in the complete loss of such records. Therefore, it is crucial to highlight that the type of zooarchaeological assemblages associated with CCMH in our study area is more comparable to those presented by Binford (1978, 1983) and Frison (1978) than to those associated with bison jumps or mass hunts in North America (Speth 2017). The few cases where regional preservation allowed the recovery of guanaco bone remains (K116, K205, and CP2 OA) provided excellent information about high frequencies and represented age groups.

A relevant point to mention is that the archaeological record of CCMH should not be addressed from a single line of evidence. Thus, in our case, the extensive spatial provisioning in the form of stone structures on the plateaus, particularly hunting blinds, can be associated with the development of multiple hunting strategies (Cassiodoro et al. 2014; Belardi et al. 2013, 2017; Flores Coni et al. 2021). Among this variability, large sets of structures stand out (Figs. 7.11 and 7.13). Of course, the overlapping of structures in the same space may result from palimpsests. In the study area, cases such as K205, CP2, and GUI10 show evidence of structural reuse over time (Flores Coni et al. 2021). However, there is also chronological evidence supporting the near-simultaneous use of different structures (K116, CP2, CP6) (Goñi et al. 2016; Dellepiane 2019; Flores Coni et al. 2021). Thus, large structure groups cannot be understood solely as the equipment of one or two solitary hunters but also as the strategic arrangement of different structures forming a structured spatial whole for directed prey management. Furthermore, if we consider that logistics may occur in two stages—first with a small number of individuals monitoring each known transit route of guanaco groups, and then informing the rest and planning a group relocation—hunting strategies would configure a differential archaeological record between isolated and grouped structures (as in our case). The variability in the lithic record of the structures also supports this (Cassiodoro and Flores Coni 2019; Flores Coni and Cassiodoro 2023), with a high frequency of projectile points reinforcing the idea of hunting sites.

Like technology, rock art representations are also relevant in addressing this issue. The presence of a significant number of sites with hundreds of motifs, the systematic overlapping of motifs indicative of the recurrent use of refuge spaces (rock walls) chronologically and spatially corresponding with facility sites, and specific hunting representations, varied age compositions of hunters, particular anecdotes of local events, etc., support the idea that these representations communicate a variety of information to locals and outsiders (both present and future). In line with this perspective, the continuous expansion of the high-plateau hunter niche (Goñi 2010) is a feature to consider when evaluating the accumulation of motifs and facilities over time, as indicated by the cited chronologies.

CCMH in pre-European contact times in COSC:

Finally, it is necessary to specify how this series of patterns in the archaeological record (RA) is understood in our study region and its relationship with ecological and environmental information to support the existence of communal, collective, and mass hunts prior to European contact.

The expansive seasonal plateau environments provide conditions for the development of these hunting activities. The particular behavior of the primary prey, along with the high frequency of stone structures, aligns with this proposal. The associated lithic record indicates that these structures likely had variable roles within the set of tasks linked to hunting (Cassiodoro et al. 2014; Flores Coni and Cassiodoro 2023). Similarly, the zooarchaeological record reflects the dominance of a single taxon (guanaco) and processing activities. Contemporaneous chronologies between some structures indicate their potential simultaneous use within a cooperative system aimed at resource acquisition. Consequently, collective hunts could be a reflection of this cooperation. In turn, this cooperation may have resulted in the acquisition of multiple prey due to the gregarious and predictable behavior of guanacos. Thus, mass hunts may have been feasible. The chronologies from site K116 place these modalities at least within the last 1000 years BP. Additionally, the areas considered, as well as the distances between associated hunting blinds, do not allow for positive comparisons with similar features from ethnographic, ethnoarchaeological, and archaeological cases in the specific literature (see Sect. 7.3.2.1).

The archaeological record of the plateaus includes not only stone structures but also rock art on basaltic walls with associated lithic materials. Chronological relationships between these two types of sites have also been established (cases in Strobel). The same applies to surface concentrations (case Cerro Pampa 2).

At the same time, the RA accounts for a variety of activities conducted within these hunting contexts. This includes the variety of rock art motifs (as exemplified by the Strobel case) and the presence of technologies (ceramics and grinding tools) linked to secondary prey processing (as exemplified by the Cerro Pampa 2 case). This processing is particularly associated with obtaining animal fat for deferred winter use (Cassiodoro 2011; Bourlot 2012; Chaile et al. 2019), based on seasonal differences in its availability (Rindel 2013). Furthermore, the mere acquisition of a set of prey requires parallel processing for deferred or immediate use. Therefore, the presence of different segments of society interacting (including women and children, for instance) is plausible. The presence of specific processing technologies and hand negatives of different sizes on basaltic walls could reflect this interaction.

Thus, the variability of activities and potential social segments suggests that communal hunts could have also been conducted on the plateaus. If they were communal, the food requirement would be high, rejecting the idea that communal hunts could exist without being massive. It must be considered that these hunts were not necessarily annual or frequent; they could have been sporadic or temporally spaced (Binford 1983).

Related to both immediate and deferred consumption is the concept of sharing. According to Binford, in reference to sharing the product of mass hunts, the action sites differ in relation to gathering and sharing in a single location, then transporting and storing elsewhere (Binford 1984). Therefore, the archaeological evidence could be found in different areas. In our case, this deferred consumption is manifested in the lowlands (Salitroso Lake basin and Cardiel Lake basin). In the lowlands, differences in skeletal parts in the zooarchaeological record (Rindel 2009; Bourlot 2012; Dellepiane 2019), the high frequency of ceramics (Cassiodoro 2008), evidence of animal-origin residues in these ceramics, primarily fats crucial in winter (Chaile et al. 2018), and the presence of formal burial areas alongside residential camps are correlated with this (Goñi 2010; García Guraieb 2010; Cassiodoro 2011).

Meanwhile, the convergence of populations on the Strobel plateau, extensively addressed in other works (Belardi and Goñi 2006; Re et al. 2017; Flores Coni 2019), with a strong seasonal component, presents an ideal scenario for hunts involving the cooperation of different individuals to obtain multiple prey.

The entirety of the archaeological record, on a regional scale, considering its spatial and temporal distribution, is a primary source of evidence for this hunting behavior in the past, at least within the last 1000 years BP.

What is noteworthy is that similar RA patterns have also been recorded in other plateau regions of Patagonia, such as the Somuncurá plateau, Buenos Aires plateau, Cardiel Chico plateau, and Central Plateau of Santa Cruz (Gradin 1996; Belardi et al. 2013, 2017; Franco et al. 2021; Miotti et al. 2021). Therefore, these hunting modalities may have developed across much of southern Patagonia before the introduction of the horse in the twenty-seventh century (Nacuzzi 2005).

7.6 Concluding Remarks

In southern Patagonia, CCMH is well documented ethnohistorically during the post-contact equestrian period. Archaeology has contributed to addressing these hunts with a greater temporal depth (Santiago and Salemme 2010; Aschero 2018). We aim to contribute to these discussions by providing definitions, establishing necessary conditions for their development, and exemplifying particularities of the archaeological record associated with a case study in the COSC.

We propose that there is sufficient evidence to support the existence of different modalities of CCMH before European contact. This evidence is based on archaeological findings combined with ecological and environmental information.

Likewise, following Binford (1984, p. 198), the internal variability of these behaviors across time and space must be considered. Therefore, the properties of the archaeological record (RA) should not be limited to a scale of uniformly defined sites. Thus, while finding "bone bed"-type sites to evidence mass hunts is not strictly necessary—though desirable—their absence does not necessarily indicate that they did not occur. Therefore, the entirety of the archaeological record, at a regional scale, considering its spatial and temporal distribution, is a primary source of evidence for this hunting behavior in the past. It is essential to remember that sites where events involving the hunting of multiple prey took place are characteristic of settlement and mobility systems that are logistically organized (Binford 1978; Borrero 1990), as in the case under study.

It is worth emphasizing that a communal or collective modality would very likely result in a massive modality. This refers to the "when" and "why," leaving the role of technology/zooarchaeology with archaeological visibility to answer the "how." In other words, the hypothesis of mass and collective hunting in Patagonia is supported by ecological, biological, and social arguments, followed by its archaeological correlation.

Social needs determine the environmental setting in which they develop, where sufficient resources must exist to support the aggregation of people. In our case, we argue that this occurs during spring/summer in the geography of the plateaus and high-altitude plains of the COSC. Furthermore, not only zooarchaeology but also technology and rock art representations play an important role in evaluating the realization of CCMH in southern Patagonia.

Finally, we understand that the discussion unfolds on different scales of analysis—temporal and spatial—which enrich the debate rather than nullify it. Future research, designed from different methodological approaches, will allow us to better understand this social behavior that has been closely linked to animal behavior itself, which, in turn, directed and shaped significant cultural responses in Patagonia's past.

Acknowledgments Firstly, we extend our sincere gratitude to the editors of this volume for the opportunity to publish our work; their suggestions have been especially valuable in improving our research. We also wish to acknowledge the entire research team who, over the years, have contributed their efforts to these investigations. The research presented here has been funded by the Instituto Nacional de Antropología y Pensamiento Latinoamericano (Cultura de la Nación, Argentina) and the following grants: PICT 2020—3521; PICT-2019-01726; PICT-2018-03813. Proyecto UBACYT 2023-2026 20020220100180BA.

Competing Interest Declaration The authors have no conflict of interest to declare that are relevant to the content of this chapter.

References

Agnolin A, Flores Coni J, Goñi R (2019) Las bolas y el viento: análisis de la distribución de bolas en las cuencas de los lagos Cardiel y Strobel. In: Gómez Otero J, Svoboda A, Banegas A (eds) Arqueología de la Patagonia: El pasado en las arenas. CONICET–IDEAUS, Puerto Madryn, pp 634–650

Aguerre A (2000) Las Vidas de Paty en la toldería Tehuelche del Río Pinturas y el después. Facultad de Filosofía y Letras, Buenos Aires

Aguilar Meza T (2018) Estructuras para la caza y el pastoreo: Un legado prehispánico. Dev Rev Estud Sobre Patrimonio Edif 5(9):153–172

Aragone A, Cassiodoro G (2005-2006) Los parapetos de Cerro Pampa: registro arqueofaunístico y tecnológico (noroeste de la provincia de Santa Cruz). Arqueología 13:131–154

Ariztegui D, Gilli A, Anselmetti F, Goñi R, Belardi J, Espinosa S (2010) Lake level changes in Central Patagonia Argentina: crossing environmental thresholds for late glacial and Holocene human occupation. J Quat Sci 25(7):1092–1099

Arkush B (2014) Communal pronghorn hunting in the Great Basin: what have we learned over the last twenty-five years? Pacific Coast Archaeol Soc Q 49:17–35

Aschero C (2018) Hunting scenes in Cueva de las Manos, style, content and chronology Río Pinturas, Santa Cruz-Argentinian Patagonia. In: Troncoso A, Armstrong F, Nash G (eds) Archaeologies of rock art, south American perspectives. Routledge, London, pp 209–237

Aunger R (2009) Human communication as niche construction. In: Shennan S (ed) Pattern and process in cultural evolution. University of California Press, Berkeley, pp 33–43

Belardi J, Goñi R (2006) Representaciones rupestres y convergencia poblacional durante momentos tardíos en Santa Cruz (Patagonia Argentina). El caso de la meseta del Strobel. In: Fiore D, Podestá M (eds) Tramas en la Piedra: Producción y usos del arte rupestre. WAC–SAA–AINA, Buenos Aires, pp 85–94

Belardi J, Espinosa S, Cassiodoro G (2005) Un paisaje de puntas: las cuencas de los lagos Cardiel y Strobel (provincia de Santa Cruz, Patagonia Argentina). Werken 7:57–76

Belardi J, Bregliani M, Rindel D, Gómez H, Bourlot T (2007) Condiciones de preservación de conjuntos arqueofaunísticos en la meseta del Strobel (Provincia de Santa Cruz, Argentina). In: Morello F, Martinic M, Prieto A, Bahamonde G (eds) Arqueología de Fuego-Patagonia. Levantando piedras, desenterrando huesos... y develando arcanos. CEQUA, Punta Arenas, pp 411–419

Belardi J, Espinosa S, Barrientos G, Carballo Marina F, Re A, Campan P, Súnico A, Guichón F (2013) Las Mesetas de San Adolfo y Cardiel Chico: estrategias de movilidad y tácticas de caza de guanacos en el SO de Santa Cruz. In: Zangrando A, Barberena R, Gil A, Neme G, Giardina M, Luna L, Otaola C, Paulides S, Salgán L, Tivoli A (eds) Tendencias teórico-metodológicas y casos de estudio en la arqueología de la Patagonia. Museo de Historia Natural de San Rafael-SAA-INAPL, San Rafael, pp 261–272

Belardi J, Carballo M, Madrid P, Barrientos G, Campan P (2017) Late Holocene guanaco hunting grounds in southern Patagonia: blinds, tactics and differential landscape use. Antiquity 91:718–731

Benedict J, Cassells E (2000) The bob Lake game drive. In: Cassells S (ed) This land of shining mountains: archeological studies in Colorado's Indian peaks wilderness area. Center for Mountain Archeology, Ward, pp 1–17

Binford L (1978) Nunamuit etnoarchaeology. Academic, New York

Binford L (1980) Willow smoke and dogs' tails: hunter gatherer settlement systems and archaeological site formation. Am Antiq 45(1):4–20

Binford L (1983) In pursuit of the past. Thames and Hudson, London

Binford L (1984) Butchering, sharing, and the archaeological record. J Anthropol Archaeol 3:235–257

Binford L (2001) Constructing frames of reference. University of California Press, Berkeley

Borrero L (1990) Fuego-Patagonian bone assemblages and the problem of communal guanaco hunting. In: Davis L, Reeves B (eds) Hunters of the recent past. Unwin Hyman, London, pp 373–399

Borrero L (2013) Estrategias de caza en Fuego-Patagonia. Comechingonia Rev Arqueol 17(1):11–26

Bourlot T (2012) Zooarqueología de sitios a cielo abierto en el lago Cardiel, provincia de Santa Cruz: Fragmentación ósea y consumo de grasa animal en grupos cazadores-recolectores del Holoceno tardío. BAR International, Oxford

Candino M, Donadio E, Pauli J (2022) Phenological drivers of ungulate migration in South America: characterizing the movement and seasonal habitat use of guanacos. Movement Ecol 10(34):1–14

Cassells E (2000) This land of shining mountains. In: Archeological studies in Colorado's Indian southwestern Lore, peaks wilderness area. Center for Mountain Archaeology, Ward

Cassiodoro G (2008) La tecnología cerámica en cazadores recolectores de la provincia de Santa Cruz. In: Austral A, Tamagnini M (eds) Problemáticas de la Arqueología Contemporánea. Universidad Nacional de Río Cuarto, Río Cuarto, pp 227–237

Cassiodoro G (2011) Movilidad y uso del espacio de cazadores-recolectores del Holoceno tardío: estudio de la variabilidad del registro tecnológico en distintos ambientes del noroeste de la provincia de Santa Cruz. BAR International, Oxford

Cassiodoro G (2016) Paisajes mesetarios en Patagonia: tecnología de Pampa del Asador- lago Guitarra (Santa Cruz). Cuad Inst Nac Antropol Pensamiento Latinoamericano 25(2):119–140

Cassiodoro G (2021) Prospecciones en la meseta del Portezuelo (Santa Cruz): una caracterización inicial del registro arqueológico. Mundo de Antes 15:179–201

Cassiodoro G, Tessone A (2014) Análisis radiocarbónico y de isótopos estables en residuos cerámicos del centro-oeste de Santa Cruz (Patagonia). Relaciones 39(1):293–299

Cassiodoro G, Flores Coni J, Dellepiane J (2013) Cronología y asentamiento en la Meseta del Guitarra Santa Cruz: el sitio Cañadón Guitarra 3. In: Zangrando A, Barberena R, Gil A, Neme G, Giardina M, Luna L, Otaola C, Paulides S, Salgán TA (eds) Tendencias teórico-metodológicas y casos de estudio en la arqueología de la Patagonia. Museo de Historia Natural de San Rafael-SAA-INAPL, San Rafael, pp 297–305

Cassiodoro G, Re A, Rindel D (2014) Estrategias de caza en espacios altos de Patagonia meridional durante el Holoceno tardío: evidencia arqueofaunística, tecnológica y rupestre. In: Cassiodoro G, Re A, Rindel D (eds) Integración de diferentes líneas de evidencia en la arqueología Argentina. Aspha, Buenos Aires, pp 113–138

Cassiodoro G, Goñi R, Guichón F (2024) Investigaciones arqueológicas en la Meseta del Asador (Santa Cruz). In: Letelier Cosmelli J, Labarca R, Sierralta S, Urbina S (eds) Arqueología de la Patagonia. Más allá de la Distancia. Universidad Austral de Chile, Puerto Montt, pp 453–468

Chaile C, Tessone A, Cassiodoro G, Bellelli C, Belardi J (2018) δ13C and δ15N in organic residues of Patagonia pottery. Implications for studies of diet and subsistence strategies among late Holocene hunter-gatherers. J Archaeol Sci Rep 20:47–56

Chaile C, Goñi R, Cassiodoro G (2019) La cerámica del fin del mundo: el uso de contenedores cerámicos en el centro-oeste de Santa Cruz (Argentina). In: Laguens A, Bonnin M, Marconetto B (eds) Libro de Resúmenes XX Congreso Nacional de Arqueología Argentina. Universidad Nacional de Córdoba, Córdoba, pp 1515–1519

Claraz G (2008) Viaje al río Chubut (1865–1866). Ediciones Continente, Buenos Aires

Colasurdo M, Santiago F, Salemme M (2012) Guanacos: sexo y edad. El caso del sitio Las Vueltas 1 (Tierra del Fuego, Argentina). Comechingonia 16(2):195–209

Crowell A, Hitchcock R (1978) Basarwa ambush hunting in Botswana. Botswana Notes Records 10:37–51

Dellepiane J (2019) Poblamiento y uso del espacio en sectores mesetarios del centro-oeste de Santa Cruz durante el Holoceno tardío. Una aproximación zooarqueológica. PhD Thesis, Facultad de Filosofía y Letras, Universidad de Buenos Aires. Unpublished

Dellepiane J, Cassiodoro G (2019) Estructuras de caza o campamentos temporales? Registro arqueológico de parapetos del sitio Cerro Pampa 6 Santa Cruz. In: Gómez Otero J, Svoboda A, Banegas A (eds) Arqueología de la Patagonia: El pasado en las arenas. CONICET–IDEAUS, Puerto Madryn, pp 361–372

Dellepiane J, Flores Coni J (2024) Parapetos en el sector este de la meseta del Strobel (Santa Cruz, Argentina), sitio K205 La Hermosa. In: Letelier Cosmelli J, Labarca R, Sierralta S, Urbina S (eds) Arqueología de la Patagonia. Más allá de la Distancia. Universidad Austral de Chile, Puerto Montt, pp 523–539

Dellepiane J, Guichón F (2023) Prospecciones remotas de estructuras de piedra (parapetos) en el sur-oeste de la meseta del Strobel (Santa Cruz, Argentina). Rev Museo Antropol 16(3):35–46

Dincauze D (2000) Environmental archaeology: principles and practice. Cambridge University Press, Cambridge

Driver J (1990) Meat in due season: the timing of communal hunts. In: Davis L, Reeves B (eds) Hunters of the recent past. Unwin Hyman, London, pp 11–33

Espinosa S, Goñi R (1999) Viven! Una fuente de obsidiana en la Provincia de Santa Cruz. In: Soplando en el Viento. Actas de las III Jornadas de Arqueología de la Patagonia. INAPL-UN de Comahue, Neuquén-Buenos Aires, pp 177–188

Flores Coni J (2018) Poblamiento humano y uso del espacio en la meseta del Strobel provincia de Santa *Cruz. Un análisis sobre la variabilidad tecnológica durante el Holoceno.* PhD Thesis. Facultad de Filosofía y Letras, Universidad de Buenos Aires. Unpublished

Flores Coni J (2019) Population convergence in the Strobel plateau: a discussion based on the study of lithic raw materials variability. J Archaeol Sci Rep 24:473–485

Flores Coni J, Cassiodoro G (2023) Technology from hunting blinds: a study of lithic assemblages in late Holocene stone structures from Patagonia (Argentina). J Lithic Stud 10:1–25

Flores Coni J, Re A, Nuevo Delaunay A, Guichón F, Lange V (2019) Múltiples evidencias, un solo lugar: el sitio K127- Vega del Flaco (Meseta del Lago Strobel, Santa Cruz). In: Gómez Otero J, Svoboda A, Banegas A (eds) Arqueología de la Patagonia: El pasado en las arenas. CONICET–IDEAUS, Puerto Madryn, pp 385–395

Flores Coni J, Dellepiane J, Cassiodoro G, Goñi R, Agnolin A (2021) Technological strategies and guaranteed return: hunting blinds and Patagonic plateaus. In: Belardi J, Bozzuto D, Fernández P, Moreno E, Neme G (eds) Ancient hunting strategies in Argentina. Springer, New York, pp 293–309

Franco N, Vetrisano L, Gilio B, Cirigliano N, Bianchi P (2021) Hunting blinds in the southern end of the Deseado massif: collective hunting strategies during the late Holocene. In: Belardi J, Bozzuto D, Fernández P, Moreno E, Neme G (eds) Ancient hunting strategies in Argentina. Springer, New York, pp 313–342

Friesen, T. M., & Stewart, A. M. (2013). To freeze or to dry: Seasonal variability in caribou processing and storage in the Barrenlands of northern Canada. Anthropozoologica, 48(1), 89–109. https://doi.org/10.5252/az2013n1a07

Frison G (1978) Prehistoric hunters of the high plains. Academic, New York

Frison G (2004) Survival by hunting: prehistoric human predators and animal prey. University of California Press, Berkeley

Furlong C (1912) Hunting the Guanaco. Outing Magaz 61(1):3–20

García Guraieb S (2010) Bioarqueología de cazadores-recolectores del Holoceno tardío de la cuenca del lago Salitroso (Santa Cruz): aspectos paleopatológicos y paleodemográficos. PhD Thesis. Facultad de Filosofía y Letras, Universidad de Buenos Aires. Unpublished

Gilli A, Anselmetti F, Ariztegui D, Platt Bradbury J, Kelts K, Markgraf V, Mckenzie J (2001) Tracking abrupt climate change in the southern hemisphere: a seismic stratigraphic study of Lago Cardiel, Argentina (49° S). Terra Nova 13(6):443–448

Goñi R (2000) Arqueología de momentos históricos fuera de los centros de conquista y colonización: un análisis de caso en el sur de la Patagonia. In: Belardi J, Carballo Marina F, Espinosa S (eds) Desde el País de los Gigantes. Perspectivas Arqueológicas en Patagonia. UNPA, Río Gallegos, pp 283–296

Goñi R (2000-2002) Fechados radiocarbónicos y registro arqueológico en la cuenca de los lagos Salitroso/Posadas (Santa Cruz). Cuad Inst Nac Antropol Pensamiento Latinoamericano 19:666–669

Goñi R (2010) Cambio climático y poblamiento humano durante el Holoceno tardío en Patagonia meridional Una perspectiva arqueológica. PhD Thesis. Universidad de Buenos Aires, Facultad de Filosofía y Letras. Unpublished

Goñi R, Cassiodoro G, Re A, Guichón F, Flores Coni J, Dellepiane J (2010) Arqueología de la meseta del Lago Guitarra (Santa Cruz). In: Bárcena R, Chiavazza H (eds) Arqueología Argentina en el Bicentenario de la Revolución de Mayo, XVII Congreso Nacional de Arqueología Argentina V. UNCuyo, Mendoza, pp 1923–1928

Goñi R, Cassiodoro G, Rindel D (2011-2012) Poblamiento de mesetas: arqueología de Pampa del Asador y Cerro Pampa (Patagonia meridional). Cuad Inst Nac Antropol Pensamiento Latinoamericano 23(1):21–36

Goñi R, Belardi J, Cassiodoro G, Re A (eds) (2014) Arqueología de las Cuencas de los Lagos Cardiel y Strobel. Poblamiento Humano y Paleoambientes en Patagonia. Aspha, Buenos Aires

Goñi R, Cassiodoro G, Flores Coni J, Dellepiane J, Agnolin A, Guichón R (2016) Estrategias de caza y movilidad. Parapetos del sitio K116 (Meseta del Strobel, Santa Cruz). In: Mena F (ed) Arqueología de Patagonia: del mar a mar. CIEP-Ñire Negro, Coyhaique, pp 441–449

Goñi R, Re A, García Guraieb S, Cassiodoro G, Tessone A, Rindel D, Dellepiane J, Flores Coni J, Guichón F, Agnolin A (2019) Climate changes, human peopling and regional differentiation during late Holocene in Patagonia. Quat Int 505:4–20

González M (1992) Paleoambientes del Pleistoceno Tardío/Holoceno Temprano en la cuenca de los lagos Belgrano y Burmeister (47°40′/48° Sur, 72° 30’ Oeste Santa Cruz). Inf Técn 9:1–7

Gould R (1968) Living archaeology: the Ngatatjara of Western Australia. Southwest J Anthropol 24(2):101–122
Gradin C (1996) Grabados y parapetos de la zona sur de la meseta del lago Buenos Aires (Provincia de Santa Cruz). In: Goméz Otero J (ed) Arqueología. Sólo Patagonia. CENPAT-CONICET, Puerto Madryn, pp 173–184
Grønnow B (2009) Caribu hunting structures and hunting grounds of the Thule culture in Angujaartorfiup Nunaa, West Greenland. In: Grønnow B (ed) On the track of the Thule culture from Bering strait to East Greenland. National Museum, Copenhague, pp 201–210
Guichón F (2012) Representaciones rupestres del sector norte de la meseta del Strobel: escalas espaciales y tendencias temporales. Bachelor Thesis. Facultad de Ciencias Sociales, UNICEN. Unpublished
Guichón F (2018) Comunicación y circulación de información entre espacios altos y bajos en Patagonia Meridional: Representaciones rupestres del lago Cardiel y la meseta del Strobel PhD Thesis, Facultad de Filosofía y Letras, Universidad de Buenos Aires. Unpublished
Guichón F, Re A (2020) Hunter-gatherer land use strategies and hand stencils in southern Patagonia (Argentina). Cuad Arte Prehist 1:122–153
Guichón F, Cassiodoro G, Schale I, Dellepiane J (2025) Mirando al norte: arqueología del sitio Guitarra 42 (GUI42), meseta del lago Guitarra (Santa Cruz, Argentina). Arqueología 31(1):14–39
Halstead P, O'Shea J (1989) Introduction: cultural responses to risk and uncertainty. In: Halstead P, O'Shea J (eds) Bad year economics: cultural Responses to risk and uncertainty. Cambridge University Press, Cambridge, pp 1–7
Hartley R (1992) Rock art on the northern Colorado plateau: variability in content and context. Aldershot, Avebury
Hogg A, Hua Q, Blackwell P, Niu M, Buck C, Guilderson T, Heaton T, Palmer J, Reimer P, Reimer R, Turney C, Zimmerman S (2013) Shcal13 southern hemisphere calibration, 0–50,000 years Cal B.P. Radiocarbon 55(4):1889–1903
Holzer A, Avner U, Porat N, Horwitz L (2010) Desert kites in the Negev desert and Northeast Sinai: their function, chronology and ecology. J Arid Environ 74(7):806–817
Horta L, Georgieff S, Aschero C, Goñi R (2017) Paleolacustrine records from late Pleistocene-Holocene in the Perito Moreno National Park, Argentinian Patagonian Andes. Quat Int 436:8–15
Jenness D (1922) The life of the copper Eskimos. Printer to the King, Acland
Kelly R (2013) The lifeways of hunter-gatherers: the foraging Spectrum. Cambridge University Press, Cambridge
La Belle J, Pelton S (2013) Communal hunting along the continental divide of northern Colorado: results from the Olson game drive (5BL147). Quat Int 297:45–63
Lancelotti J (2009) Caracterización limnológica de lagunas de la Provincia de Santa Cruz y efectos de la introducción de Trucha Arco iris (Oncorhynchus mykiss) sobre las comunidades receptoras. PhD Thesis, Universidad Nacional del Comahue, Centro Regional Universitario Bariloche. Unpublished
Lubinski P (2013) What is adequate evidence for mass procurement of ungulates in zooarchaeology? Quat Int 297:167–175
Marlowe F (2010) The Hadza: hunter-gatherers of Tanzania. University of California Press, Berkeley
Martínez L, Re A (2024) Los colores del centro-oeste de Santa Cruz (Argentina). Primera caracterización al registro arqueológico de pigmentos y en soportes móviles con sustancias colorantes. Andes Antropol Hist 35(2):391–425
Mazzoni E, Rabassa J (2010) Inventario y clasificación de manifestaciones basálticas de Patagonia mediante imágenes satelitales y SIG, Provincia de Santa Cruz. Rev Asoc Geol Argent 66(4):608–618
Mengoni Goñalons G (1999) Cazadores de guanacos de la estepa patagónica. Sociedad Argentina de Antropología, Buenos Aires

Miotti L, Salemme M, Hermo D, Magnín L, Rabassa J (2004) Yamnago 137 años después: otro lenguaje para la misma región. In: Civalero MT, Fernández P, Guraieb G (eds) Contra viento y marea. Arqueología de la Patagonia. SAA-AINA, Buenos Aires, pp 775–796

Miotti L, Marchionni L, Hermo D, Terranova E, Magnin L (2021) Changes and continuities of hunting practices from the late Pleistocene to the late Holocene among nomadic societies of the Patagonian plateaus. In: Belardi J, Bozzuto D, Fernández P, Moreno E, Neme G (eds) Ancient hunting strategies in southern South America. Springer, New York, pp 259–291

Morin E, Bird D, Winterhalder B, Bliege Bird R (2024) Why do humans hunt cooperatively? Ethnohistoric data reveal the contexts, advantages, and evolutionary importance of communal hunting. Curr Anthropol 65(5):876–921

Musters G (2005) Vida entre los Patagones. El Elefante Blanco, Buenos Aires

Nacuzzi L (2005) Identidades impuestas. Tehuelches, aucas y pampas en el norte de la Patagonia. Sociedad Argentina de Antropología, Buenos Aires

Oliva G, González L, Rial P, Livraghi E (2001) El ambiente en la Patagonia Austral. In: Borreli P, Oliva G (eds) Ganadería Ovina Sustentable en la Patagonia Austral. Tecnologías de Manejo Extensivo. INTA, Buenos Aires, pp 19–82

Onelli C (1904) Trepando los Andes. Compañía S.A. de Billetes de Banco, Buenos Aires

Paruelo J, Aguilar M, Golluscio R, León R (1992) La Patagonia extraandina. Análisis de la estructura y el funcionamiento de la vegetación a distintas escalas. Ecol Austral 2:123–136

Quade J, Kaplan M (2017) Lake-level stratigraphy and geochronology revisited at Lago (Lake) Cardiel, Argentina, and changes in the southern hemispheric westerlies over the last 25 ka. Quat Sci Rev 177:173–188

Ramos V (2002) El magmatismo neógeno de la cordillera patagónica. In: Haller M (ed) Geología y Recursos Naturales de Santa Cruz. Relatorio del XV Congreso Geológico Argentino. CGA, El Calafate, pp 187–199

Ratto N, Orgaz M (2002-2004) Cacería comunal de camélidos en los Andes: el caso de macroestructuras La Lampaya y El Matambre en Cazaderos Grande (Chascuil, Dpto. Tinogasta, Catamarca, Argentina). Arqueología 12:75–99

Re A (2010) Representaciones rupestres en mesetas altas de la provincia de Santa Cruz. Circulación de información en espacios de uso estacional. PhD Thesis, Facultad de Filosofía y Letras, Universidad de Buenos Aires. Unpublished

Re A (2016) Las escenas de caza del Holoceno tardío en Patagonia centro-meridional. In: Mena F (ed) Arqueología de Patagonia: de mar a mar. CIEP-Ñire Negro, Coyhaique, pp 486–495

Re A, Goñi R, Flores Coni J, Guichón R, Dellepiane J, Umaño M (2017) Arqueología de la meseta del Strobel Patagonia meridional: 15 años después. Relaciones 42(1):33–158

Re A, Guichón F, Ahets Etcheberry E, Rouan Sirolli M (2019) Representaciones rupestres y jerarquización del espacio al interior de la meseta del Strobel (Santa Cruz, Argentina). Sociedades de Paisajes Áridos y Semi-Áridos. Rev Cien Lab Arqueol Etnohist 12(2):90–113

Re A, Guichón F, Espinosa M, Martínez L (2021) Los puntos claves para la comunicación por medios materiales durante el Holoceno tardío en el centro-oeste de Santa Cruz (Patagonia meridional, Argentina). Bol Museo Chileno Arte Precolombino 26(2):107–131

Re A, Guichón F, Lange V, Martínez L, Criado A (2024) Disponibilidad de agua y representaciones rupestres: nuevos relevamientos en el sector este de la meseta del lago Strobel (Santa Cruz, Argentina). In: Letelier Cosmelli J, Labarca R, Sierralta S, Urbina S (eds) Arqueología de la Patagonia. Más allá de la Distancia. Universidad Austral de Chile, Puerto Montt, pp 523–539

Rindel D (2009) Arqueología de momentos tardíos en el noroeste de la Provincia de Santa Cruz Argentina: una perspectiva faunística. PhD thesis. Facultad de Filosofía y Letras, Universidad de Buenos Aires. Unpublished

Rindel D (2013) Marcos de referencia y frecuencia de partes esqueletarias de guanaco en sitios de Patagonia meridional: el caso del Índice de Médula Insaturada. In: Zangrando A, Barberena R, Gil A, Neme G, Giardina M, Luna L, Otaola C, Paulides S, Salgán L, Tivoli A (eds) Tendencias teórico-metodológicas y casos de estudio en la arqueología de la Patagonia. Museo de Historia Natural de San Rafael-SAA-INAPL, San Rafael, pp 515–522

Rindel D, Cassiodoro G, Aragone A (2007) La utilización de las mesetas altas durante el Holoceno tardío: el sitio Cerro Pampa 2 Ojo de Agua (Santa Cruz). In: Morello F, Martinic M, Prieto A, Bahamonde G (eds) Arqueología de Fuego-Patagonia. Levantando piedras, desenterrando huesos... y develando arcanos. CEQUA, Punta Arenas, pp 649–662

Rindel D, Goñi R, Belardi J, Bourlot T (2017) Climatic changes and hunter-gatherer populations: Archaeozoological trends in southern Patagonia. In: Monks G (ed) Climate change and human responses. Springer, Dordrecht, pp 153–172

Santiago F, Salemme M (2010) A guanaco kill site in Tierra del Fuego, Argentina. The case of Las Vueltas 1. Before Farm 2:1–17

Santiago F, Salemme M (2016) Guanaco hunting strategies in the northern plains of Tierra del Fuego, Argentina. J Anthropol Archaeol 43:110–127. https://doi.org/10.1016/j.jaa.2016.07.002

Speth J (1997) Communal bison hunting in western North America: background for the study of Paleolithic bison hunting in Europe. In: Patou-Mathis M, Otte M (eds) L'Alimentation des Hommes du Paléolithique: Approche Pluridisciplinaire. Université de Liège, Liège, pp 23–57

Speth J (2017) 13,000 years of communal bison hunting in western North America. In: Albarella U (ed) The Oxford handbook of zooarchaeology. Academic, Oxford, pp 525–540

Speth J (2019) Communal bison hunting in Western North America: current understandings and unresolved issues. In: Goldfus H, Gruber M, Yona S, Fabian P (eds) Studies in archaeology and ancient cultures in honor of Isaac Gilead. Archaeopress, Oxford, pp 278–294

Spier L (1928) Havasupai ethnography. The Trustees, Los Ángeles

Stine S (1994) Extreme and persistent drought in California and Patagonia during medieval time. Nature 369:546–549

Stine S, Stine M (1990) A record from Lake Cardiel of climate Change in southern America. Nature 345:705–708

Stuiver M, Reimer P (1993) Extended 14C database and revised CALIB radiocarbon calibration program. Radiocarbon 35:215–230

Thomas D (1983) The archaeology of Monitor Valley. Anthropol Pap AMNH 58:1–159

Ugan A (2005) Does size matter? Body size, mass collecting, and their implications for understanding prehistoric foraging behavior. Am Antiq 70(1):75–89

Whallon R (2006) Social networks and information: non-utilitarian mobility among hunter gatherers. J Anthropol Archaeol 25(2):259–270

Chapter 8
Pumas Between the Forest and the River, Between the Domestic and the Symbolic. Prehispanic Record of *Puma concolor* in the Site Las Marías (Magdalena, Buenos Aires, Argentina)

Fernanda Day Pilaría

> *…con los colmillos prendidos a las venas, el puma.*
> *había vaciado a sus víctimas sorbiéndoles toda la sangre.*
> *Yo vi las cabras al pasar, y les aseguro, chiquitos, que me encendí también en ira.*
> *al ver las pobres cuatro cabras sacrificadas por la bestia sedienta de sangre.*
> *El puma, al igual que el hurón, deja de lado cualquier manjar por la sangre tibia.*
> *Los bebedores de sangre. Horacio Quiroga.*
> *...with its fangs stuck in its veins, the cougar had emptied its victims by sucking all their blood. I saw the goats as I passed by, and I assure you, little ones, that I was also inflamed with rage at the sight of the poor four goats slaughtered by the bloodthirsty beast. Like the ferret, the puma gives up any delicacy for warm blood.*
>
> *The blood drinkers, Horacio Quiroga. (Free translation by the author).*

8.1 Introduction

The species *Puma concolor* (Linnaeus, 1771) has scarce identification in the archaeological records of the Argentine Pampas region, despite its historical presence in almost the entire national territory, except Tierra del Fuego and other South Atlantic islands (Cabrera 1961). Interpretations are highlighted for some sites in the northern area of Buenos Aires stand out, where the species would be linked to symbolic

F. Day Pilaría (✉)
Laboratorio de Análisis Cerámico, Facultad de Ciencias Naturales y Museo, Universidad Nacional de La Plata, La Plata, Argentina

R. Frontini et al. (eds.), *Non-human Animals and People: Interactions, Agents and Materialities*, The Latin American Studies Book Series,
https://doi.org/10.1007/978-3-031-97495-3_8

activities (Acosta et al. 2010). In a similar sense, the puma specimens from the Calera site (Olavarría, Buenos Aires province), in association with other elements, allow us to postulate the ritual or ceremonial character of the site (Politis et al. 2005). Other sites such as Nutria Mansa I with the presence of carnivores and felids suggest a link to the belief system and symbolic value (e.g., Bonomo 2006). In none of these cases is exploitation for "food or subsistence" purposes inferred. In this context, the remains of *P. concolor* with evidence of anthropic exploitation recovered at the Las Marías site (Day Pilaría 2018) stand out. This paper aims to interpret the strategies related to the acquisition, use, and consumption of the *P. concolor* at the Las Marías site (Magdalena district, Buenos Aires province, Argentina) and thus investigate the role they may have played in the sociocultural contexts of the prehispanic past of the Río de la Plata area. To achieve this, a zooarchaeological approach to the puma assemblage is implemented, including a morphometric study with estimations of body mass, age, and sex, and an analysis of the ethnographic and ethnozoological literature available for the species.

8.2 Archaeology of the Las Marías Site

Las Marías is an archaeological site located 1 km from the coast of the Río de La Plata, in the Magdalena district of Buenos Aires province, Argentina (Fig. 8.1). It is located in a high area, on a system of shell ridges with elevations of 2 and 2.5 m (Cavallotto 1995) with permeable soil where a xerophytic tala forest developed, composed mainly of species such as tala (*Celtis tala*), coronillo (*Scutia buxifolia*), sombra de toro (*Jodina rhombifolia*), and molle (*Schinus longifolius*) (Fig. 8.2a, b).

Fig. 8.1 Location of the Las Marías site (Magdalena district, Buenos Aires province, Argentina)

The site has three radiocarbon dates that place it in the Late Holocene, indicating that it may have had a prolonged occupation or repeated occupations at different times (Table 8.1).

Numerous studies by the research team on the various materials recovered suggest that different environmental management practices were carried out at the site, such as fishing, hunting, as well as plant gathering, and likely horticultural activities (Paleo and Pérez Meroni 2007; Day Pilaría 2018; Auge et al. 2021; Auge and Day Pilaría 2023). Likewise, the local manufacture of pottery and the domestic use of vessels for cooking by exposing the pottery directly to fire or embers have been proposed (Paleo and Pérez Meroni 2005-2006, 2007, 2008; Pérez Meroni et al. 2013; Paleo et al. 2018). Globular forms dominate, with three morpho-functional categories identified: processing and service containers, storage vessels, and transference instruments known as tubular potteries (Paleo and Pérez Meroni 2005-2006; Ghiani Echenique and León 2014). The lithic assemblage is composed of flaked stone tools such as scrapers, knives, and burins made from fine-grained orthoquartzite and chalcedony, and other pieces made by abrasion and polishing such as grinding stones, *boleadora* balls, and milling hand stones on granites, gneisses, coarse quartzites, diorites, and volcanic rocks. The raw materials are of allochthonous origin, possibly from the Sierras Septentrionales (Tandilia), Sierras Australes, and secondary deposits of the Uruguay River, evidence that highlights the participation in exchange networks involving long distances (Paleo and Pérez Meroni 2007; García et al. 2011).

Pollen, organic residues, and microbotanical analyses in ceramics (Paleo and Pérez Meroni 2007; Pérez Meroni et al. 2010; Paleo et al. 2018; Auge et al. 2021; Auge 2024) identified the use of native plants such as sombra de toro (*Jodina rhombifolia*), tala (*Celtis tala*), and brotal (*Anredera cordifolia*). Domesticated plants of broad macro-regional circulation, such as maize (*Zea mays*) and possibly manioc (aff. *Manihot esculenta*), were also identified (Pérez Meroni et al. 2010; Auge et al. 2021; Auge 2024).

The archaeozoological record consists of 8500 remains, 65% of which were identified at the taxonomic level. Highlights include fish such as black sea bass (*Pogonias courbina*), catfish (*Pterodoras granulosus*), small- and medium-sized birds and rhea (*Rhea americana*), and different species of mammals such as marsh deer (*Blastocerus dichotomus*), pampas deer (*Ozotoceros bezoarticus*), puma (*Puma concolor*), dog (*Canis familiaris*), and rodents such as cuis (*Cavia aperea*), tucu tucu (*Ctenomys talarum*), and coypu (*Myocastor coypus*) (Day Pilaría et al. 2013, 2014, 2024; Day Pilaría 2018). The assemblage is well-preserved and shows various modifications of natural and cultural origin on the bone surfaces, including intentional fractures, thermal alterations, and cut marks. Various activities related to animal capture, processing, and cooking were interpreted, as well as the integral consumption of several of the animal species mentioned, including manufacturing of bone tools (Day Pilaría et al. 2015; Day Pilaría 2018). There are 10 bone tools from Las Marías, including needles, smoothers, plano-convex points, awls, harpoons, tembetá, decorated bones, with manufacturing, and use features (e.g., polishing, burnt, cut marks), and they were made on fish and pampas deer bones (Day

Fig. 8.2 Modern landscape of Las Marías site (Magdalena district, Buenos Aires province, Argentina). (**a**) Coast. (**b**) Native forest

Table 8.1 Radiocarbon dating of the Las Marías site (Magdalena district, Buenos Aires province, Argentina)

Years BP-δC[13]	Dated materials	Laboratory reference	References
1122 ± 25 δ[13]-18.7‰	Bone of *Canis familiaris*	AA109894 X31810	Pérez Meroni et al. (2013)
1590 ± 40 δ[13]-19.0%	Bone of *Lama guanicoe*	NSRL-12552 CURL-6072	Paleo and Pérez Meroni (2007)
1820 ± 50 δ[13]-12.32%	Bone of *Pogonias courbina*	NSRL-12553 CURL-6073	Paleo and Pérez Meroni (2007)

Pilaría et al. 2015). In addition, the record contains mineralized bone remains of marsh deer (*B. dichotomus*), which have a very distinctive black luster and coloring (Day Pilaría 2018). These unique elements would have distinctive values, and meanings within the cultural repertory of the Las Marías site.

Terrestrial carnivores are represented by three specimens of fox (*Pseudalopex gymnocercus*), one specimen of prehispanic dog (*Canis familiaris*), five remains assigned to the family Felidae, and 14 teeth for the Order Carnivora (Day Pilaría 2018). In this way, and in relation to the representation of the species in the sites of the Pampas region, the 43 remains determined for *P. concolor* recovered at the site of Las Marías stand out.

Therefore, the numerous lines of work implemented in Las Marías suggest that it was a residential camp where multiple activities were carried out in close connection with the surrounding local areas, such as the coast of the Río de la Plata, minor watercourses, forest and grassland surrounding the site, and with a participation in interaction and exchange networks at regional level (Paleo and Pérez Meroni 2007; García et al. 2011; Day Pilaría 2018; Auge et al. 2021; among others).

8.3 The Species *Puma concolor*

Puma concolor (Linnaeus, 1771) is a carnivore of the family Felidae of considerable size, with males typically weighing between 50 and 65 kg (although some studies report weights reaching up to 80 or even 105 kg), and females between 34 and 48 kg, showing a marked sexual dimorphism (Currier 1983; Chebez and Nigro 2010; Perera and Aprile 2012; Segura 2018; De Angelo et al. 2019). The length of specimens, including the tail, can reach 2.20 m in males and 2 m in females (Currier 1983). The body is elongated and has short, muscular legs. The coat is short, and of a uniform tan color that turns reddish or grayish according to the environment in which it lives. The young have black spots, which disappear around 4 months of age (Chebez and Nigro 2010; Perera and Aprile 2012; Segura 2018), probably around the time they begin to accompany their mothers on hunts (Currier 1983). They remain with their mothers until they are 2 years old, and the age of sexual maturity in females begins between 2 and 3 years of age (Currier 1983).

Another interesting aspect of their behavior is that they make loud vocalizations but cannot roar like other big cats (Currier 1983). Pumas have a carnivorous diet based on hunting and scavenging, and are solitary, with crepuscular and nocturnal habits. Forests, grasslands, and shrublands, among other environments, are optimal habitats for their presence. In the Pampas ecoregion it plays a role as a top predator (De Angelo et al. 2019).

Regarding the ranges of action of the species in South America, variations of between 24 and 211 km^2 are estimated for both sexes (De Angelo et al. 2019 and bibliography cited therein). However, for a true range expansion to occur, females must disperse and establish populations beyond the known geographic distribution of the species. Other studies indicate that females disperse to a lesser extent than males, and when they do disperse, they travel shorter distances (Chimento and De Lucca 2014; and literature cited therein).

The species has a wide distribution throughout the American continent, where it inhabits different environments. In Argentina, it was historically found in a great portion of the territory though currently its distribution is restricted due to the pressure of agricultural, livestock, and industrial activities, the advance of urbanization, habitat loss and fragmentation, the reduction of natural prey, poaching, because of the threat to livestock, and fires that cause mortality and displacement (Chimento and De Lucca 2014; Brancatelli and Yezzi 2017). Biogeographical studies postulate that the species has been recolonizing the Pampas region in recent decades, evidenced by the existence of pregnant females and cubs, although its distribution is still restricted. Its presence in urban and peri-urban areas, close to dwellings or intensively used spaces, is infrequent and occasional. Recently, the species has been recorded practically in the entire Pampas ecoregion corresponding to the province of Buenos Aires, in the southeast of Córdoba, and south of Santa Fe (Argentina), including the districts of Magdalena and Punta Indio, the archaeological area of study. In this sense, the environments of the east of Buenos Aires province, where tala forests, grasslands, wetlands, and numerous watercourses are frequent, provide favorable conditions for the establishment of a breeding population of pumas (for an update on the geographical distribution of the species see Chimento and De Lucca 2014; De Lucca and Chimento 2020).

8.4 Materials and Methods

Thirty-two remains were determined for the species *P. concolor* in Day Pilaría (2018); harvesting activities such as chipping and marrow consumption were interpreted. In this opportunity, the zooarchaeological study is expanded to specifically determine new specimens and deepen the observation and characterization of anthropic modifications. In addition, morphometric and sex estimations are included. In this way, the zooarchaeological materials were taxonomically and anatomically determined (Salemme et al. 1991; Mengoni Goñalons 1999, 2006-2010), consulting the specimens of the Vertebrate Zoology Division of the La Plata

Museum. Taxonomic abundance calculations, such as Minimum Number of Specimens (NISP) and Minimum Number of Individuals (MNI), and of skeletal parts, Minimum Number of Elements and Minimum Anatomical Units (MNE, MAU, and MAU%, respectively) (Grayson 1984; Mengoni Goñalons 1999).

Different taphonomic variables of natural (e.g., root marks, mineral precipitation, among others), and anthropogenic origin, such as cut marks, intentional fractures, and thermal alterations were observed and recorded with a 15× low-power microscope with the aim of inferring the processes of formation of the assemblage and the human activities carried out with the animal resource (e.g., processing, use of skeletal parts, cooking methods, discarding, manufacture of tools, among others) (Binford 1981; Miotti 1990; Mengoni Goñalons 1999, 2006-2010). In the cut marks, attributes such as location, and orientation concerning the bone axis, length and depth, number of incisions measured with a Vernier caliper, and internal micro-striations and grooves were observed (e.g., Binford 1981; Bromage and Boyde 1984; Shipman 1983; Shipman and Rose 1983; Mengoni Goñalons 2006-2010). The photographs were taken with a digital camera coupled to a binocular magnifying glass with a magnification of 6.3×. Intentional fractures were evaluated for type, shape, and length and classified according to the shape and texture, angle and direction, surface aspect and color, presence of bone flakes, and presence and location of impact marks (Mengoni Goñalons 2006-2010; Miotti 1990). Thermal alterations were evaluated based on the coloration criterion to infer whether they were directly or indirectly exposed to fire during cooking, subsequently discarded in the hearths, or used as combustion material (Mengoni Goñalons 2006-2010).

Morphometric studies were performed by Dr. Lucio González Venanzi (CONICET, Universidad Nacional de La Plata). The following measurements were taken from eight specimens: anteroposterior length and labiolingual width of the upper third premolar (P3), lower canine (c1), lower third premolar (p3), lower fourth premolar (p4), and lower first molar (m1) (Thackeray and Kieser 1992; Van Valkenburg 1990), following the nomenclature proposed in Prevosti (2023). Estimates of body mass were made from two fragments of mandibles belonging to two different specimens, which, in addition to canines, premolars, and molars, preserve the m1 necessary to take the measurements. The formulas Log mass = 3.05*log m1AP-2.15; %PE 29.72 (Van Valkenburg 1990) and Log mass = 3.37*log m1AP-2.57; %PE 27.73 (Thackeray and Kieser 1992) were applied. These measurements also allowed for the sex estimation of the specimens. Relative age was determined by assessing tooth wear (Cougar Management Guidelines Working Group 2007).

For the interpretations, biological and ethnoecological studies of the species were taken into account, as well as the information provided by the ethnographic and ethnozoological literature available on carnivores, especially felids and in particular, *P. concolor* (e.g., Arenas 2003; Villar 2015; Manzano-García and Martínez 2017; Tamburini and Cáceres 2017; Wajner 2018). Mentions and examples are recovered that explore the ways in which humans relate to pumas, capture strategies, exploitation practices, and the sociocultural connotations attributed to them. The collection of this information allows us to generate questions of archaeological interest about the relationships between humans and animals and about the ways of inhabiting the landscapes in the past.

8.5 Results

The NISP identified for the species *P. concolor* is 43, including isolated teeth or tooth fragments. There are 32 bone elements, mostly skeletal parts of the appendicular skeleton, both forelimb and hind limb (Table 8.2). All elements belong exclusively to the autopodium: metatarsals, metacarpals, and phalanges.

For the skull, two hemimandibles were identified as belonging to the same specimen, along with a smaller mandible fragment from another individual (Fig. 8.3a). There were two canines, four lower first molars, one lower fourth premolar, and one upper third premolar, some of them complete and others fragmented. Based on the four lower molars of right laterality, an MNI of four individuals for the species *P. concolor* from the site Las Marías is consigned. The observations on dental wear indicate that these are individuals within an adult age range. Additionally, body mass estimates could be made in the two specimens with the m1 to measure (Table 8.3).

Applying both formulas (Van Valkenburg 1990; Thackeray and Kieser 1992) corroborates and strengthens the estimates. From these results, it is observed that the estimated body mass corresponds to that of adult female pumas, ranging between

Table 8.2 Representation of specimens and skeletal parts of *Puma concolor* at the Las Marías site (Magdalena district, Buenos Aires province, Argentina)

Element	NISP	MNE	MNI
Skull (jaw)	3	2	4
Skull (isolated dental pieces: canine, m1, PM3, pm4)	9	2	
Radius	1	1	1
Metacarpus	3	3	2
Metatarsus	2	2	1
Metapodium	6	6	2
Pelvis	1	1	1
Patella	1	1	1
Tibia	3	3	2
Calcaneus	1	1	1
Astragalus	1	1	1
Tarsal/carpal	2	2	1
Proximal phalanx	2	2	1
Medial phalanx	3	3	1
Distal phalanx	4	4	1
Claw	1	1	1
Total	43	34	4

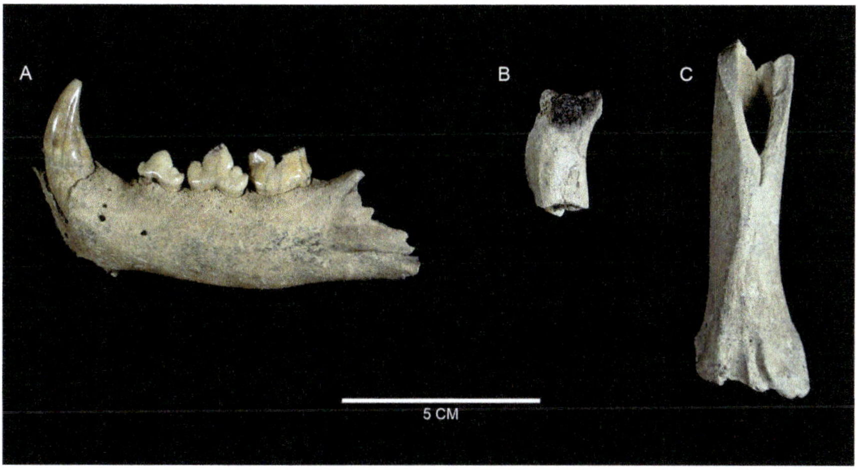

Fig. 8.3 Jaw (**a**), radius with thermal alteration (**b**) and intentional fracture in the tibia (**c**) of Puma concolor from the Las Marías site (Magdalena district, Buenos Aires province, Argentina)

Table 8.3 Estimation of body mass and sex in two individuals of *Puma concolor* from the Las Marías site (Magdalena district, Buenos Aires province, Argentina)

Specimen/ sample	Formula	Measurement (mm)	Value (kg)	Estimated sex
Specimen 1. Low left first molar	Log masa = 3.05*log m1AP-2.15; %PE 29.72 (Van Valkenburg 1990)	16.83	38.86481043	Female
	Log masa = 3.37*log m1AP-2.57; %PE 27.73 (Thackeray and Kieser 1992)	16.83	36.46728833	Female
Specimen 1. Low right first molar	Log masa = 3.05*log m1AP-2.15; %PE 29.72 (Van Valkenburg 1990)	17.23	41.751309	Female
	Log masa = 3.37*log m1AP-2.57; %PE 27.73 (Thackeray and Kieser 1992)	17.23	39.47129589	Female

34 and 48 kg (Currier 1983; Chebez and Nigro 2010; Perera and Aprile 2012; Segura 2018; De Angelo et al. 2019). These values do not overlap with the lower weight range of males (50 kg). Finally, it is worth mentioning that, attending to geographical differences in size, a recent record for Buenos Aires province reported a male of 80 kg (De Angelo et al. 2019), i.e., almost double the values obtained for the archaeological samples.

The assemblage is well-preserved and unaffected by weathering. Root marks ($n = 22$) and manganese precipitation ($n = 18$) are the most common modifications of natural origin on the bone surfaces.

The modifications of cultural origin observed correspond to fractures, cut marks, and thermal alterations. Fresh helical fractures were recorded in the tibia, metatarsus, and metapodial. Fractures in the distal fragments of these elements would indicate the intention to access the bone marrow and the possibility of manufacturing bone instruments (Fig. 8.3c). This type of fracture is recurrently recorded in other archaeological specimens of medium and large mammals, such as cervids (Day Pilaría et al. 2014; Day Pilaría 2018). Transverse and longitudinal dry fractures were also observed on the assemblage.

Thermal alterations were recorded in three specimens in a localized manner: black coloration is observed over the medial glenoid cavity of the proximal epiphysis of a radius (Fig. 8.3b), black and grayish tones over the distal trochlea of an astragalus and, finally, black coloration is observed in the area of the proximal insertion tuberosity of a third phalanx.

Cut marks were observed on five specimens: tibia, metatarsus, fourth metapodial, talus, and third phalanx (Fig. 8.4). On the distal tibia fragment, two parallel cut marks were recorded, which appear to be continuous around the perimeter of the element, extending to the posterior face of the specimen (Fig. 8.4a). On the metapodial, cut marks can be observed on the muscle insertion notch of the proximal portion, while on the fourth metatarsal, two 0.8 mm incisions are located on the tuberosity (Fig. 8.4b). On the talus, seven short incisions of variable depths were

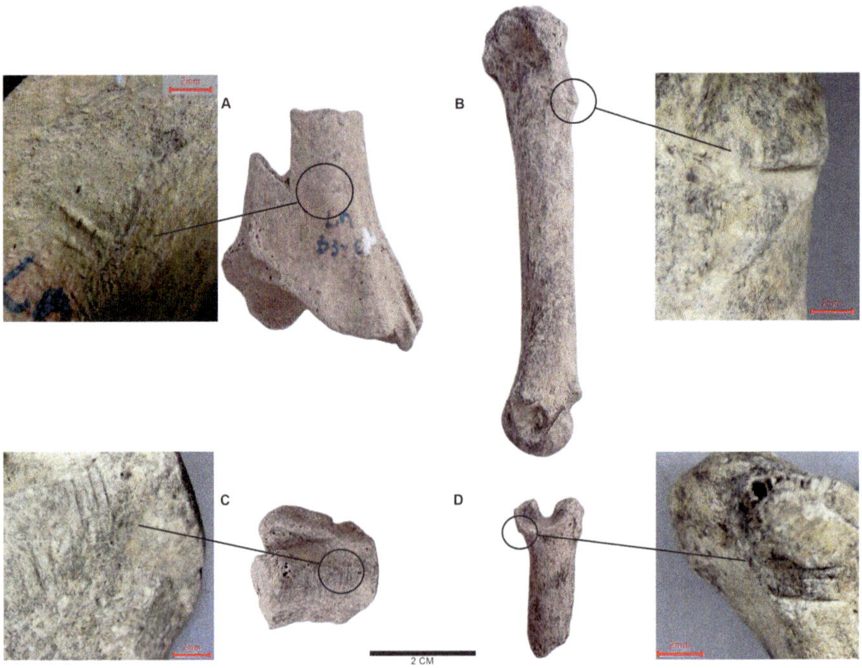

Fig. 8.4 Cut marks on Puma concolor elements from Las Marías (Magdalena district, Buenos Aires province, Argentina): distal tibia (**a**), metatarsal (**b**), astragalus (**c**) and phalanx (**d**)

recorded, located on the tuberosity of the element (Fig. 8.4c). Finally, on the third phalanx two 0.3 mm incisions were observed on the left side of the proximal insertion tuberosity (Fig. 8.4d).

8.6 Discussion

The zooarchaeological assemblage of the *P. concolor* species recorded at the Las Marías site (Magdalena district, Buenos Aires province, Argentina) is regionally relevant since its identification is scarce or null in other contexts of the Pampas region.

It is estimated that four specimens are represented in the assemblage, and 34 bone elements whose representation of skeletal parts suggests the complete entrance of the species to the site. Cranial elements were identified: three mandible fragments and nine dental pieces, including canines, molars, and premolars. So far, no surface modifications (e.g., polishing or perforations) as those identified in carnivore teeth from other archaeological sites (Politis 2000; Acosta 2005; Laporte 2014; Acosta et al. 2015) were recognized on any of the pieces. Probably other elements of the axial skeleton could be determined in nonspecific categories (e.g., medium-large Mammalia). The remaining identified elements belong to the appendicular skeleton, where three tibia fragments and 29 elements of the autopodium (carpals, tarsals, metacarpals, metatarsals, and phalanges) stand out in terms of quantity.

The processing of *P. concolor* is a novel aspect compared to what has been interpreted for the carnivore record of the Pampas region. The puma remains from the Las Marías site show fresh, possibly intentional fractures, cut marks, and thermal alterations. There are helicoidal fractures in the tibia, metatarsus, and metapodial bones that would indicate the intention to access bone marrow or to use the bones for manufacturing artifacts, although so far, no tools made from puma remains have been identified. Regarding the consumption of marrow, it is worth mentioning that within the belief system of the southern Tehuelche ethnic group, pumas would have had a very important role, and "with the purpose of acquiring of the strength and the courage of this traditional enemy, they would heat the bones and absorbed the bone marrow" (Llaras Samitier 1950 in Bonomo 2006, p. 99. Free translation by the author).

The thermal alterations are recorded in a localized manner, limited to proximal portions of the radius, talus, and phalanx, with the latter elements also showing cut marks. The areas where thermal alterations are observed are exposed tuberosities of the bones, which may explain the localized burning of their surfaces as a product of exposure to fire in some cooking events by roasting or as part of the treatment to harden the bone in the process of making instruments (Pérez Jimeno 2004; Day Pilaría et al. 2015; Pérez Jimeno and del Papa 2016).

The recorded cut marks turn out to be numerous taking into account the few identified in the zooarchaeological assemblages of the site and the study area (Day Pilaría 2019). They were observed on the tibia, metatarsus, metapodium, talus, and phalanx. The marks on the distal portion of the tibia may be related to the skinning process or be linked to the preparation of the surface to fracture and access the bone marrow (Binford 1981; Miotti 1990) and/or to a stage in the preparation of bone instruments since the parallel and continuous marks could be a perimeter sawing (Acosta 2000; del Papa et al. 2013). For their part, the marks observed on the metapodium, talus, and phalanx are interpreted as a product of leathering to obtain the skin of the specimen and/or dismemberment (Binford 1981; Miotti 1990). In this sense, the recurring representation of the skeletal parts of the zeugopodium and autopodium, which can be deliberately left attached to the hides (Aguerre 2000) would indicate the intention to take advantage of the skins. Leather is one of the materials used since ancient times in domestic and everyday activities (e.g., clothing, protection, housing, containers), which is also configured as a reference for cultural meanings and symbolic values.

The results of the morphometric and sex estimates indicate the presence of two adult females with a body mass between 36 and 41 kg. They could be mothers teaching their cubs to hunt, who stay with them from the age of 4 months to about 2 years, or they could be solitary adult females. Its capture could have been made in the vicinity of the camp, located in the native forest, surrounded by grasslands and bushes, a favorable environment for the species. Also, females have a smaller dispersal range than males, and travel shorter distances (Chimento and De Lucca 2014), which could point to a certain proximity to the Las Marías environment. These felines have crepuscular, nocturnal and solitary habits, behaviors that could have influenced the possibilities of encounter and capture, for instance, in an individual hunting event. On the other hand, they have a carnivorous diet based on hunting, but also scavenging, an aspect that could have encouraged them to approach the camp in search of food waste.

As a hunting strategy for these felines, the ethnographic work of Arenas (2003) among the Wichi and Toba populations of the Central Chaco (Argentina) mentions the use of clubs or maces, which are sticks made of different woods, with heads and holes at the ends, of variable length (0.50 m tied to the waist and functions as a throwing weapon, and 1 m used for striking). This strategy was used to hunt animals that were chased with dogs. It is worth remembering that the Las Marías site has a record of a prehispanic dog dated 1122 ± 25 years BP (Day Pilaría and Pérez Meroni 2021). Spears, rod-shaped wooden sticks with pointed ends, were also used as hunting tools for large animals. Arenas (2003) notes that this species does not arouse fear among hunters, who define it as "easy" or "weak," that with a blow to the head or snout it can fall because it is "not very hard." He also points out that hunters extract the skin and viscera, which are given to the dogs. It is also a very fatty animal, and its meat, highly regarded, is prepared roasted, stewed, boiled, or fried in fat or oil. The paws are discarded, and the heads are boiled or cooked over embers for the dogs (Arenas 2003). Likewise, regarding hunting strategies, an ethnobiological study in the Chaco Seco region of Córdoba province (Argentina) indicates that, for dangerous species such as the puma or the peccary (*Pecari tajacu*), or when long distances must be traveled, hunters organized in small groups of two or three people made up of relatives or friends (Tamburini and Cáceres 2017). These authors highlight that puma meat is valued for its flavor and quality because "it is a very clean animal (…) it eats fresh meat, does not eat a dead animal, it always hunts and eats" (Tamburini and Cáceres 2017, p. 10. Free translation by the author). They also mention that currently it is not hunted for food but rather when it causes harm and that in those cases some farmers use it to eat. For the Sierras Grandes area (province of Córdoba, Argentina), Wajner (2018) points out that some residents hunt pumas with dogs and lassoes, ad that the meat is considered the richest in the area, used in different preparations such as breaded cuts, pickles, empanadas, and roast. He also mentions that hunting pumas is a very difficult task because it is a very agile animal, which is why it is an activity that provides reputation and prestige. It can take up to 2 days to track one, requiring hunters to go out in groups, know where the puma is sleeping, and "then coordinate between some of the residents of the area to converge on the location from different cardinal points, as it is said that if one arrives late to the place, the puma escapes through that spot" (Wajner 2018, p. 41; free translation by the author). Tamburini and Cáceres (2017) also highlight that the hunters categorize the puma, and other animals, as very dangerous, unlike what Arenas (2003) points out in his ethnography of the Central Chaco. For this reason, they are considered valuable prizes or trophies and in general, their skins, skulls, limbs, and teeth are displayed in their homes. In their interviews with the farmers of the region, they recover that "it is difficult to catch it," "there are puma hunters," "boar, puma, I am not afraid of those." They also record the "endurance" that hunters must have to walk long distances and return with their prey. The knowledge to follow the trail, the handling of the dogs and the resistance, among other qualities that they learn, "fills them with pride and satisfaction" and "legitimizes the symbolic status of the hunter" (Tamburini and Cáceres 2017).

Villar (2015) highlights similar aspects in his work on an ethnographic and ethnoarchaeological perspective in the Cajón Valley (Catamarca, Argentina). He notes that hunting is mainly a male activity, which can be carried out in groups or individually, and it is learned from an early age. It is also an occupation that involves a display of practical knowledge, requires great physical effort, and uses dogs. Another interesting aspect that he records is the preservation of certain parts of the pumas such as skins, skulls, and claws that can be used as pendants, and that the fat is preserved and then used to relieve bruises. Generally, local residents do not consume its meat. Villar also points out that the processing is carried out at the place of capture, which implies a careful selection of the parts that will be preserved and those that will be discarded. The latter can also receive different treatments, such as simply leaving them abandoned at the hunting place or being buried as an offering, which highlights the interrelation of the economic and symbolic spheres (Villar 2015).

Another ethnozoological study in protected areas of the mountains of the central east Córdoba province (Argentina), records that currently the puma is hunted especially due to the threat it poses to small-scale livestock producers, that its meat is consumed by humans, while abdominal fat is used for different health conditions, for example, joints issues and asthma (Manzano-García and Martínez 2017). Another study records the use of puma fat to treat rheumatism problems (Wajner 2018). These studies highlight that in the area pumas are labeled as harmful species, especially when "they teach their young to hunt, killing up to 20 lambs and not eating them, it is just to teach the young how to kill; the stallholders get crazy, the puma has to eat" (Manzano-García and Martínez 2017, p. 39; free translation by the author). Villar (2015) mentions that, in connection with the productive activity of raising goats and sheep in the area, it is considered that "the puma stands out for killing animals excessively and without eating them entirely." This behavior is seen with some disdain, as the residents express "that it kills for pleasure or satisfaction," which is why pumas are locally known as "the damage," also called "the lion" (Villar 2015, p. 36; free translation by the author).

Recent studies on hunting activity in Argentina provide interesting elements regarding puma hunting practices. Brancatelli and Yezzi (2017) highlight that in leisure hunting or sport hunting, "pumas are hunted by ambush" (generally by chance, when the hunter is lying waiting for another prey and a puma appears) and by stalking (actively searching for them in areas where they are abundant). They note that "the size of the animal is not taken into account but rather the measurements of its skull," which are valued as trophies, making the hunting selective. For both national and foreign hunters, collecting an adult puma in good law constitutes an important trophy due to the difficulty of finding them. The commercialization of their skin and other parts of their body, such as claws, obtained by persecution due to attacks on livestock or sport hunting, can be casual, illegal and of little importance.

The *P. concolor* species has a scarce representation in the archaeological sites of the Argentine Pampas region. The following provides a survey of the findings present in the regional archaeological literature.

For the northern area (Lower Paraná Wetland), the presence of carnivores is extremely low, however, the remains of canids and, to a lesser extent, felids, show clear evidence of anthropic action (Acosta et al. 2010). For the *P. concolor* species, a canine perforated in the root was identified at the La Argentina site, which suggests its use as a pendant (Acosta et al. 2015), and a phalanx at the Anahí site (Acosta 2005). In general terms, emphasize that these findings of canids and felids remains would not be related to subsistence activities, given that they were mostly used as artifacts or associated with funerary contexts, indicating symbolic and ideological significance (Acosta et al. 2010). For these sites, and others in the northern area of the Pampas region, teeth and other remains of other carnivores such as maned wolf (*Chrysocyon brachyurus*), extinct fox (*Dusicyon avus*), domestic dog (*Canis familiaris*), gray fox (*Lycalopex gymnocercus*), and jaguar (*Panthera onca*) have been documented (Acosta et al. 2015).

For the Laguna de los Pampas site (Lincoln, Buenos Aires province), a specimen of puma is recorded (Politis et al. 2012). Meanwhile, in the basins of the Calera site located in the Sierras Bayas (Olavarría, Buenos Aires province) (ca. 3400 and 1750 years BP), remains of the autopodium, metapodium, and phalanges of the species *P. concolor* were recovered (Politis et al. 2005; Álvarez 2009). In addition, remains of other species of carnivores were found, some of which show an intentional arrangement. Based on different lines of analysis, it is interpreted that the Calera site was formed as a result of rituals or ceremonies, including the burial of objects as offerings, conducted in the northwest sector of Tandilia hilly area during a large part of the Late Holocene (Politis et al. 2005; Kaufmann and Álvarez 2007). However, at the Las Marías site, there is no deliberate or sectorized arrangement of puma remains, which were recovered in different grids and artificial levels during the excavations.

At Tapera Moreira site (La Pampa province), a puma phalanx was found although without anthropogenic modifications evidence. Here, as well as in other archaeological contexts in the province, bone elements assigned to the Felidae family were recovered (*pers. comm.* Cabrera 1961).

For the middle valley of the Río Negro (Río Negro province), two puma specimens exhibiting signs of thermal alteration were identified at the Pomona and Colforta 1 sites. They consist of autopodium elements that would have been introduced with the skin and not for meat consumption. For this area, numerous chronicles by travelers and naturalists mention that pumas were consumed in Northern Patagonia in the nineteenth century (Mange 2019).

Other species of carnivores identified in archaeological contexts close to the study area are: the presence of *Panthera onca* (jaguar or yaguareté) at the Cañada Rocha site, in the northern area of Buenos Aires province, canid skull fragment from Río Luján I (Salemme 1988), and at the Don Santiago site, in the Paraná Delta (Caggiano 1984). For the La Maza I site (Berisso, Buenos Aires) (Salemme et al. 1985), the presence of a recovered canid metacarpal fragment is mentioned, which could be assigned to *Chrysocyon brachyurus* or a large breed of *Canis familiaris*. In the same region, Austral (1977) recorded remains of *Panthera onca palustris* and *P. concolor* in the archaeological record of the El Ceibo site, while Brunazzo (1999) reported *Canis* sp. remains at the La Norma site (Berisso, Buenos Aires). In addition, for the Nutria Mansa I site (Gral. Alvarado and Lobería counties, Buenos Aires province), carnivore remains were found, including canids (*Dusicyon gymnocercus* pampas fox, *Dusicyon avus* extinct fox, and *C. brachyurus* maned wolf) and felids (*P. onca* yaguareté), without evidence of cultural modifications. It is postulated that their presence at the site would not be related to subsistence purposes, but rather could be related to beliefs and the symbolic value of objects in different everyday settings (Bonomo 2006).

In their survey of archaeofaunas for the Pampas region, Martínez and Gutiérrez (2004) mention the presence of *Felis concolor* at the Fortín Necochea site (Interserrana Area) and Cueva Tixi (Tandilia Hilly Area) and *Felis* sp. at the Ostrich site (Ventania Hilly Area). In addition, 11 elements were recovered at the La Toma site (Coronel Pringles district, Buenos Aires province) with evidence of thermal alteration and cut marks indicating their exploitation (Álvarez and Salemme 2015).

This review of the bibliography shows, on the one hand, the scarce presence of *P. concolor* remains in the zooarchaeological records of the Pampas region. On the other hand, in the contexts where the species is present, its special place and role in ceremonial, ritual, funerary, or even domestic spaces is highlighted, without showing signs to infer its use for "food or subsistence" purposes. In the case of the Las Marías site, characterized as a multipurpose site, anthropic modifications observed (intentional fractures, thermal alterations, and cut marks) suggest the consumption of meat and bone marrow of *P. concolor*, the potential use of bones to manufacture instruments and the use of the skins of the specimens. These interpretations, together with the minimum number of four recorded individuals, could imply a differential cultural consumption compared to other species, granting them particular valuations and meanings, as suggested by other authors for the prehispanic Pampa-Patagonian populations (Politis et al. 2005; Bonomo 2006). The archaeofaunal record from Las Marías shows other uses, consumption and particular relationships between fauna and human populations, such as the case of the mandible remains of *C. familiaris* (Day Pilaría 2018; Day Pilaría and Pérez Meroni 2021), numerous fish remains that show, on the one hand, a prominent link with Río de la Plata and, on the other, the use of their bones as a support for making tools (Day Pilaría et al. 2015; Day Pilaría 2018), mineralized bones, deer antlers. and a bone plate of *Tolypeutes matacus* that could have been collected in the vicinity of the camp due to their notable characteristics (Day Pilaría 2018). Likewise, microbotanical analyses on ceramics provide evidence of the preferences, beliefs, and knowledge about plants held by past riverine populations (Auge 2024). Meanwhile, studies on ceramic technology indicate, in addition to their role in the domestic sphere, their communicational intention, and their social dimension (Paleo and Pérez Meroni 2005-2006).

8.7 Final Considerations

Various archaeological studies indicate that among hunter-gatherer groups, the symbolic and everyday aspects are clearly intertwined (Bonomo 2006), and that, in addition, the relationships they establish with the animal world are unique, where the differences between humans and fauna are a matter of degree, which is why they do not employ categories to distinguish themselves from them. For the case study, this connection can be found in the representation of the skeletal parts, in the consumption of marrow and meat to incorporate the strength and vigor of the pumas, and in the use of skins as significant elements of prestige and differentiation. At the Las Marías camp, inhabited by hunters, gatherers, fishermen, potters and, possibly, horticulturists, the use of pumas could have operated as an encounter between the domestic experience and the symbolic universe. This was a residential site in which multiple activities were carried out and which in turn was inserted in interaction and exchange networks at the regional level (Paleo and Pérez Meroni 2007; García et al. 2011; Day Pilaría 2018; Auge et al. 2021).

In this way, pumas could have occupied a place and a differential role for the populations of the Las Marías site, constituting symbolic references, along with other elements and practices (e.g., mineralized bones, *boleadora* balls, and decorated bones, see Bonomo 2006; Paleo and Pérez Meroni 2007; Day Pilaría et al. 2015; Day Pilaría 2018), and, at the same time, intervene in the belief system and in the ways of inhabiting the landscape and constructing everyday space (Villar 2015). This contribution provides new data to understand human-fauna relationships in the Río de la Plata area during the Late Holocene and allows us to interpret the significant social value that pumas had in the lives of the people who inhabited the Las Marías site.

Acknowledgments Specially to Lucio González Venanzi for carrying out the morphometric estimates, Luis del Papa, Juana Fuertes, Naiquen Ghiani Echenique, Anabella Sfeir to the teammates of the Arqueología rioplatense. To Romina Frontini, Paula Escosteguy, and Mónica Salemme, editors of the special volume for the invitation and their editorial work. This article is part of the project *Archaeological investigations in the districts of Magdalena, Punta Indio and the coastal area of Berisso, Province of Buenos Aires. Social dynamics and landscape construction* (11/995) from the UNLP, directed by M. C. Paleo.

Competing Interest Declaration The authors have no conflict of interest to declare that are relevant to the content of this chapter.

References

Acosta A (2000) Huellas de corte relacionadas con la manufactura de artefactos óseos en el nordeste de la provincia de Buenos Aires. Relac Soc Argentina Antropol 25:159–178

Acosta A (2005) Zooarqueología de cazadores-recolectores del extremo nororiental de la provincia de Buenos Aires (Humedal del río Paraná inferior, Región Pampeana, Argentina). Ph.D. Dissertation, Facultad de Ciencias Naturales y Museo, Universidad Nacional de La Plata, La Plata. https://sedici.unlp.edu.ar/handle/10915/4476

Acosta A, Buc N, Mucciolo L (2010) Linking evidences: from carcass processing to bone technology. The case of the lower Paraná wetlands (late Holocene, Argentina). In: Legrand-Pineau A, Sidéra I, Buc N, David E, Scheinsohn V (eds) Ancient and modern bone artefacts from America to Russia cultural, technological and functional signature. BAR International, Sonipat, pp 303–314

Acosta A, Buc N, Ramírez M, Prevosti F, Loponte D (2015) Producción y uso de objetos ornamentales elaborados sobre dientes de carnívoros en contextos arqueológicos del Humedal del Paraná Inferior. Rev Museo Antropol 8(2):33–46. https://doi.org/10.31048/1852.4826.v8.n2.11853

Aguerre A (2000) Las vidas de Pati en la toldería tehuelche del Río Pinturas y el después. Universidad de Buenos Aires, Buenos Aires

Álvarez MC (2009) Análisis de los restos faunísticos del sitio Calera (Sierras Bayas, partido de Olavarría). Un aporte a su funcionalidad a través del estudio de los mamíferos. Relac Soc Argentina Antropol 34:29–51

Álvarez MC, Salemme M (2015) Tendencias en la subsistencia en el Holoceno tardío en la región pampeana, Argentina: el caso del sitio La Toma (partido de Coronel Pringles, provincia de Buenos Aires). Relac Soc Argentina Antropol 60(1):123–148

Arenas P (2003) Etnografía y Alimentación Entre los Tobas Ñachilamoleek y Wichi-Lhukutas del Chaco Central (Argentina)

Auge MA (2024) Las interacciones entre el entorno vegetal y las poblaciones que habitaron los partidos de Magdalena y Punta Indio desde el Holoceno tardío hasta el siglo XIX. Ph.D. Dissertation, Facultad de Ciencias Naturales y Museo, Universidad Nacional de La Plata, Argentina. https://sedici.unlp.edu.ar/handle/10915/168354

Auge MA, Day Pilaría F (2023) De animales y plantas a la producción de comidas en el sitio Las Marías (Magdalena, Buenos Aires). Rev Folia Hist Nordeste 47:195–212. https://doi.org/10.30972/fhn.0476787

Auge MA, Andreoni DF, Pérez Meroni MM, Paleo MC (2021) Plantas en la olla: análisis de microrrestos vegetales aplicado a cerámica del sitio arqueológico Las Marías, provincia de Buenos Aires. Comechingonia 25(2):59–81

Austral A (1977) Arqueología del contacto hispano-indígena en el área Platense Meridional. Obra del Centenario del Museo de La Plata II (Antropología). Museo de La Plata, La Plata, pp 69–96

Binford LR (1981) Bones: ancient men and modern myths. Academic, New York

Bonomo M (2006) Un acercamiento a la dimensión simbólica de la cultura material en la región pampeana. Relac Soc Argentina Antropol 31:89–116

Brancatelli GIE, Yezzi AL (2017) Análisis de la explotación de puma en Argentina. Tellus BioScriba 8(1):1–8

Bromage T, Boyde A (1984) Microscopics criteria for the determination of directionality of cutmarks on bone. Am J Phys Anthropol 65:359–366

Brunazzo GA (1999) Investigaciones arqueológicas en el sitio La Norma (Partido de Berisso, Provincia de Buenos Aires, Argentina). Actas del XII Congreso Nacional de Arqueología Argentina, pp 101–106

Cabrera AL (1961) Los félidos vivientes de la República Argentina. Rev Museo Argentino Cien Nat Bernardino Riv Zool 6(5):161–247

Caggiano MA (1984) Prehistoria del Noreste Argentino, sus vinculaciones con la República Oriental del Uruguay y sur de Brasil. Pesquisas Antropol 38:1–109

Cavallotto JL (1995) Evolución geomorfológica de la llanura costera ubicada en el margen sur del Río de la Plata. Ph.D. Dissertation, Facultad de Ciencias Naturales y Museo, Universidad Nacional de La Plata, La Plata. https://sedici.unlp.edu.ar/handle/10915/4804

Chebez JC, Nigro NA (2010) Aportes preliminares para un plan de conservación y manejo del puma (*Puma concolor*) en la República Argentina. Fundación de Historia Natural Félix de Azara, Buenos Aires, p 21

Chimento N, De Lucca E (2014) *Puma concolor* recoloniza el centro y este del ecosistema de las pampas. Fundación de Historia Natural Félix de Azara. Tercera Serie 4(2):13–51

Cougar Management Guidelines Working Group (2007) Cougar management guidelines, 1st edn. Wildfutures Press, Washington

Currier MJP (1983) Felis concolor. Mammalian Species 200:1–7

Day Pilaría F (2018) Gestión de los recursos faunísticos en sociedades cazadoras, recolectoras y pescadoras. Análisis arqueozoológico en sitios del litoral del Río de la Plata (partidos de Magdalena y Punta Indio, provincia de Buenos Aires). Ph.D. Dissertation, Facultad de Ciencias Naturales y Museo, Universidad Nacional de La Plata, Argentina. http://sedici.unlp.edu.ar/handle/10915/73571

Day Pilaría F (2019) Pocos cortes, ¿buenos procesadores? El caso de los conjuntos arqueofaunísticos de los sitios del litoral del Río de la Plata. Comechingonia Rev Arqueol 23(2):59–79

Day Pilaría F, Pérez Meroni M (2021) Un perro en la orilla del Río de la Plata. Presencia de *Canis familiaris* en el sitio Las Marías (Magdalena, Buenos Aires, Argentina). In: Quintana C, Álvarez MC, Bonnat F, Mazzanti D, Barros MP, Puente V, Bonomo M (eds) Libro de resúmenes del IX Congreso de Arqueología de la Región Pampeana Argentina, 1st ed, p 26

Day Pilaría F, Merino M, Gambaro R (2013) Explotación y consumo de cérvidos en el litoral fluvial bonaerense durante el Holoceno tardío final: análisis de los sitios San Clemente VI y Las Marías. Rev Museo La Plata Sección Antropol 13(87):153–166

Day Pilaría F, Paleo MC, Merino M (2014) Prácticas de procesamiento de cérvidos en el sitio Las Marías (Partido de Magdalena, Buenos Aires, Argentina). Rev Mexicana Etnobiol 12(2):50–63. https://revistaetnobiologia.mx/index.php/etno/article/view/266

Day Pilaría F, Paleo MC, Pérez Meroni M (2015) Hasta los huesos. Análisis de los artefactos óseos del litoral fluvial del Río de La Plata. Resúmenes VI Encuentro de Arqueología del Nordeste Argentino, Gualeguaychú, p 62

Day Pilaría F, García Lerena MS, Ghiani Echenique N (2024) Venados junto al río. Historia de la relación entre el venado de las pampas (*Ozotoceros bezoarticus*) y las sociedades humanas en la costa rioplatense. Rev Museo La Plata 9(1):50–77. https://doi.org/10.24215/25456377e189

De Angelo C, Llanos R, Guerisoli MM, Varela D, Valenzuela A.E, Pía MV, Monteverde M, Reppucci JI, Lucherini M, D'Agostino R, Bolgeri MJ, Quiroga VA (2019) Puma concolor. In: SAyDS–SAREM (eds.) Categorización 2019 de los mamíferos de Argentina según su riesgo de extinción. Lista Roja de los mamíferos de Argentina. https://doi.org/10.31687/SaremLR.19.150

De Lucca E, Chimento N (2020) El puma (*Puma concolor*) en las pampas de la provincia de Buenos Aires: Una actualización sobre distribución geográfica y conflicto con el hombre; Fundación de Historia Natural "Félix de Azara". Historia Nat 10(2):53–79

del Papa LM, Togo J, De Santis L (2013) Primera aproximación a la tafonomía de la región Chaco-Santiagueña. Sitio Maquijata, Santiago del Estero. In Izeta A, Mengoni Goñalons G (eds) De la Puna a las Sierras: Avances y Perspectivas en Zooarqueología Andina. BAR International series 2564, pp 17–38

García MS, Paleo MC, López L (2011) Análisis del material lítico del sitio Las Marías, Holoceno tardío, partido de Magdalena, Buenos Aires. Arqueol Rosarina Hoy 3:115–125

Ghiani Echenique NM, León PN (2014) Metodología 3D para la reconstrucción de formas cerámicas en contextos de cazadores-recolectores. Sitio Las Marías (partido de Magdalena, provincia de Buenos Aires). In: Castro Esnal A, Funes ML, Grosso M, Kuperszmit N, Murgo A, Romero G (eds) Entre pasados y presentes IV: Estudios contemporáneos en Ciencias Antropológicas. Asociación Amigos del Instituto Nacional de Antropología, Buenos Aires, pp 579–595

Grayson DK (1984) Quantitative zooarchaeology: topics in the analysis of archaeological faunas. Academic, New York

Kaufmann C, Álvarez MC (2007) La arqueofauna del sitio Calera (Sierras Bayas, región pampeana): un abordaje a los aspectos rituales del descarte de huesos de animales. In: Bayón C, Flegenheimer N, González M, Pupio A, Frère M (eds) Arqueología en las pampas II. Universidad Nacional del Sur, Bahía Blanca, pp 745–764

Laporte L (2014) Estudio del ajuar funerario de los entierros humanos. In: Politis G, Gutiérrez M, Scabuzzo C (eds) Estado actual de las Investigaciones en el Sitio Arqueológico Arroyo Seco 2. Universidad Nacional del Centro de la Provincia de Buenos Aires, Tandil, pp 393–415

Mange E (2019) Investigaciones arqueológicas en la margen sur del valle medio-superior del río Negro (provincia de río Negro). Ph.D. Dissertation, Facultad de Ciencias Naturales y Museo, Universidad Nacional de La Plata, Argentina. https://sedici.unlp.edu.ar/handle/10915/80578

Manzano-García J, Martínez GJ (2017) Percepción de la fauna silvestre en áreas protegidas de la provincia de Córdoba, Argentina: un enfoque etnozoológico. Rev Etnobiol 15(2):32–45. https://revistaetnobiologia.mx/index.php/etno/article/view/134

Martínez G, Gutiérrez M (2004) Tendencias en la explotación humana de la fauna durante el Pleistoceno final y Holoceno en la Región Pampeana (Argentina). In: Mengoni Goñalons G (ed) Zooarchaeology of South America. Archaeopress, Bicester, pp 81–98

Mengoni Goñalons G (1999) Cazadores de guanacos de la estepa patagónica. Sociedad Argentina de Antropología, Buenos Aires

Mengoni Goñalons G (2006-2010) Zooarqueología en la práctica: algunos temas metodológicos. Xama 19-23:83–113

Miotti L (1990-1992) La experimentación simulativa de fracturas y marcas óseas y sus implicancias arqueológicas. Arqueol Contemporánea 3: 39- 61

Paleo MC, Pérez Meroni M (2005-2006) Dimensión social de la tecnología cerámica en sociedades cazadoras-recolectoras. Rev Museu Arqueol Etnol 15:73–85

Paleo MC, Pérez Meroni M (2007) Primeros resultados del sitio "Las Marías" partido de Magdalena, Provincia de Buenos Aires. In Oliva F, de Grandis N, Rodriguez J (eds) Arqueología Argentina en los inicios de un nuevo siglo, pp 275–286

Paleo MC, Pérez Meroni M (2008) Relación forma-función de un conjunto alfarero del partido de Magdalena, provincia de Buenos Aires: una aproximación metodológica. In: Austral A, Tamagnini MM (eds) Problemáticas de la arqueología contemporánea. Universidad Nacional de Río Cuarto, Río Cuarto, pp 73–85

Paleo MC, Pérez Meroni M, Menestrina F, Romero L (2018) Indagando sobre prácticas alimentarias en la costa del Río de la Plata a través del estudio de ácidos grasos. Comechingonia Rev Arqueol 22(2):135–165

Perera J, Aprile G (2012) Guía de felinos de Sudamérica. Londaiz Laborde Ediciones, pp 34–39

Pérez Jimeno L (2004) Análisis comparativo de dos conjuntos de artefactos óseos procedentes de la llanura aluvial del Paraná y la pampa bonaerense. In: Martínez G, Gutiérrez MA, Curtoni R, Berón M, Madrid P (eds) Aproximaciones Contemporáneas a la Arqueología Pampeana. Perspectivas teóricas, metodológicas, analíticas y casos de estudio. UNCPBA, Olavarría, pp 319–334

Pérez Jimeno L, del Papa L (2016) Presencia del grupo morfológico definido como tubo-ave e hipótesis de uso. El humedal del Paraná medio y el Chaco seco como caso de estudio. Rev Antropol Museo Entre Ríos 2(1):100–118

Pérez Meroni M, Paleo MC, Pochettino ML, Lema VS (2010) Procesamiento y consumo de vegetales por grupos cazadores-recolectores del Holoceno tardío en los Partidos de Magdalena y Punta Indio, Provincia de Buenos Aires. In: Berón M, Luna L, Bonomo M, Montalvo C, Aranda C, Carrera Aizpitarte M (eds) Mamül Mapu: pasado y presente desde la arqueología pampeana. Editorial Libros del Espinillo, Ayacucho, pp 215–226

Pérez Meroni M, Paleo MC, López L, Ghiani Echenique N (2013) Caracterización de pastas cerámicas del sitio Las Marías, Partido de Magdalena, Provincia de Buenos Aires. Integración de distintas líneas de trabajo. Cuad Inst Nac Antropol Pensamiento Latinoamericano 1(2):95–106

Politis G (2000) Los cazadores de la Llanura. In: Tarragó M (ed) Nueva Historia Argentina, l. Los pueblos originarios y la conquista. Sudamericana, Buenos Aires, pp 61–103

Politis G, Messineo P, Kaufmann C, Barros MP, Álvarez MC, Di Prado V, Scalise R (2005) Persistencia ritual entre cazadores- recolectores de la llanura pampeana. Bol Arqueol 9:67–90

Politis G, Messineo PG, González ME, Álvarez MC, Favier Dubois CM (2012) Primeros resultados de las investigaciones en el sitio Laguna de los Pampas (partido de Lincoln, provincia de Buenos Aires). Relac Soc Argentina Antropol 37(2):463–472. https://saantropologia.com.ar/relaciones-37-2/

Prevosti FJ (2023) Sistemática de los grandes cánidos (Mammalia, Carnivora, Canidae) fósiles de América del Sur. Publ Electr Asoc Paleontol Argentina 23(1):78–192. https://doi.org/10.5710/PEAPA.28.10.2022.417

Salemme M (1988) Paleoetnozoología del sector bonaerense de la región pampeana, con especial atención a los mamíferos. Ph.D. Dissertation, Facultad de Ciencias Naturales y Museo, Universidad Nacional de La Plata

Salemme M, Tonni EP, Ceruti CN (1985) Los materiales faunísticos del sitio arqueológico "La Maza I" (partido de Berisso, prov. Buenos Aires, Argentina): revisión crítica. Rev Asoc Cien Nat Litoral 16(2):169–178

Salemme M, Miotti L, Tonni E (1991) La determinación sistemática de los mamíferos en el análisis arqueofaunístico. In: Ratto N, Haber A (eds) De Procesos, Contextos y Otros Huesos. ICA-Sección Prehistoria, Facultad de Filosofía y Letras, Universidad de Buenos Aires, Buenos Aires, pp 65–73

Segura V (2018) Puma, león, Puma concolor. In Scrocchi G, Szumik C (eds) Universo Tucumano. Fundación Miguel Lillo

Shipman P (1983) Early hominid lifestyle: hunting and gathering or foraging and scavenging? In: Clutton-Brock J, Grigson C (eds) Animals and archaeology, 1: hunters and their prey. BAR International Series, Oxford, pp 31–49

Shipman P, Rose J (1983) Early hominid hunting, butchering, and carcass-processing behaviors: approaches to the fossil record. J Anthropol Archaeol 2:57–98

Tamburini DM, Cáceres DM (2017) Estrategias de uso de la fauna silvestre por las comunidades campesinas de Argentina central. Rev Etnobiol 15(3):5–23. https://revistaetnobiologia.mx/index.php/etno/article/view/124

Thackeray J, Kieser J (1992) Body mass and carnassial length in modern and fossil carnivores. Ann Transv Mus 35(24):337–341

Van Valkenburg B (1990) Skeletal and dental predictors of body mass in carnivores. In: Damuth J, Macfadden BJ (eds) Body size in mammalian Paleobiology: estimation and biological implication. Cambridge University Press, Cambridge, pp 181–205

Villar P (2015) El lugar del puma. Representaciones sociales y cultura material en el Valle del Cajón. KULA. Antropólogos del Atlántico Sur, Catamarca, pp 30–40

Wajner M (2018) Conocimiento ecológico local sobre animales silvestres y sus redes de interacción etnobiológicas, por parte de pobladores rurales de los alrededores de la Cuchilla Nevada, Sierras Grandes, Córdoba. Undergraduate Thesis, Biological Sciences, Facultad de Ciencias Exactas, Físicas y Naturales Universidad Nacional de Córdoba, p 85. https://rdu.unc.edu.ar/handle/11086/6219?show=full

Index

R. Frontini et al. (eds.), *Non-human Animals and People: Interactions, Agents and Materialities*, The Latin American Studies Book Series,
https://doi.org/10.1007/978-3-031-97495-3